Study Guide/Solutions Manual

to accompany

Genetics:
From Genes to Genomes

Fourth Edition

Leland H. Hartwell
Fred Hutchison Cancer Research Center

Leroy Hood
The Institue for Systems Biology

Michael L. Goldberg
Cornell University

Ann E. Reynolds
University of Washington

Lee M. Silver
Princeton University

Prepared by
Debra Nero

Study Guide/Solutions Manual
GENETICS: FROM GENES TO GENOMES, FOURTH EDITION
LELAND H. HARTWELL, LEROY HOOD, MICHAEL L. GOLDBERG, ANN E. REYNOLDS, AND LEE M. SILVER

Published by McGraw-Hill Higher Education, an imprint of The McGraw-Hill Companies, Inc., 1221 Avenue of the Americas, New York, NY 10020. Copyright © 2011, 2008, and 2004 by The McGraw-Hill Companies, Inc. All rights reserved.

This book is printed on acid-free paper.

1 2 3 4 5 6 7 8 9 0 QDB/QDB 1 0 9 8 7 6 5 4 3 2 1

ISBN: 978–0–07–729511–0
MHID: 0–07–729511–0

www.mhhe.com

Table of Contents

Introduction

Chapter 2   Mendel's Breakthrough: Patterns, Particles and Principles of Heredity

Synopsis:

Chapter 2 covers the basic principles of inheritance, first described by Mendel, that form the foundation of the Laws of Segregation and Independent Assortment. You will see in chapter 4 how these laws relate to chromosome segregation during meiosis. Chapter 2 contains most of the essential terminology used to describe inheritance. You should become very familiar with and fluent in the use of these terms because they will be used in increasingly sophisticated ways in subsequent chapters. A good way to assure that you have a solid grasp of the meanings of the new terms is to pretend you are describing each word or phenomenon to a friend or relative who is not a science major. Often giving an example of each term is useful. The first problem at the end of the chapter is also a useful gauge of how well you know these terms.

A few of the terms defined in this chapter are very critical yet often misunderstood. Learn to be precise about the way in which you use these terms:

genes and *alleles* of genes - a gene determines a trait; and there are different alleles or forms of a gene. The color gene in pea has two alleles: the yellow allele and the green allele;

genotype and *phenotype* - genotype is the genetic makeup of an organism (written as alleles) and phenotype is what the organism looks like;

homozygous and *heterozygous* - when both alleles of a gene are the same, we say the organism is homozygous for that gene; if the two alleles are different, the organism is heterozygous;

dominant and *recessive* - the dominant allele is the one that controls the phenotype in the heterozygous genotype.

Significant Elements:

After reading the chapter and thinking about the concepts you should be able to:

♦ Remember there are two alleles of each gene when describing genotypes of individuals. If you are describing gametes remember there is only one allele of each gene per gamete.

♦ Recognize if a trait is dominant or recessive by considering the phenotype of the F_1 generation.

♦ Recognize ratios:

(i) in the Study Guide monohybrid ratio refers to any ratio involving one gene, e.g. the phenotypic monohybrid ratio from a cross between two individuals heterozygous for one gene (3 dominant :1 recessive);

(ii) in the Study Guide dihybrid ratio refers to any ratio involving two genes, for e.g. a phenotypic dihybrid ratio from a cross between two individuals heterozygous for two genes (9:3:3:1).

♦   Recognize the need for and be able to set up a test cross.

♦   Determine probabilities using the basic rules of probability:

(i) *Product rule:* If two outcomes must occur together, the probability of one outcome AND the other occurring is the product of the two individual probabilities. (The final outcome is the result of two independent events.) So, the probability of getting a 4 on one die AND a 4 on the second die is the product of the two individual probabilities.

(ii) *Sum rule:* If there is more than one way in which an outcome can be produced, the probability of either one OR the other occurring is the sum of the individual probabilities. In this case, the outcomes are mutually exclusive.

♦   Draw and interpret pedigrees as in **Figures 2.21 and 2.22** using the helpful information and hints found in **Figure 2.20 and Table 2.2**.

♦   Set up Punnett squares by determining the gametes produced by the parents and the probabilities of the potential offspring as in **Figures 2.11 and 2.15**.

♦   Use and interpret branched line diagrams as in **Figure 2.17** and Problem 2-24a.

Problem Solving - How to Begin:

♦   As you work on problems in the first few chapters, you will begin to recognize some similarities in the type of problem. The majority of problems in the first few chapters involve genetic crosses. It is best to BE CONSISTENT in your approach and formatting to solve such problems. To begin, rewrite the information given in a format that will be useful in solving the problem: in other words, learn to DIAGRAM THE CROSS.

  phenotype of one parent x phenotype of other parent → phenotype(s) of progeny

The goal is to assign genotypes to the parents then use these predicted genotypes to generate genotypes, phenotypes and ratios of progeny. If the predicted progeny match the observed data you were given in the problem then the genetic explanation you have created is correct.

♦   Use the THREE ESSENTIAL QUESTIONS to determine basic information about the genotypes and/or phenotypes of the parents and/or offspring. These questions are distilled from many years of helping students figure out how to solve problems. They are designed to force you to focus on the underlying genetic basis of the information in a problem. Each of the questions has an identifying characteristic that helps you answer the question - see the Hints for an idea of what sort of information allows you to formulate specific answers. As you go through further chapters the Hints will be further refined.

THREE ESSENTIAL QUESTIONS (3EQ):

1. How many genes are involved in the cross?

2. For **each gene** involved in the cross: what are the phenotypes associated with the gene? Which phenotype is the dominant one and why? Which phenotype is the recessive one and why?

[3. For **each** gene involved in the cross: is it X-linked or autosomal?]

At this point, only questions 3EQ #1 and 3EQ #2 may be applied. The material that is the basis of question 3EQ #3 will be covered in Chapter 4.

Hints:

For 3EQ #1 look for the number of different phenotypes or phenotypic classes in the progeny. In Chapter 2 each gene has only 2 phenotypes.

For 3EQ #2 if the parents of a cross are true-breeding, look at the phenotype of the F_1 individuals. Their genotype must be heterozygous, and their phenotype is thus controlled by the dominant allele of the gene. Also, look at the F_2 progeny – the 3/4 portion of the 3:1 phenotypic monohybrid ratio is the dominant one.

Solutions to Problems:

Vocabulary

2-1. a. **4**; b. **3**; c. **6**; d. **7**; e. **11**; f. **13**; g. **10**; h. **2**; i. **14**; j. **9**; k. **12**; l. **8**; m. **5**; n. **1**.

Section 2.1 - Background

2-2. People held **two basic misconceptions** about inheritance. Firstly, it was believed that **one parent contributes the most** to an offspring's inherited features. Second was the idea of **blended inheritance** - the parental traits become mixed and forever changed in the offspring.

2-3. There are **several advantages to using peas** for the study of inheritance. **(1) Peas have a fairly rapid generation time** (at least two generations per year if grown in the field, three or four generations per year if grown in greenhouses). **(2) Peas can either self-fertilize or be artificially crossed** by an experimenter. **(3) Peas produce large numbers of offspring** (hundreds per parent). **(4) Peas can be maintained as pure-breeding lines**, simplifying the ability to perform subsequent crosses. **(5) Because peas have been maintained as inbred stocks, two easily distinguished, discrete forms of many phenotypic traits are known. (6) Peas are easy and inexpensive to grow.**

In contrast, studying **genetics in humans has several disadvantages. (1) The generation time of humans is long** (roughly 20 years). **(2) There is no self-fertilization in humans, and it is not ethical to manipulate crosses. (3) Humans produce only a small number of offspring per mating** (usually one) or per individual (almost always less than 20). **(4) Although people that are homozygous for a trait (analogous to pure-breeding) exist, homozygosity cannot be maintained** because mating with another individual is needed to produce the next generation. **(5) Because human populations are not inbred, most human traits show a continuum of phenotypes**; only a few traits have two very distinct forms. **(6) People require a lot of expensive care to "grow".**

There is nonetheless one advantage to the study of genetics in humans. Because many inherited traits result in disease syndromes, and because the world's population now exceeds 6 billion, **a very large number of individuals with variant phenotypes can be recognized. Thus, the number of genes identified in this way is rapidly increasing.**

Section 2.2 – Genetic Analysis According to Mendel

2-4.

a. Two phenotypes are seen in the second generation of this cross, normal and albino. Thus there is **one gene** controlling the phenotypes in this cross (3EQ#1). To answer 3EQ#2 note that the phenotype of the first generation progeny is normal color and that in the second generation there is a ratio of 3 normal : 1 albino. Both of these prove that allele controlling the **normal** phenotype **is dominant to** the allele controlling the **albino** phenotype.

b. Test crosses are crosses between an organism in which there is interest and an organism that is homozygous recessive for all the genes of interest. The **male parent is** albino and must have the *aa* genotype. The **normally colored offspring** must receive an *A* allele from the mother, so their genotype is *Aa*; **the albino offspring** must receive an *a* allele from the mother, so their genotype is *aa*. Therefore **the female parent is heterozygous *Aa*.**

2-5. Because two phenotypes result from the mating of two cats of the same phenotype, the short-haired parent cats must have been heterozygous. The phenotype expressed in the heterozygotes (the parent cats) is the dominant phenotype. Therefore, **short-hair is dominant to long-hair.**

2-6.

a.   3EQ #1 is answered in the problem (1 gene); 3EQ #2 is the question! Two affected individuals have an affected child and a normal child. This is not possible if the affected individuals were homozygous for a recessive allele conferring piebald spotting. Therefore, **the piebald trait must be the dominant phenotype.**

b.   If the trait is dominant, the piebald parents could be either homozygous (*PP*) or heterozygous (*Pp*). However, because the two affected individuals have an unaffected child *(pp)*, **they both must be heterozygous (*Pp*).**

Diagram the cross:

Spotted x spotted → 1 spotted : 1 normal

Pp   x   *Pp*→   1 *Pp*   : 1 *pp*

2-7. Do a **test cross** between your normal winged fly (*W-*) and a short winged fly that must be homozygous recessive (*ww*). The possible results are diagrammed here; the first genotype in each cross is that of the normal winged fly whose genotype was originally unknown:

WW x *ww* → all *W-* (normal wings)

or   *Ww* x *ww* → 1/2 *W-* (normal wings) : 1/2 *ww* (short wings).

2-8. Diagram the crosses:

closed x open → F$_1$ open → F$_2$ 145 open : 59 closed

F$_1$ open x closed → 81 open : 77 closed

The results of the crosses all fit the pattern of inheritance of a single gene, with the closed trait being recessive. The first cross is a cross like Mendel did with his pure-breeding plants, although we don't know from the information provided if the starting plants were pure-breeding or not. The F$_1$ result of this cross shows that open is dominant. **The closed parent must be homozygous for the recessive allele. Because only one phenotype is seen the F$_1$ the open parent must be the homozygous dominant genotype.** Thus the parental cucumber plants were indeed true-breeding homozygotes. The self-fertilization of the F$_1$ resulting in **a ratio of 3 open : 1 closed ratio shows that these phenotypes are controlled by one gene and that the F$_1$ plants are all heterozygous.** The second cross is a test cross. It confirms that the F$_1$ plants from the first cross are heterozygous hybrids, **because there is a 1:1 ratio of open and closed progeny.** Thus, all the data is consistent with one gene with two alleles and open being the dominant trait.

2-9. The **dominant trait (short tail) is easier to eliminate** from the population by selective breeding. **You can recognize every animal that has inherited the short tail allele,** because only one such dominant allele is needed to see the phenotype. **Short-tailed mice can be prevented from mating.** The recessive dilute coat color allele, on the other hand, can be passed unrecognized from generation to generation in heterozygous mice (carriers). The heterozygous mice do not express the phenotype, so they cannot be distinguished from homozygous dominant mice with normal coat color. Only the homozygous recessive mice express the dilute phenotype and could be prevented from mating; the heterozygotes could not.

2-10.

a. In this problem, the first of the 3 Essential Questions (3EQ #1) is answered for you (there is only 1 gene), as is 3EQ # 2 (the dominant allele is dimple D and the recessive allele is nondimple d; this dominance relationship will be symbolized in the form dimple D > nondimple d from now on.) Next, diagram the cross. In all cases, the male parent is written first for consistency.

$$\text{nondimple} \ \times \ \text{dimpled} \ \rightarrow \ \text{proportion } F_1 \text{ with dimple?}$$

Therefore: $dd \ \times \ Dd^* \ \rightarrow \ \textbf{1/2 dimple} : \text{1/2 nondimple}$

(*Note that the dimpled woman in this cross had a dd [nondimple] mother, so the woman's genotype MUST be heterozygous.)

b. dimpled x nondimple → nondimple F_1

$D? \ \times \ dd \ \ \rightarrow \ \ dd$

Because they have a nondimple child (dd), the husband must also have a d allele to contribute to the offspring. **The husband has genotype Dd.**

c. dimpled x nondimple → eight F_1, all dimpled

$D? \ \times \ dd \ \ \rightarrow \ \ 8 \ D\text{-}$

The D allele in the children must come from their father. The father could be either DD or Dd but **it is more probable that his genotype is DD.** We cannot rule out the heterozygous genotype (Dd). However, the probability that all 8 children would inherit the D allele from a Dd parent is only $(1/2)^8$ or 1/256.

2-11.

a. The only unambiguous cross is:

homozygous recessive x homozygous recessive → all homozygous recessive

The only cross that fits this criteria is: dry x dry → all dry. Therefore, **dry is the recessive phenotype (ss) and sticky is the dominant phenotype (S-).**

b. A 1:1 ratio comes from a testcross of heterozygous sticky (*Ss*) x dry (*ss*). However, **the sticky x dry matings here include both the *Ss* x *ss* AND the homozygous sticky (*SS*) x dry (*ss*)**. A 3:1 ratio comes from crosses between two heterozygotes, *Ss* x *Ss*. However, the *SS* individuals are also sticky. Thus the sticky x sticky matings in this human population are a mix of matings between two heterozygotes (*Ss* x *Ss*), between two homozygotes (*SS* x *SS*) and between a homozygote and heterozygote (*SS* x *Ss*). **The 3:1 ratio of the heterozygote cross is therefore obscured by being combined with results of the two other crosses.**

2-12. Diagram the cross:

black x red → 1 black : 1 red

No, **you cannot tell** how coat color is inherited from the results of this one mating. In effect, this was a test cross – a cross between animals of different phenotypes resulting in offspring of two phenotypes. This does not indicate whether red or black is the dominant phenotype. To determine which phenotype is dominant, remember that an animal with a recessive phenotype must be homozygous. Thus, **if you mate several red horses to each other and also mate several black horses to each other, the crosses that always yield only offspring with the parental phenotype must have been between homozygous recessives.** For example, if all the black x black matings result in only black offspring, black is recessive. Some of the red x red crosses (that is, crosses between heterozygotes) would then result in both red and black offspring in a ratio of 3:1. To establish this point, you might have to do several red x red crosses, because some of these crosses could be between red horses homozygous for the dominant allele. You could of course ensure that you were sampling heterozygotes by using the progeny of black x red crosses (such as that described in the problem) for subsequent black x black or red x red crosses.

2-13.

a. **1/6** because a die has 6 different sides.

b. There are three possible even numbers (2, 4, and 6). The probability of obtaining any one of these is 1/6. Because the 3 events are mutually exclusive, use the sum rule: 1/6 + 1/6 + 1/6 = 3/6 = **1/2**.

c. You must roll either a 3 or a 6, so 1/6 + 1/6 = 2/6 = **1/3**.

d. Each die is independent of the other, thus the product rule is used: 1/6 x 1/6 = **1/36**.

e. The probability of getting an even number on one die is 3/6 = 1/2 (see part b). This is also the probability of getting an odd number on the second die. This result could happen either of 2 ways – you could get the odd number first and the even number second, or vice-versa. Thus the probability of both occurring is 1/2 x 1/2 x 2 = **1/2**.

f. The probability of any specific number on a die = 1/6. The probability of the same number on the other die =1/6. The probability of both occurring at same time is 1/6 x 1/6 = 1/36. The same probability is true for the other 5 possible numbers on the dice. Thus the probability of any of these mutually exclusive situations occurring is 1/36 + 1/36 + 1/36 + 1/36 + 1/36 + 1/36 = 6/36 = **1/6**.

g. The probability of getting two numbers both over four is the probability of getting a 5 or 6 on one die (1/6 + 1/6 = 1/3) and 5 or 6 on the other die (1/3). The results for the two dice are independent events, so 1/3 x 1/3 = **1/9**.

2-14. The **probability of drawing a face card** = 12 face cards / 52 cards = **0.231**. The **probability of drawing a red card** = 26 / 52 = **0.5**. The **probability of drawing a red face card** = probability of a red card x probability of a face card = 0.231 x 0.5 = **0.116**.

2-15.

a. The *Aa bb CC DD* woman can produce 2 genetically different eggs that vary in their allele of the first gene (*A* or *a*). She is homozygous for the other 3 genes and can only make eggs with the *b C D* alleles for these genes. Thus, using the product rule (because the inheritance of each gene is independent), she can make 2 x 1 x 1 x 1 = **2 different types of gametes**: (*A b C D* and *a b C D*).

b. Using the same logic, an *AA Bb Cc dd* woman can produce 1 x 2 x 2 x 1 = **4 different types of gametes**: *A* (*B* or *b*) (*C* or *c*) *d*.

c. A woman of genotype *Aa Bb cc Dd* can make 2 x 2 x 1 x 2 = **8 different types of gametes**: (*A* or *a*) (*B* or *b*) *c* (*D* or *d*).

d. A woman who is a quadruple heterozygote can make 2 x 2 x 2 x 2 = **16 different types of gametes**: (*A* or *a*) (*B* or *b*) (*C* or *c*) (*D* or *d*).

2-16.

a. The probability of any phenotype in this cross depends only on the gamete from the heterozygous parent alone. The probability that a child will resemble the quadruply heterozygous parent is thus 1/2*A* x 1/2*B* x 1/2*C* x 1/2*D* = 1/16. The probability that a child will resemble the quadruply homozygous recessive parent is 1/2*a* x 1/2*b* x 1/2*c* x 1/2*d* = 1/16. **The probability that a child will resemble either parent is then 1/16 + 1/16 = 1/8**. This cross will produce 2 different phenotypes for each gene or 2 x 2 x 2 x 2 = **16 potential phenotypes**.

b. The probability of a child resembling the recessive parent is 0; the probability of a child resembling the dominant parent is 1 x 1 x 1 x 1 = 1. The probability that a child will resemble one of the two parents is 0+1=**1. Only 1 phenotype is possible** in the progeny (dominant for all 4 genes), as $(1)^4 = 1$.

c. The probability that a child would show the dominant phenotype for any one gene is 3/4 in this sort of cross (remember the 3/4 : 1/4 monohybrid ratio of phenotypes), so the probability of resembling the parent for all four genes is $(3/4)^4 = $ **81/256**. There are 2 phenotypes possible for each gene, so $(2)^4 = $ **16 different kinds of progeny.**

d. All progeny will resemble their parents because all of the alleles from both parents are identical, so the **probability = 1**. There is only 1 phenotype possible for each gene in this cross; because $(1)^4=1$, the child can have only **one possible phenotype** when considering all four genes.

2-17.

a. The combination of alleles in the egg and sperm allows only one genotype: ***aa Bb Cc DD Ee***.

b. Because the inheritance of each gene is independent, you can use the product rule to determine the number of different types of gametes that are possible: 1 x 2 x 2 x 1 x 2 = 8 (as in <u>problem 2-15</u>). To figure out the types of gametes, consider the possibilities for each gene separately and then the possible combinations of genes in a consistent order. For each gene the possibilities are: a, $(B : b)$, $(C : c)$, D, and $(E : e)$. The possibilities can be determined using the product rule. Thus for the first 2 genes [a] x [$B : b$]gives [$a B : a b$] x [$C : c$] gives [$a B C : a B c : a b C : a b c$] x [$D$] gives [$a B C D : a B c D : a b C D : a b c D$] x [$E : e$] gives [***a B C D E : a B C D e : a B c D E : a B c D e : a b C D E : a b C D e : a b c D E : a b c D e***]. This problem can also be visualized with a branch diagram:

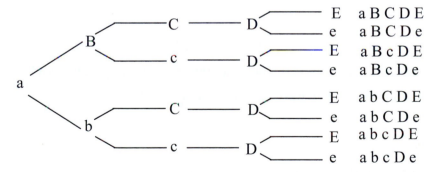

2-18. The first two parts of this problem involve the probability of occurrence of two independent traits: the sex of a child and galactosemia. The parents are heterozygous for galactosemia, so there is a 1/4 chance that a child will be affected (that is, homozygous recessive). The probability that a child is a girl is 1/2. The probability of an affected girl is therefore 1/2 x 1/4 = 1/8.

a. Fraternal (non-identical) twins result from two independent fertilization events and therefore the probability that both will be girls with galactosemia is the product of their individual probabilities (see above); 1/8 x 1/8 = **1/64**.

b. For identical twins, one fertilization event gave rise to two individuals. The probability that both are girls with galactosemia is **1/8**.

For parts c-g, remember that each child is an independent genetic event. The sex of the children is not at issue in these parts of the problem.

c. Both parents are carriers (heterozygous), so the probability of having an unaffected child is 3/4. The probability of 4 unaffected children is 3/4 x 3/4 x 3/4 x 3/4 = **81/256**.

d. The probability that at least one child is affected is all outcomes except the one mentioned in part c. Thus, the probability is 1 - 81/256 = **175/256**.

e. The probability of an affected child is 1/4 while the probability of an unaffected child is 3/4. Therefore 1/4 x 1/4 x 3/4 x 3/4 = **9/256**.

f. The probability of 2 affected and 1 unaffected in any one particular birth order is 1/4 x 1/4 x 3/4 = 3/64. There are 3 mutually exclusive birth orders that could produce 2 affecteds and 1 unaffected – unaffected child first born, unaffected child second born, and unaffected child third born. Thus, there is a 3/64 + 3/64 + 3/64 = **9/64** chance that 2 out of 3 children will be affected.

g. The phenotype of the last child is independent of all others, so the probability of an affected child is **1/4**.

2-19. Diagram the cross, where *P* is the normal pigmentation allele and *p* is the albino allele:

      normal x normal → albino

       P?   x   *P?* → *pp*

An albino must be homozygous recessive *pp*. The parents are normal in pigmentation and therefore could be *PP* or *Pp*. Because they have an albino child, **they must both be carriers (*Pp*). The probability that their next child will have the *pp* genotype is 1/4.**

2-20. Diagram the cross:

      yellow round x yellow round → 156 yellow round : 54 yellow wrinkled

The monohybrid ratio for seed shape is 156 round : 54 wrinkled = 3 round : 1 wrinkled. The parents must therefore have been heterozygous (*Rr*) for the pea shape gene. All the offspring are yellow and therefore have the *Yy* or *YY* genotype. The parent plants were **Y- *Rr* x *YY Rr*** (that is, you know at least one of the parents must have been *YY*).

2-21. Diagram the cross:

smooth black ♂ x rough white ♀ → F₁ rough black

→ F₂ 8 smooth white : 25 smooth black : 23 rough white : 69 rough black

a. Since only **one phenotype was seen in the first generation of the cross, we can assume that the parents were true breeding**, and that the F₁ generation consists of heterozygous animals. **The phenotype of the F₁ progeny indicates that rough and black are the dominant phenotypes. Four phenotypes are seen in the F₂ generation so there are two genes controlling the phenotypes in this cross.** Therefore, *R* = rough, *r* = smooth; *B* = black, *b* = white. In the F₂ generation, consider each gene separately. For the coat texture, there were 8 + 25 = 33 smooth : 23 + 69 = 92 round, or a ratio of ~1 smooth : 3 round. For the coat color, there were 8 + 23 = 31 white : 25 + 69 = 94 black, or about ~1 white : 3 black, so the F₂ progeny support the conclusion that the F₁ animals were heterozygous for both genes.

b. An F₁ male is heterozygous for both genes, or *Rr Bb*. The smooth white female must be homozygous recessive; that is, *rr bb*. Thus, *Rr Bb* x *rr bb* → 1/2 *Rr* (rough) : 1/2 *rr* (smooth) and 1/2 *Bb* (black) : 1/2 *bb* (white). The inheritance of these genes is independent, so apply the product rule to find the expected phenotypic ratios among the progeny, or **1/4 rough black : 1/4 rough white : 1/4 smooth black : 1/4 smooth white**.

2-22. Diagram the cross:

YY rr X *yy RR* → all *Yy Rr* → 9/16 *Y- R-* (yellow round) : 3/16 *Y- rr* (yellow wrinkled) : 3/16 *yy R-* (green round) :1/16 *yy rr* (green wrinkled).

Each F₂ pea results from a separate fertilization event. The probability of 7 yellow round F₂ peas is $(9/16)^7 = 4{,}782{,}969/268{,}435{,}456 = $ **0.018**.

2-23.

a. First diagram the cross, and then figure out the monohybrid ratios for each gene:

Aa Tt x *Aa Tt* → 3/4 *A-* (achoo) : 1/4 *aa* (non-achoo) and 3/4 *T-* (trembling) : 1/4 *tt* (non-trembling).

The probability that a child will be *A-* (and have achoo syndrome) is independent of the probability that it will lack a trembling chin, so the probability of a child with achoo syndrome but without trembling chin is 3/4 *A-* x 1/4 *tt* = **3/16**.

b. The probability that a child would have neither dominant trait is 1/4 *aa* x 1/4 *tt* = **1/16**.

2-24. The F$_1$ must be heterozygous for all the genes because the parents were pure-breeding (homozygous). The appearance of the F$_1$ establishes that the dominant phenotypes for the four traits are tall, purple flowers, axial flowers and green pods.

a.   From a heterozygous F$_1$ x F$_1$, both dominant and recessive phenotypes can be seen for each gene. Thus, you expect 2 x 2 x 2 x 2 = 16 different phenotypes when considering the four traits together. The possibilities can be determined using the product rule with the pairs of phenotypes for each gene, because the traits are inherited independently. Thus: [tall : dwarf] x [green : yellow] gives [tall green : tall yellow : dwarf green : dwarf yellow] x [purple : white] gives [tall green purple : tall yellow purple : dwarf green purple : dwarf yellow purple : tall green white : tall yellow white : dwarf green white : dwarf yellow white] x [terminal : axial] which gives **tall green purple terminal : tall yellow purple terminal : dwarf green purple terminal : dwarf yellow purple terminal : tall green white terminal : tall yellow white terminal : dwarf green white terminal : dwarf yellow white terminal : tall green purple axial : tall yellow purple axial : dwarf green purple axial : dwarf yellow purple axial : tall green white axial : tall yellow white axial : dwarf green white axial : dwarf yellow white axial**. The possibilities can also be determined using the branch method shown below, which might in this complicated problem be easier to track.

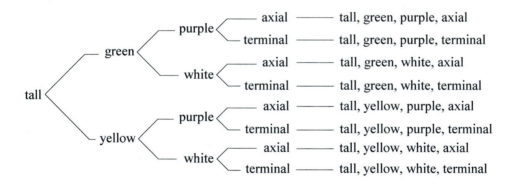

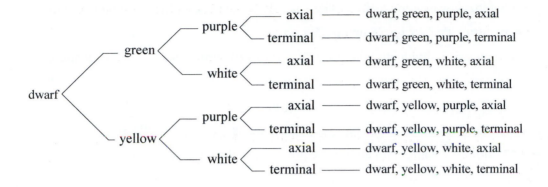

b. Designate the alleles: T = tall, t = dwarf; G = green; g = yellow; P = purple, p = white; A = axial, a = terminal. The cross $Tt\ Gg\ Pp\ Aa$ (an F$_1$ plant) x $tt\ gg\ pp\ AA$ (the dwarf parent) will produce 2 phenotypes for the tall, green and purple genes, but only 1 phenotype (axial) for the fourth gene or 2 x 2 x 2 x 1 = **8 different phenotypes**. The first 3 genes will give a 1/2 dominant : 1/2 recessive ratio of the phenotypes (for example 1/2 T : 1/2 t) as this is in effect a test cross for each gene. Thus, the proportion of each phenotype in the progeny will be 1/2 x 1/2 x 1/2 x 1 = 1/8.

Using either of the methods described in part a, the progeny will be **1/8 tall green purple axial : 1/8 tall yellow purple axial : 1/8 dwarf green purple axial : 1/8 dwarf yellow purple axial : 1/8 tall green white axial : 1/8 tall yellow white axial : 1/8 dwarf green white axial : 1/8 dwarf yellow white axial**.

2-25. Apply the 3EQ to each cross separately. Remember that 4 phenotypic classes in the progeny means there are 2 genes controlling the phenotypes. Determine the phenotypic ratio for each gene separately. A 3:1 monohybrid ratio tells you which phenotype is dominant and that both parents were heterozygous for the trait; in contrast, a 1:1 ratio results from a test cross where the dominant parent was heterozygous.

a. 3EQ#1 - there are 2 genes in this cross (4 phenotypes). 3EQ#2 - one gene controls purple : white with a monohybrid ratio of 94 + 28 = 122 purple : 32 + 11 = 43 white or ~3 purple : 1 white. The second gene controls spiny : smooth with a monohybrid ratio of 94 + 32 =1 26 spiny : 28 + 11 = 39 smooth or ~3 spiny : 1 smooth. Thus, designate the alleles P = **purple**, p = **white**; S = **spiny**, s = **smooth**. Therefore, this is a straightforward dihybrid cross: $Pp\ Ss$ x $Pp\ Ss$ → **9 P- S- : 3 P- ss : 3 pp S- : 1 pp ss**.

b. The 1 spiny : 1 smooth ratio indicates a test cross for the pod shape gene. Because all progeny were purple, at least one parent plant must have been homozygous for the P allele of the flower color gene. The cross was **either $PP\ Ss$ x P- ss or P- Ss x $PP\ ss$.**

c. This is similar to part b, except that here all the progeny were spiny so at least one parent must have been homozygous for the *S* allele. The 1 purple : 1 white test cross ratio indicates that the parents were **either *Pp S-* x *pp SS* or *Pp SS* x *pp S-*.**

d. Looking at each trait individually, there are 89 + 31 = 120 purple : 92 + 27 = 119 white. A 1 purple:1 white monohybrid ratio denotes a test cross. For the other gene, there are 89 + 92 = 181 spiny : 31 + 27 =5 8 smooth, or a 3 spiny : 1 smooth ratio indicating that the parents were both heterozygous for the *S* gene. The genotypes of the parents were ***pp Ss* x *Pp Ss*.**

e. There is a 3 purple : 1 white ratio among the progeny, so the parents were both heterozygous for the *P* gene. All progeny have smooth pods so the parents were both homozygous recessive *ss*. The genotypes of the parents are ***Pp ss* x *Pp ss*.**

f. There is a 3 spiny : 1 smooth ratio, indicative of a cross between heterozygotes (*Ss* x *Ss*). All progeny were white so the parents must have been homozygous recessive *pp*. The genotypes of the parents are ***pp Ss* x *pp Ss*.**

2-26. Three characters (genes) are being analyzed in this cross. While we can usually tell which alleles are dominant from the phenotype of the heterozygote, we are not told the phenotype of the heterozygote (that is, the original pea plant that was selfed). Instead, use the monohybrid phenotypic ratios to determine which allele is dominant and which is recessive for each gene. Consider height first. There are 272 + 92 + 88 + 35 = 487 tall plants and 93 + 31 + 29 + 11 = 164 dwarf plants. This is a ratio of ~3 tall : 1 dwarf, indicating that **tall is dominant**. Next consider pod shape, where there are 272 + 92 + 93 + 31 = 488 inflated pods and 88 + 35 + 29 + 11 = 163 flat pods, or approximately 3 inflated : 1 flat, so **inflated is dominant**. Finally, consider flower color. There were 272 + 88 + 93 + 29 + 11 = 493 purple flowers and 92 + 35 + 31 + 11 = 169 white flowers, or ~3 purple : 1 white. Thus, **purple is dominant**.

2-27. Apply 3EQ#1 and 3EQ#2 to each cross to help you diagram these crosses. Remember that you are told that tiny wings = t, normal wings = T, narrow eye = n and oval (normal) eye = N.

In cross 1 there are 2 phenotypes that differ in the offspring, so there is one gene controlling the oval and narrow eyes (3EQ#1). All of the parents and offspring show the tiny wing phenotype so there is no variability in the gene controlling this trait, and all flies in this cross are *tt*. In considering 3EQ#2 note that the eye phenotypes in the offspring are seen in a ratio of 3 oval : 1 narrow. This phenotypic monohybrid ratio means that both parents are heterozygous for the gene (*Nn*). Thus the genotypes for the parents in **cross 1** are: ***tt Nn* x *tt Nn*.**

In cross 2 consider the wing trait first. The female parent is tiny (*tt*) so this is a test cross for the wings. The offspring show both tiny and normal in a ratio of 82 : 85 or a ratio of 1 tiny : 1 normal.

Therefore the normal male parent must be heterozygous for this gene (*Tt*). For eyes the narrow parent is homozygous recessive (*nn*) so again this is a test cross for this gene. Again both eye phenotypes are seen in the offspring in a ratio of 1 oval : 1 narrow, so the oval female parent is a *Nn* heterozygote. Thus the genotypes for the parents in **cross 2** are: ***Tt nn* x *tt Nn*.**

Consider the wing phenotype in the offspring of cross 3. Both wing phenotypes are seen in a ratio of 64 normal flies : 21 tiny or a 3 normal : 1 tiny. Thus both parents are *Tt* heterozygotes. The male parent is narrow (*nn*), so cross 3 is a test cross for eyes. Both phenotypes are seen in the offspring in a 1 normal : 1 narrow ratio, so the female parent is heterozygous for this gene. The genotypes of the parents in **cross 3** are: ***Tt nn* x *Tt Nn*.**

When examining cross 4 you notice a monohybrid phenotypic ratio of 3 normal : 1 tiny for the wings in the offspring. Thus both parents are heterozygous for this gene (*Tt*). Because the male parent has narrow eyes (*nn*), this cross is a test cross for eyes. All of the progeny have oval eyes, so the female parent must be homozygous dominant for this trait. Thus the genotypes of the parents in **cross 4** are: ***Tt nn* x *Tt NN*.**

2-28.

a. Analyze each gene separately: *Tt* x *Tt* will give 3/4 *T*- (normal wing) offspring. The cross *nn* x *Nn* will give 1/2 *N*- (normal eye) offspring. To calculate the probability of the normal offspring apply the product rule to the normal portions of the monohybrid ratios by cross multiplying these two fractions (and their gene symbols): 3/4 *T*- x 1/2 *N*- = 3/8 *T*- *N*-. Thus **3/8** of the offspring of this cross will have **normal wings and oval eyes**.

b. Diagram the cross:

Tt nn x Tt Nn → ?

Find the phenotypic monohybrid ratio separately for each gene in the offspring. Then cross multiply these monohybrid ratios to find the phenotypic dihybrid ratio. A cross of *Tt* x *Tt* → 3/4 *T*- (normal wings) : 1/4 *tt* (tiny wings). For the eyes the cross is *nn* x *Nn* → 1/2 *N*- (oval) : 1/2 *nn* (narrow). Applying the product rule gives 3/8 *T*- *N*- (normal oval) : 3/8 *T*- *nn* (normal narrow) : 1/8 *tt N*- (tiny oval) : 1/8 *tt nn* (tiny narrow). When you multiply each fraction by 200 progeny you will see **75 normal oval : 75 normal narrow : 25 tiny oval : 25 tiny narrow**.

Section 2.3 – Mendelian Inheritance in Humans

2-29.

a. **Recessive** - two unaffected individuals have an affected child (*aa*). Therefore the parents involved in the consanguineous marriage must both be carriers (*Aa*).

b. **Dominant** - the trait is seen in each generation and every affected person (*A-*) has an affected parent. Note that III-3 is unaffected (*aa*) even though both his parents are - this would not be possible for a recessive trait. The term carrier is not applicable, because everyone with a single *A* allele shows the trait.

c. **Recessive** - two unaffected, carrier parents (*Aa*) have an affected child (***aa***), as in part a.

2-30.

a. Cutis laxa must be a recessive trait because affected child II-4 has normal parents. Because II-4 is affected she must have received an affected allele (*CL*) from both parents. The mother (I-3) is heterozygous (CL^+ *CL*). If this trait was X linked then the father (I-4) would be hemizygous *CL* on his X chromosome and he would be affected. Because he is normal he must be heterozygous (CL^+ *CL*) and the trait is **autosomal recessive**.

b. You are told that this trait is rare, so unrelated people in the pedigree, like I-2, are homozygous normal (CL^+ CL^+). Diagram the cross that gives rise to II-2: *CL CL* (I-1) x CL^+ CL^+ (I-2) → CL^+ *CL*. Thus the **probability that II-2 is a carrier is 100%**.

c. As described in part a both parents in this cross are carriers: CL^+ *CL* x CL^+ *CL*. II-3 is <u>not</u> affected so he can NOT be the *CL CL* genotype. Therefore there is a 1/3 probability that he is the CL^+ CL^+ genotype and a **2/3 probability that he is a carrier** (CL^+ *CL*).

d. As shown in part b II-2 must be a carrier (CL^+ *CL*). In order to have an affected child II-3 must also be a carrier. The probability of this is 2/3 as shown in part c. The probability of two heterozygous parents having an affected child is 1/4. Apply the product rule to these probabilities: 1 probability that II-2 is CL^+ *CL* x 2/3 probability that II-3 is CL^+ *CL* x 14 probability of an affected child = 2/12 = **1/6**.

2-31. Diagram the cross! In humans, this is usually done as a pedigree. Remember that the affected siblings must be *CF CF*.

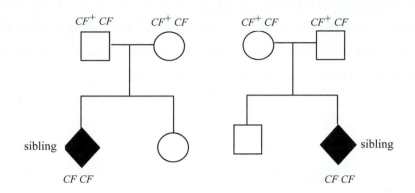

a. The probability that II-2 is a carrier is **2/3**. Both families have an affected sibling, so both sets of parents (that is, all the people in generation I) must have been carriers. Thus, the expected genotypic ratio in the children is 1/4 affected : 1/2 carrier : 1/4 homozygous normal. II-2 is NOT affected, so she cannot be *CF CF*. Of the remaining possible genotypes, 2 are heterozygous. There is therefore a 2/3 chance that she is a carrier.

b. The probability that II-2 x II-3 will have an affected child is 2/3 (the probability that the mother is a carrier as seen in part a) x 2/3 (the probability the father is a carrier using the same reasoning) x 1/4 (the probability that two carriers can produce an affected child) = **1/9**.

c. The probability that both parents are carriers and that their child will be a carrier is 2/3 x 2/3 x 1/2 = 2/9 (using the same reasoning as in part b, except asking that the child be a carrier instead of affected). However, it is also possible for $CF^+ CF^+$ x $CF^+ CF$ parents to have children that are carriers. Remember that there are 2 possible ways for this particular mating to occur – homozygous father x heterozygous mother or vice versa. Thus the probability of this sort of mating is 2 x 1/3 (the probability that a particular parent is $CF^+ CF^+$) x 2/3 (the probability that the other parent is $CF^+ CF$ x 1/2 (the probability such a mating could produce a carrier child = 2/9. The probability that a child could be carrier from either of these two scenarios (where both parents are carriers or where only one parent is a carrier) is the sum of these mutually exclusive events, or 2/9 + 2/9 = **4/9**.

2-32.

a. Because the disease is rare the affected father is most likely to be heterozygous (*Hh*). There is a **1/2** chance that the son inherited the *H* allele from his father and will thus develop the disease.

b.   The probability of an affected child is: 1/2 (the probability that Joe is *Hh*) x 1/2 (the probability that the child inherits the *H* allele if Joe is *Hh*) = **1/4.**

2-33. The trait is **recessive** because pairs of unaffected individuals (I-1 x I-2 as well as II-3 x II-4) had affected children (II-1, III-1, and III-2). There are also two cases in which an unrelated individual must have been a carrier (II-4 and either I-1 or I-2), so the disease allele appears to be **common** in the population.

2-34.

a.   The inheritance pattern seen in **Figure 2.21** could be caused by a **rare dominant mutation**. In this case, the affected individuals would be heterozygous (*Hh*) and the normal individuals would be *hh*. Any mating between an affected individual and an unaffected individual would give 1/2 normal (*hh*) : 1/2 affected (*Hh*) children. However, the same pattern of inheritance could be seen if the disease were caused by a **common recessive mutation**. In the case of a common recessive mutation, all the affected individuals would be *hh*. Because the mutant allele is common in the population, most or even all of the unrelated individuals could be assumed to be carriers (*Hh*). Matings between affected and unaffected individuals would then also yield phenotypic ratios of progeny of 1/2 normal (*Hh*) : 1/2 affected (*hh*).

b.   **Determine the phenotype of the 14 children of III-6 and IV-8.** If the disease is due to a **recessive allele**, then III-6 and IV-8 must be homozygotes for this recessive allele, and **all their children must have the disease.** If the disease is due to a **dominant mutation**, then III-6 and IV-8 must be heterozygotes (because they are affected but they each had one unaffected parent), and **1/4 of their 14 children would be expected to be unaffected.**

Alternatively, you could **look at the progeny of matings between unaffected individuals** in the pedigree such as III-1 and an unaffected spouse. If the disease were due to a **dominant mutation**, these matings would all be homozygous recessive x homozygous recessive and would **never give affected children.** If the disease is due to a **recessive mutation**, then many of these individuals would be carriers, and if the trait is common then at least some of the spouses would also be carriers, so such matings **could give affected children.**

2-35. Diagram the cross by drawing a pedigree.

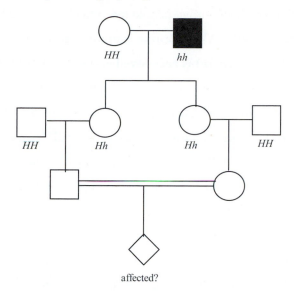

affected?

a. Assuming the disease is very rare, the first generation is *HH* unaffected (I-1) x *hh* affected (I-2). Thus, both of the children (II-2 and II-3) must be carriers (*Hh*). Again assuming this trait is rare in the population, those people marrying into the family (II-1 and II-4) are homozygous normal (*HH*). Therefore, the probability that III-1 is a carrier is 1/2; III-2 has the same chance of being a carrier. Thus the probability that a child produced by these two first cousins would be affected is 1/2 (the probability that III-1 is a carrier) x 1/2 (the probability that III-2 is a carrier) x 1/4 (the probability the child of two carriers would have an *hh* genotype) = **1/16** = 0.0625.

b. If 1/10 people in the population are carriers, then the probability that II-1 and II-4 are *Hh* is 0.1 for each. In this case an affected child in generation IV can only occur if III-1 and III-2 are both carriers. III-1 can be a carrier as the result of 2 different matings: (i) II-1 homozygous normal x II-2 carrier or (ii) II-1 carrier x II-2 carrier. (Note that II-2 must be a carrier because of the normal phenotype and the fact that one parent was affected.) The probability of III-1 being a carrier is thus the probability of mating (i) x the probability of generating a *Hh* child from mating (i) + the probability of mating (ii) x the probability of generating an *Hh* child from mating (ii) = 0.9 (the probability II-1 is *HH*, which is the probability for mating [i]) x 1/2 (the probability that III-1 will inherit *h* in mating [i]) + 0.1 (the probability II-1 is *H*, which is the probability for mating [ii]) x 2/3 (the probability that III-1 will inherit *h* in mating [ii]; remember that III-1 is known not to be *hh*) = 0.45 + 0.067 = 0.517. The chance that III-2 will inherit *h* is exactly the same. Thus, the probability that IV-1 is *hh* = 0.517 (the probability III-1 is *Hh*) x 0.517 (the probability that III-2 is *Hh*) X 1/4 (the probability the child of two carriers will be *hh*) = **0.067**. This number is slightly higher than the answer to part a, which was 0.0625, so the increased likelihood that II-1 or II-4 is a carrier makes it only slightly more likely that IV-1 will be affected.

2-36.

a. Both diseases are known to be rare, so normal people marrying into the pedigree are assumed to be homozygous normal. **Nail-patella (*N*) syndrome is dominant** because all affected children have an affected parent. **Alkaptonuria (*a*) is recessive** because the affected children are the result of a consanguineous mating between 2 unaffected individuals (III-3 x III-4). Genotypes: **I-1 *nn Aa*; I-2 *Nn AA* (or I-1 *nn AA* and I-2 *Nn Aa*); II-1 *nn AA*; II-2 *nn Aa*; II-3 *Nn A-*; II-4 *nn A-*; II-5 *Nn Aa*; II-6 *nn AA*; III-1 *nn AA*; III-2 *nn A-*; III-3 *nn Aa*; III-4 *Nn Aa*; III-5 *nn A-*; III-6 *nn A-*; IV-1 *nn A-*; IV-2 *nn A-*; IV-3 *Nn A-*; IV-4 *nn A-*; IV-5 *Nn aa*; IV-6 *nn aa*; IV-7 *nn A-*.**

b. The cross is *nn A-* (IV-2) x *Nn aa* (IV-5). The ambiguity in the genotype of IV-2 is due to the uncertainty of her father's genotype (III-2). His parents' genotypes are *nn AA* (II-1) x *nn Aa* (II-2) so there is a 1/2 chance III-2 is *nn AA* and a 1/2 chance he is *nn Aa*. Thus, for each of the phenotypes below you must consider both possible genotypes for IV-2. For each part below, calculate the probability of the child inheriting the correct gametes from IV-2 x the probability of obtaining the correct gametes from IV-3 to give the desired phenotype. If both the possible IV-2 genotypes can produce the needed gametes, you will need to sum the two probabilities.

   - for the child to have both syndromes (*N- aa*), IV-2 would have to contribute an *n a* gamete. This could only occur if IV-2 were *nn Aa*. The probability IV-2 is *nn Aa* is 1/2, and the probability of receiving an *n a* gamete from IV-2 if he is *nn Aa* is also 1/2. The probability that IV-5 would supply an *N a* gamete is also 1/2. Thus, 1/2 x 1/2 x 1/2 = 1/8. There is no need to sum probabilities in this case because IV-2 cannot produce an *n a* gamete if his genotype is *nn AA*.

   - for the child to have only nail-patella syndrome (*N- A-*), IV-2 would have to provide an *n A* gamete and IV-5 an *N a* gamete. This could occur if IV-2 were *nn Aa*; the probability is 1/2 (the probability IV-2 is *Aa*) x 1/2 (the probability of an *A* gamete if IV-2 is *Aa*) x 1/2 (the probability of an *N a* gamete from IV-5] = 1/8. This could also occur if IV-2 were *nn AA*. Here, the probability is 1/2 (the probability IV-2 is *nn AA*) x 1 (the probability of an *n A* gamete if IV-2 is *nn AA*) x 1/2 (the probability of an *N a* gamete from IV-5] = 1/4. Summing the probabilities for the two mutually exclusive IV-2 genotypes, 1/8 + 1/4 = **3/8**.

   - for the child to have just alkaptonuria (*nn aa*), IV-2 would have to contribute an *n a* gamete. This could only occur if IV-2 were *nn Aa*. The probability IV-2 is *nn Aa* is 1/2, and the probability of receiving an *n a* gamete from IV-2 if he is *nn Aa* is also 1/2. The probability that IV-5 would supply an *n a* gamete is also 1/2. Thus, 1/2 x 1/2 x 1/2 = **1/8**. There is no need to sum probabilities in this case because IV-2 cannot produce an *n a* gamete if his genotype is *nn AA*.

- the probability of neither defect is 1 – (sum of the first 3) = 1 - (1/8 + 3/8 + 1/8) = 1 - 5/8 = **3/8**. You can make this calculation because there are only the four possible outcomes and you have already calculated the probabilities of three of them.

2-37. Diagram the cross(es):

midphalangeal x midphalangeal → 1853 midphalangeal : 209 normal

$M?$ x $M?$ → $M?$: mm

The following crosses are possible:

MM x MM → all MM

Mm x MM → all $M-$

MM x Mm → all $M-$

Mm x Mm → 3/4 $M-$: 1/4 mm

The 209 normal children must have arisen from the last cross, so approximately 3 x 209 = 630 children should be their $M-$ siblings. Thus, about 840 of the children or **~40% came from the last mating** and the other 60% of the children were the result of one or more of the other matings. This problem illustrates that much care in interpretation is required when the results of many matings in mixed populations are reported.

Chapter 3   Extensions to Mendel: Complexities in Relating Genotype to Phenotype

Synopsis:

This chapter builds on the principles of segregation and independent assortment that you learned in Chapter 2. An understanding of those basic principles will help you understand the more complex inheritance patterns in Chapter 3. While the basic principles for the inheritance of alleles of one or more genes hold true, the expression of the corresponding phenotypes is more complicated.

Chapter 3 describes several examples of single gene inheritance in which phenotypic monohybrid ratios are different from the complete dominance examples in Chapter 2 (see Significant Elements on next page and **Table 3.1**). These variant phenotypic monohybrid ratios may be caused by:

♦ incomplete dominance;

♦ codominance;

♦ dominance series of multiple alleles

♦ lethal alleles;

♦ pleiotropy.

Also introduced in this chapter are examples in which **two or more interacting genes** determine the phenotype. Remember from Chapter 2 that crosses involving two independently assorting genes have a 9 *A-B-* : 3 *A- bb* : 3 *aa B-* : 1 *aa bb* dihybrid F_2 phenotypic ratio. If two genes interact then the dihybrid ratio is a modification of the 9:3:3:1 ratio. This modification involves the same genotypes for the progeny. Because of the effect of an allele of one gene on the other gene (**epistasis**), the four classes can be added together in different combinations. Multigene inheritance (>2 genes involved) leads to even more phenotypic classes. With more genes controlling a trait you see a continuous range of phenotypes instead of discrete traits, as in **Figure 3.21**.

Penetrance and **expressivity** are terms used to describe some cases of variable phenotypic expression in different individuals. Penetrance describes the fraction of individuals with a mutant genotype who are affected while expressivity describes the extent to which individuals with a mutant genotype are affected.

Significant Elements:

After reading the chapter and thinking about the concepts you should be able to:

♦ Understand that novel phenotypes arise when there is codominance or incomplete dominance. The novel phenotype will appear in the F_1 generation. In the F_2 generation, this same phenotype <u>must</u> be the largest component of the 1:2:1 monohybrid ratio.

♦ Realize that if you see a series of crosses involving different phenotypes for a certain trait, for example coat color, and each individual cross gives a monohybrid ratio, then all the phenotypes are controlled by one gene with many alleles. In other words, the problem involves an **allelic series** as in **<u>Figure 3.6</u>**. It is important to write a dominance hierarchy for the alleles of the gene, e.g. $a = b > c$. Thus, a is codominant or incompletely dominant to b and both a and b are completely dominant to c.

♦ Understand that **lethal mutations** are almost always recessive alleles, as shown in **<u>Figure 3.9</u>**. If there is a recessive lethal allele present in a cross you can never make that allele homozygous. Therefore the cross must have involved parents heterozygous for the lethal allele. Instead of the expected 1:2:1 ratio in the progeny, one of the 1/4 classes is lethal, so the monohybrid phenotypic ration will be 2/3 heterozygous phenotype : 1/3 other (viable) homozygous phenotype caused by homozygosity for the other allele. The recessive lethal allele may be pleiotropic and show a different, dominant phenotype, as in **<u>Figure 3.9</u>**.

♦ Remember that **epistasis involves two genes**. In epistasis, <u>none of the progeny die</u>. All are present, but instead of four phenotypic classes in a 9:3:3:1 phenotypic dihybrid ratio, you will see an epistatic variation where 2 or 3 of the phenotypes have been summed together, for example 9:3:4 or 9:7 or 12:3:1.

Problem Solving Tips:

♦ Solve enough problems so you can distinguish single and two gene traits on the basis of inheritance patterns. Look for the number of classes in the F_2 generation to identify single gene inheritance (3:1, 1:2:1, 1:1 or 2:1) versus 2 gene inheritance (9:3:3:1 or an epistatic variation). In the Study Guide 'monohybrid ratio' is used in a more general sense than in the text to refer to any ratio that is based on the segregation of the alleles of a single gene (3:1, 1:2:1, 1:1 or 2:1). Likewise, in the Study Guide dihybrid ratio refers to any ratio based on the segregation of the alleles of 2 genes (9:3:3:1, 1:3:4, 15:1, etc).

♦ Be able to derive the monohybrid phenotypic ratios for incomplete dominance/codominance and lethal alleles involving inheritance of a single gene.

♦ It is critical that you understand the 9:3:3:1 phenotypic dihybrid ratio involves the 4 classes 9/16 *A- B-* : 3/16 *A- bb* : 3/16 *aa B-* : 1/16 *aa bb*, where a dashed line (-) indicates either a dominant or recessive allele.

♦ If the phenotype involves 2 genes, be able to propose ways in which two genes interact based on offspring ratios. Do not merely memorize the altered ratios (**Table 3.2**); instead think through what the combinations of alleles mean.

♦ Remember the product rule of probability and use it to determine proportions of genotypes or phenotypes for independently assorting genes.

Problem Solving - How to Begin:

THREE ESSENTIAL QUESTIONS (3EQ):

#1. How many genes are involved in the cross?

#2. For **each gene** involved in the cross: what are the phenotypes associated with the gene? Which phenotype is the dominant one and why? Which phenotype is the recessive one and why?

[#3. For **each** gene involved in the cross: is it X-linked or autosomal?]

At this point, only questions #1 and #2 may be applied. The material that is the basis of question #3 will be covered in Chapter 4.

Hints:

BE CONSISTENT. Always diagram the crosses and write out the genotypes. Set the problems up the same way. Note the repetitive approach to many of the problems in this chapter. Make sure you always distinguish between genotypes and phenotypes when working the problems.

To answer **3EQ**#1, look for the number of phenotypic classes in the F_2 progeny. Two phenotypes usually means 1 gene, 4 phenotypes MUST be due to 2 genes. Three phenotypic classes is an ambiguous result - this could result from 1 gene with codominance or incomplete dominance, or from 2 genes with epistasis. Use the ratio of phenotypes to distinguish between these possibilities. One gene with codominance/incomplete dominance MUST give 1:2:1 while 2 genes with 3 phenotypes will be an epistatic variation of a 9:3:3:1.

For **3EQ**#2 when one gene is involved, look at the phenotype of the F_1 individuals. If the phenotype of the F_1 progeny is like one of the parents, then that phenotype is the dominant one. Also, examine the F_2 progeny – the 3/4 portion of the 3:1 phenotypic monohybrid ratio is the dominant one. If the phenotype of the F1 progeny is unlike either parent, then it may be the alleles of the gene are codominant or incompletely dominant. In this case, the novel F_1 phenotype will be seen again in the largest class of F_2 progeny.

After you answer 3EQ#1 and #2 to the best of your ability, use the answers to assign genotypes to the parents of the cross. Then follow the cross through, figuring out the <u>expected phenotypes and genotypes</u> in the F$_1$ and F$_2$ generations. Remember to assign the expected phenotypes in a manner consistent with those initially assigned to the parents. Next, compare your predicted results to the observed data you were given. If the 2 sets of information match, then your initial genotypes were correct! In many cases there may be two possible set of genotypes for the parents. If your predicted results do not match the data given, try the other set of genotypes for the parents. See <u>problem 3-24</u> for an illustration of this issue.

Solutions to Problems:

<u>Vocabulary</u>

3-1. a. **2**; b. **6**; c. **11**; d. **8**; e. **7**; f. **9**; g. **12**; h. **3**; i. **5**; j. **4**; k. **1**; l. **10**.

<u>Section 3.1 – Single Gene Extensions to Mendel</u>

3-2. The problem states that the intermediate pink phenotype is caused by incomplete dominance for the alleles of a single gene. We suggest that you employ genotype symbols that can show the lack of complete dominance; the obvious *R* for red and *r* for white does not reflect the complexity of this situation. In such cases we recommend using a base letter as the gene symbol and then employing superscripts to show the different alleles. To avoid any possible misinterpretations, it is always advantageous to include a separate statement making the complexities of the dominant/recessive complications clear. Designate the two alleles f^r = red and f^w = white, so the possible genotypes are $f^r f^r$ = red; $f^r f^w$ = pink; and $f^w f^w$ = white. Note that the phenotypic ratio is the same as the genotypic ratio in incomplete dominance.

a. Diagram the cross: $f^r f^w$ x $f^r f^w$ → **1/4 $f^r f^r$ (red) : 1/2 $f^r f^w$ (pink) : 1/4 $f^w f^w$ (white)**.

b. $f^w f^w$ x $f^r f^w$ → **1/2 $f^r f^w$ (pink) : 1/2 $f^w f^w$ (white)**.

c. $f^r f^r$ x $f^r f^r$ → **1 $f^r f^r$ (red)**.

d. $f^r f^r$ x $f^r f^w$ → **1/2 $f^r f^r$ (red) : 1/2 $f^r f^w$ (pink)**.

e. $f^w f^w$ x $f^w f^w$ → **1 $f^w f^w$ (white)**.

f. $f^r f^r$ x $f^w f^w$ → **1 $f^r f^w$ (pink)**.

The cross shown in **part f** is the most efficient way to produce pink flowers, because all the progeny will be pink.

3-3. Diagram the cross: yellow x yellow $\rightarrow$ 38 yellow : 22 red : 20 white

Three phenotypes in the progeny show that the yellow parents are not true breeding. The ratio of the progeny is close to 1/2 : 1/4 : 1/4. This is the result expected for crosses between individuals heterozygous for incompletely dominant genes. Thus:

$c^r c^w$ x $c^r c^w$ $\rightarrow$ **1/2 $c^r c^w$ (yellow) : 1/4 $c^r c^r$ (red) : 1/4 $c^w c^w$ (white).**

3-4.

a. Diagram the cross: $e^+ e^+$ x $e^+ e$ $\rightarrow$ 1/2 $e^+ e^+$: 1/2 $e^+ e$. The trident marking is only found in the heterozygotes, so the probability is **1/2**.

b. The offspring with the trident marking are $e^+ e$, so the cross is $e^+ e$ x $e^+ e$ $\rightarrow$ 1/4 ee : 1/2 $e^+ e$: 1/4 $e^+ e^+$. Therefore, of 300 offspring, **75 should have ebony bodies, 150 should have the trident marking and 75 should have honey-colored bodies.**

3-5. The cross is: white long x purple short $\rightarrow$ 301 long purple : 99 short purple : 612 long pink : 195 short pink : 295 long white : 98 short white

Deconstruct this dihybrid phenotypic ratio for two genes into separate constituent monohybrid ratios for each of the 2 traits, flower color and pod length. For flower color note that there are 3 phenotypes: 301 + 99 purple : 612 +195 pink : 295 + 98 white = 400 purple : 807 pink : 393 white = 1/4 purple : 1/2 pink : 1/4 white. This is a typical monohybrid ratio for an incompletely dominant gene, so **flower color is caused by an incompletely dominant gene with c^p giving purple when homozygous, c^w giving white when homozygous, and the $c^p c^w$ heterozygotes giving pink.** For pod length, the phenotypic ratio is 301 + 612 + 295 long : 99 + 195 + 98 short = 1208 long : 392 short = 3/4 long : 1/4 short. This 3:1 ratio is that expected for a cross between individuals heterozygous for a gene in which one allele is completely dominant to the other, so **pod shape is controlled by 1 gene with long (L) completely dominant to short (l).**

3-6. A cross between individuals heterozygous for an incompletely dominant gene give a ratio of 1/4 (one homozygote) : 1/2 (heterozygote with the same phenotype as the parents) : 1/4 (other homozygote). Because the problem already states which genotypes correspond to which phenotypes, you know that the color gene will give a monohybrid phenotypic ratio of 1/4 red : 1/2 purple : 1/4 white, while the shape gene will give a monohybrid phenotypic ratio of 1/4 long : 1/2 oval : 1/4 round. Because the inheritance of these two genes is independent, use the product rule to generate all the possible phenotype combinations (note that there will be 3 x 3 = 9 classes) and their probabilities, thus generating the dihybrid phenotypic ratio for two incompletely dominant genes: **1/16 red long : 1/8 red oval : 1/16 red round : 1/8 purple long : 1/4 purple oval : 1/8 purple round : 1/16 white**

long : 1/8 white oval : 1/16 white round. As an example, to determine the probability of red long progeny, multiply 1/4 (probability of red) x 1/4 (probability of long) = 1/16. If you have trouble keeping track of the 9 possible classes, it may be helpful to list the classes in the form of a branch diagram.

Phenotype	Probability of phenotype
red, long	$1/4 \times 1/4 = 1/16$
red, oval	$1/4 \times 1/2 = 1/8$
red, round	$1/4 \times 1/4 = 1/16$
purple, long	$1/2 \times 1/4 = 1/8$
purple, oval	$1/2 \times 1/2 = 1/4$
purple, round	$1/2 \times 1/4 = 1/8$
white, long	$1/4 \times 1/4 = 1/16$
white, oval	$1/4 \times 1/2 = 1/8$
white, round	$1/4 \times 1/4 = 1/16$

3-7.

a. The most likely mode of inheritance is a **single gene with incomplete dominance** such that $f^{n}f^{n}$ = normal (<250 mg/dl), $f^{n}f^{a}$ = intermediate levels of serum cholesterol (250-500 mg/dl) and $f^{a}f^{a}$ homozygotes = elevated levels (>500 mg/dl). Some of the individuals in the pedigrees do not fit this hypothesis. In t 2 of the families, two normal parents have a child with intermediate levels of serum cholesterol: **Family 2 –I-2 $f^{n}f^{n}$ x I-3 $f^{n}f^{n}$ $\rightarrow$ 3 $f^{a}f^{n}$ children; and Family 4 - I-1 $f^{n}f^{n}$ x I-2 $f^{n}f^{n}$ $\rightarrow$ 2 $f^{n}f^{a}$ children.**

b. **Factors other than just the genotype are involved in the expression of the phenotype. Such factors could include diet, level of exercise, and other genes.**

3-8.

a. A person with sickle-cell anemia is a homozygote for the sickle-cell allele: $Hb^{S}Hb^{S}$.

b. The child must be homozygous $Hb^{S}Hb^{S}$ and therefore must have inherited a mutant allele from each parent. Because the parent is phenotypically normal, he/she must be a **carrier with genotype $Hb^{S}Hb^{A}$**.

c. Each individual has two alleles of every gene, including the β-globin gene. If an individual is heterozygous, he/she has two different alleles. Thus, if each parent is heterozygous for different alleles, there are **four possible alleles** that could be found in the five children. This is the maximum number of different alleles possible (barring the very rare occurrence of a new, novel mutation in a gamete that gave rise to one of the children). If one or both of the parents were homozygous for any one allele, the number of alleles distributed to the children would of course be less than four.

3-9. Remember that the gene determining ABO blood groups has 3 alleles and that $I^A = I^B > i$.

a. The O phenotype means the girl's genotype is ii. Each parent contributed an i allele, so her parents could be **ii (O) or I^Ai (A) or I^Bi (B)**.

b. A person with the B phenotype could have either genotype I^BI^B or genotype I^Bi. The mother is A and thus could not have contributed an I^B allele to this daughter. Instead, because the daughter clearly does not have an I^A allele, the mother must have contributed the i allele to this daughter. The mother must have been an I^Ai heterozygote. The father must have contributed the I^B allele to his daughter, so he could be **either I^BI^B, I^Bi, or I^BI^A**.

c. The genotypes of the girl and her mother must both be I^AI^B. The father must contribute either the I^A or the I^B allele, so there is only one phenotype and genotype which would exclude a man as her father - **the O phenotype (genotype ii)**.

3-10. To approach this problem, look at the mother/child combinations to determine what alleles the father must have contributed to each child's genotype.

a. The father had to contribute I^B, N, and Rh^- alleles to the child. The only male fitting these requirements is **male d** whose phenotype is B, MN, and Rh^+ (note that the father must be Rh^+Rh^- because the daughter is Rh^-).

b. The father had to contribute i, N, and Rh^- alleles. The father could be either male c (O MN Rh^+) or male d (B MN Rh^+). As we saw previously, male c is the only male fitting the requirements for the father in part a. Assuming one child per male as instructed by the problem, the father in part b must be **male c**.

c. The father had to contribute I^A, M, and Rh^- alleles. Only **male b** (A M Rh^+) fits these criteria.

d. The father had to contribute either I^B or i, M, and Rh^-. Three males have the alleles required: these are male a, male c, and male d. However, of these three possibilities, only **male a** remains unassigned to a mother/child pair.

3-11. Designate the alleles: p^m (marbled) $> p^s$ (spotted) $= p^d$ (dotted) $> p^c$ (clear).

a. Diagram the crosses:

    1. p^mp^m (homozygous marbled) x p^sp^s (spotted) $\rightarrow$ p^mp^s (marbled F_1)

    2. p^dp^d x p^cp^c $\rightarrow$ p^dp^c (dotted F_1)

    3. p^mp^s x p^dp^c $\rightarrow$ 1/4 p^mp^d (marbled) : 1/4 p^mp^c (marbled) : 1/4 p^sp^d (spotted dotted) :

       1/4 p^sp^c (spotted) = **1/4 spotted dotted : 1/2 marbled : 1/4 spotted**.

b.   The F_1 from cross 1 are **marbled ($p^m p^s$)** from the first cross **and dotted ($p^d p^c$)** from the second cross as shown in part a.

3-12. Designate the gene p (for pattern; this is a different p gene than that in the previous <u>problem 3-11</u> because the two problems involve different plant species). There are 7 alleles, p^1-p^7, with p^7 being the allele that codes for absence of pattern and $p^1 > p^2 > p^3 > p^4 > p^5 > p^6 > p^7$.

a.   There are **7 different patterns** possible. These are associated with the following genotypes: p^1-, $p^2 p^a$ (where $p^a = p^2$, p^3, $p^4 ... p^7$), $p^3 p^b$ (where $p^b = p^3$, p^4, $p^5 ... p^7$), $p^4 p^c$ (where $p^c = p^4$, p^5, p^6, and p^7), $p^5 p^d$ (where $p^d = p^5$, p^6, and p^7), $p^6 p^e$ (where $p^e = p^6$ and p^7), and $p^7 p^7$.

b.   The phenotype dictated by the allele p^1 has the greatest number of genotypes associated with it = **7** ($p^1 p^1$ $p^1 p^2$ $p^1 p^3$, etc.). **The absence of pattern** is caused by just one genotype, $p^7 p^7$.

c.   This finding suggests that **the allele determining absence of pattern (p^7) is very common** in these clover plants with the $p^7 p^7$ genotype is the most frequent in the population. The other alleles are present, but are much less common in this population.

3.13.

a.   First analyze these crosses for the answers to the 3EQ (see Problem Solving Hints above). All of the crosses have results that can be explained by one gene - either a 3:1 phenotypic monohybrid ratio showing that one allele is completely dominant to the other, or a 1:1 ratio showing that a test cross was done for a single gene, or all progeny with the same phenotype as the parents. You can thus conclude that all of the coat colors are controlled by the alleles of **one gene, with chinchilla (C) > himalaya (c^h) > albino (c^a)**.

b.   1. $c^h c^a$ x $c^h c^a$
   2. $c^h c^a$ x $c^a c^a$
   3. $C c^h$ x $C (c^h$ or $c^a)$
   4. CC x c^h $(c^h$ or $c^a)$
   5. $C c^a$ x $C c^a$
   6. $c^h c^h$ x $c^a c^a$
   7. $C c^a$ x $c^a c^a$
   8. $c^a c^a$ x $c^a c^a$
   9. $C c^h$ x $c^h (c^h$ or $c^a)$ or $C c^a$ x $c^h c^h$
   10. $C c^a$ x $c^h c^a$.

c. $C c^h$ **(from cross 9) x $C c^a$ (from cross 10)** $\rightarrow$ 1/4 CC (chinchilla) : 1/4 $C c^a$ (chinchilla) : 1/4 $C c^h$ (chinchilla) : 1/4 $c^h c^a$ (himalaya) = **3/4 chinchilla : 1/4 himalaya**, or $C c^a$ **(cross 9) x $C c^a$ (cross 10)** $\rightarrow$ **3/4 C- chinchilla : 1/4 $c^a c^a$ albino.**

3.14. You know from the incompatibility system that each plant must be heterozygous for two different alleles of the incompatibility gene. A '-' in the chart indicates no seeds were produced in a particular mating, which means the parents have an incompatibility allele in common. A '+' in the chart means the parents do not have <u>any</u> alleles in common so that they can produce seeds. Thus, looking at the results in the table, plants 2, 3, and 5 must have one (or both) of the alleles from plant 1 since they are incompatible with plant 1. Arbitrarily designate two different alleles for plant 1 as i^1 and i^2. Plants 2, 3 and 5 must have either allele i^1 or i^2 or both. Plant 4 has neither allele present in plant 1. Let's give plant 4 alleles i^3 and i^4. Furthermore, plant 2 does not share alleles with plants 3, 4 or 5 so it must have the allele found in plant 1 that plant 5 does not contain. Designate i^1 as the allele shared between plants 1 and 2 and designate i^2 as the allele shared between plants 1 and 3 and 5. Plant 3 shares an allele with plant 5, i^2, but it does not share an allele with plant 4. Call the second allele in plant 3 i^5. Plants 4 and 5 share an allele, say i^3. We still don't know a second allele for plant 2, but it is not any of those carried by plants 3, 4, or 5, so it must be another allele i^6. Thus, **there are six alleles total**.

plant	genotype
1	$i^1 i^2$
2	$i^1 i^6$
3	$i^2 i^5$
4	$i^3 i^4$
5	$i^2 i^3$

3-15.

a.   This ratio is approximately **2/3 Curly : 1/3 normal**.

b.   The expected result for this cross is: $Cy^+Cy \times Cy^+Cy \rightarrow$ 1/4 $CyCy$ (?): 1/2 Cy^+Cy (Curly) : 1/4 Cy^+Cy^+ (normal). If the **Cy/Cy** genotype **is lethal** then the expected ratio will match the observed data.

c.   The cross is $Cy^+Cy \times Cy^+Cy^+ \rightarrow$ 1/2 Cy^+Cy : 1/2 Cy^+Cy^+, so there would be approximately **90 Curly winged and 90 normal winged flies**.

3-16.

a.   There are actually **two different phenotypes** mentioned in this problem. One phenotype is the shape of the erythrocytes. All people with the genotype Sph^-Sph^- have **spherical erythrocytes**. Therefore this phenotype **is fully penetrant and shows no variation in expression**. The second **phenotype is anemia. Here the expressivity among anemic patients varies from severe to mild**. There are even **some people with the Sph^-Sph^- genotype (150/2400) with no symptoms**

of anemia at all. Thus the penetrance of the anemic phenotype is 2,240/2,400 or 0.94.

b.   The disease causing phenotype is the anemia and the severity of the anemia is greatly reduced when the spleen functions poorly and does not "read" the spherical erythrocytes. Therefore **treatment might involve removing the spleen (an organ which is not essential to survival). The more efficiently the spleen functions the earlier in a patient's life it should be removed**.

Sph^-Sph^- individuals with no symptoms of anemia should not be subjected to this drastic treatment.

3-17.

a.   The 2/3 montezuma : 1/3 wild type phenotypic ratio, and the statement that montezumas are never true-breeding, together suggest that there is a **recessive lethal allele** of this gene. When there is a recessive lethal, crossing two heterozygotes results in a 1:2:1 genotypic ratio, but one of the 1/4 classes of homozygotes do not survive. **The result is the 2:1 phenotypic ratio as seen in this cross. Both the montezuma parents were therefore heterozygous, Mm.** The *M* allele must confer the montezuma coloring in a dominant fashion, but homozygosity for *M* is lethal.

b.   Designate the alleles: *M* = montezuma, *m* = greenish; *F* = normal fin, *f* = ruffled. Diagram the cross: ***MmFF* x *mmff* →** expected monohybrid ratio for the *M* gene alone: 1/2 *Mm* (montezuma) : 1/2 *mm* (wild type); expected monohybrid ratio for the *F* gene alone: all *Ff*. The expected dihybrid ratio = **1/2 *Mm Ff* (montezuma) : 1/2 *mm Ff* (greenish, normal fin)**.

c.   ***MmFf* x *Mm Ff* →** expected monohybrid ratio for the *M* gene alone: 2/3 montezuma (*Mm*) : 1/3 greenish (*mm*); expected monohybrid ratio for the *F* gene alone: 3/4 normal fin (*F*-) : 1/4 ruffled (*ff*). The expectations when considering both genes together is: **6/12 montezuma normal fin : 2/12 montezuma ruffled fin : 3/12 green normal fin : 1/12 green ruffled fin.**

3-18.

a.   The pattern in both families **looks like a recessive trait** since unaffected individuals have affected progeny and the trait skips generations. For example, in the Smiths II-3 must be a carrier, but in order for III-5 to be affected II-4 must also be a carrier. **If the trait is rare (as is this one) you wouldn't expect two heterozygotes to marry by chance as many times as required by these pedigrees. The alternative explanation is that the trait is dominant but not 100% penetrant.**

b.   Assuming this is a dominant but not completely penetrant trait, **individuals II-3 and III-6 in the Smiths' pedigree individual and II-6 in the Jeffersons' pedigree** must carry the dominant allele but not express it in their phenotypes.

c.   If the trait were common, **recessive inheritance** is the more likely mode of inheritance.

d. **None**; in cases where two unaffected parents have an affected child, both parents would be carriers of the recessive trait.

3-19. If polycystic kidney disease is dominant, then the child is *Pp* and inherited the *P* disease allele from one parent or the other, yet phenotypically the parents are *pp*. Perhaps one of the parents is indeed *Pp,* but this parent does not show the disease phenotype for some reason. As we will see in the next chapter, such situations are not uncommon: the unexpressed dominant allele is said to have **incomplete penetrance** in these cases. Alternatively, it could be that both parents are indeed *pp* and the *P* allele inherited by the child was due to a **spontaneous mutation** during the formation of the gamete in one of the parents; again, we will discuss this topic in the next chapter. It is also possible that the **father of the child is not the male parent of the couple**. In this case the biological father must have the disease.

Section 3.2 – Multifactorial Extensions to Mendel

3-20. The cross is: walnut x single → F_1 walnut x F_1 walnut → 93 walnut : 29 rose : 32 pea : 11 single

a. 3EQ#1 - four F_2 phenotypes means there are 2 genes, *A* and *B*. Both genes affect the same structure, the comb. The F_2 phenotypic dihybrid ratio among the progeny is close to 9:3:3:1, so there is no epistasis. Because walnut is the most abundant F_2 phenotype, it must be the phenotype due to the *A- B-* genotype. Single combs are the least frequent class, and are thus *aa bb*. Now assign genotypes to the cross. If the walnut F_2 are *A- B-*, then the original walnut parent must have been *AA BB*:

> *AA BB* **x** *aa bb* → *Aa Bb* **(walnut)** → **9/16** *A- B-* **(walnut) : 3/16** *A- bb* **(rose) : 3/16** *aa B-* **(pea) : 1/6** *aa bb* **(single).**

b. Diagram the cross, recalling that the problem states the parents are homozygous:

> *AA bb* **(rose) x** *aa BB* **(pea)** → *Aa Bb* **(walnut)** → **9/16** *A- B-* **(walnut) : 3/16** *A- bb* **(rose) : 3/16** *aa B-* **(pea) : 1/6** *aa bb* **(single).** This F_2 is in identical proportions as the F_2 generation in part a.

c. Diagram the cross: *A- B-* (walnut) x *aa B-* (pea) → 12 *A- B-* (walnut) : 11 *aa B-* (pea) : 3 *A- bb* (rose) : 4 *aa bb* (single). Because there are pea and single progeny, you know that the walnut parent must be *Aa*. The 1 *A-* : 1 *aa* monohybrid ratio in the progeny also tells you the walnut parent must have been *Aa*. Because some of the progeny are single, you know that both parents must be *Bb*. In this case, the monohybrid ratio for the *B* gene is 3 *B-* : 1 *bb*, so both parents were *Bb*. The original cross must have been *Aa Bb* x *aa Bb*. You can verify that this cross would yield the observed ratio of progeny by multiplying the probabilities expected for each gene alone. For example, you anticipate that 1/2 the progeny would be *Aa* and 3/4 of the progeny would be *Bb,* so 1/2 x 3/4 = 3/8 of the progeny should be walnut; this is close to the 12 walnut chickens seen among 30 total progeny.

d.   Diagram the cross: *A- B-* (walnut) x *A- bb* (rose) → all *A- B-* (walnut). The progeny are all walnut, so **the walnut parent must be BB**. No pea progeny are seen, so **both parents cannot be Aa, so one of the two parents must be AA.** This could be either the walnut or the rose parent or both.

3-21. black x chestnut → bay → black : bay : chestnut : liver

Four phenotypes in the F$_2$ generation means there are two genes determining coat color. The F$_1$ bay animals produce four phenotypic classes, so they must be doubly heterozygous, *Aa Bb*. Crossing a liver colored horse to either of the original parents resulted in the parent's phenotype. The liver horse's alleles do not affect the phenotype, suggesting the recessive genotype *aa bb*. Though it is probable that the original black mare was *AA bb* and the chestnut stallion was *aa BB,* each of these animals only produced 3 progeny, so it cannot be definitively concluded that these animals were homozygous for the dominant allele they carry. Thus, **the black mare was A- bb, the chestnut stallion was aa B-, and the F$_1$ bay animals are Aa Bb. The F2 horses were: bay (A- B-), liver (aa bb), chestnut (aa B-), and black (A- bb).**

3-22.

a.   Because unaffected individuals had affected children, the **trait is recessive**. From affected individual II-1, you know the mutant allele is present in this generation. The trait was passed on through II-2 who was a carrier. All children of affected individuals III-2 x III-3 are affected, as predicted for a recessive trait. However, generation V seems inconsistent with recessive inheritance of a single gene. This result is consistent with two different genes involved in hearing with a defect in either gene leading to deafness. **The two family lines shown contain mutations in two separate genes, and the mutant alleles of both genes determining deafness are recessive.**

b.   Individuals in generation V are doubly heterozygous (***Aa Bb***), having inherited a dominant and recessive allele of each gene from their parents (*aa BB* x *AA bb*). The people in generation V are not affected because **the product of the dominant allele of each gene is sufficient for normal function.** This is an example of the complementation of two genes.

3-23. green x yellow → green → 9 green : 7 yellow

a.   Two phenotypes in the F$_2$ generation could be due to one gene or to two genes with epistasis. If this is one gene, then *GG* x *gg* → *Gg* → 3/4 *G-* (green) : 1/4 *gg* (yellow). The actual result is a

9:7 ratio, not a 3:1 ratio. 9:7 is an epistatic variant of the 9:3:3:1 phenotypic dihybrid ratio, so there are **2 genes** controlling color. The genotypes are:

AA BB (green) x *aa bb* (yellow) $\rightarrow$ *Aa Bb* (green) $\rightarrow$ 9/16 *A- B-* (green) : 3/16 *A- bb* (yellow) : 3/16 *aa B-* (yellow) : 1/16 *aa bb* (yellow).

b.   **F$_1$** *Aa Bb* x *aa bb* $\rightarrow$ 1/4 *Aa Bb* (green) : 1/4 *aa Bb* (yellow) : 1/4 *Aa bb* (yellow) : 1/4 *aabb* (yellow) = 1/4 green : 3/4 yellow.

3-24.

a.   white x white $\rightarrow$ F$_1$ white $\rightarrow$ 126 white : 33 purple

3EQ#1 - At first glance this cross seems to involve only one gene, as true breeding white parents give white F$_1$s. However, if this were true, then the F$_2$ MUST be totally white as well! The purple F$_2$ plants show that this cross is NOT controlled by 1 gene. These results must be **due to 2 genes**. What ratio is 126 : 33? Usually when you are given raw numbers of individuals for the classes, you divide through by the smallest number, yielding in this case 3.8 white : 1 purple. This is neither a recognizable monohybrid nor dihybrid ratio. Dividing through by the smallest class is NOT the correct way to convert raw numbers to a ratio. In crosses controlled by 2 genes there must be 16 genotypes in the F$_2$ progeny, even though the phenotypes may not be distributed in the usual 9/16 : 3/16 : 3/16 : 1/16 ratio. If the 159 F$_2$ progeny are divided equally into 16 genotypes, then there are 159/16 = ~10 F$_2$ plants/genotype. The 126 white F$_2$s therefore represent 126/10 = 13 of these genotypes. Likewise the 33 purple plants represent 33/10 = 3 genotypes. The correct F$_2$ dihybrid phenotypic ratio is thus 13 white : 3 purple.

You can now assign genotypes to the parents in the cross. Because the parents are homozygous (true-breeding) and there are 2 genes controlling the phenotypes, there are two possible ways to set up the genotypes of the parents. One option is: *AA BB* (white) x *aa bb* (white) $\rightarrow$ *Aa Bb* (white, same as *AA BB* parent) $\rightarrow$ 9 *A- B-* (white) : 3 *A- bb* (unknown phenotype) : 3 *aa B-* (unknown phenotype) : 1 *aa bb* (white, same as *aa bb* parent). If you assume that *A- bb* is white and *aa B-* is purple (or vice versa), then this is a match for the observed data presented in the cross above (9 + 3 + 1 = 13 white : 3 purple).

Alternatively, you could try to diagram the cross as *AA bb* x *aa BB* $\rightarrow$ *Aa Bb* (whose phenotype is unknown as this is NOT a genotype seen in the parents) $\rightarrow$ 9 *A-B-* (same unknown phenotype as in the F$_1$) : 3 *A- bb* (white like the *AA bb* parent) : 3 *aa B-* (white like the *aa BB* parent) : 1 *aa bb* (unknown phenotype). Such a cross cannot give an F$_2$ phenotypic ratio of 13 white : 3 purple. The only F$_2$ classes that could be purple are *A- B-*, but this is impossible because it is too large and because the F$_1$ flies must then have been purple, or the *aa bb* class which is too small. Therefore, the first set of possible genotypes (written in bold above) is the best fit for the observed data.

b. white F$_2$ x self white F$_2$ → 3/4 white : 1/4 purple. Assume that the *aa B-* class is purple in part a above. A 3:1 monohybrid ratio means the parents are both heterozygous for one gene with purple due to the recessive allele. The second gene is not affecting the ratio, so both parents must be homozygous for the same allele of that gene. Thus the cross must be: ***Aa BB*** (white) x *Aa BB* (white self cross) → 3/4 *A- BB* (white) : 1/4 *aa BB* (purple).

c. purple F$_2$ x self → 3 purple : 1 white. Again, the selfed parent must be heterozygous for one gene and homozygous for the other gene. Because purple is *aa B-*, the genotypes of the purple F$_2$ plant must be ***aa Bb***.

d. white F$_2$ x white F$_2$ (not a self cross) → 1/2 purple : 1/2 white. The 1:1 monohybrid ratio means a test cross was done for one of the genes. The second gene is not altering the ratio in the progeny, so the parents must be homozygous for that gene. If purple is *aa B-*, then the genotypes of the parents must be ***aa bb* (white)** x ***Aa BB* (white)** → 1/2 *Aa Bb* (white) : 1/2 *aa Bb* (purple).

3-25. Dominance relationships are between <u>alleles of the same gene</u>. **Only one gene is involved when considering dominance relationships. Epistasis involves two genes.** The alleles of one gene affect the phenotypic expression of the second gene.

3-26.

a. *Aa Bb* x *Aa Bb* → 9 *A- B-* : 3 *A- bb* :3 *aa B-* : 1 *aa bb*

   Since the defect in enzyme is only seen if <u>both</u> genes are defective, only *aa bb* will result in abnormal progeny, giving a phenotypic dihybrid ratio of **15 normal : 1 abnormal**.

b. *Aa Bb Cc* x *Aa Bb Cc* → the dihybrid ratio for *A* and *B* is 9 *A- B-* : 3 *A- bb* :3 *aa B-* : 1 *aa bb*; while the monohybrid ratio for *C* is 3 *C-* : 1 *cc*. Use the product rule to generate the expected phenotypic trihybrid ratio. Remember that the <u>only</u> abnormal genotype will be *aa bb cc*, which will occur with a probability of 1/16 x 1/4 = 1/64. The expected phenotypic trihybrid ratio is thus **63/64 normal : 1/64 abnormal**.

3-27. $I^A I^B$ *Ss* x $I^A I^A$ *Ss* → expected monohybrid ratio for the *I* gene of 1/2 $I^A I^A$: 1/2 $I^A I^B$; expected ratio for the *S* gene considered alone of 3/4 *S-* : 1/4 *ss*. Use the product rule to generate the phenotypic ratio for both genes considered together and then assign phenotypes, remembering that all individuals with the *ss* genotype look like type O. The phenotypic ratio for both genes is: 3/8 $I^A I^A$ *S-* : 3/8 $I^A I^B$ *S-* : 1/8 $I^A I^A$ *ss* : 1/8 $I^A I^B$ *ss* = 3/8 A : 3/8 AB : 1/8 O : 1/8 O = **3/8 Type A : 3/8 Type AB : 2/8 Type O**.

3-28. The difference between pleiotropic mutations and traits determined by several genes would be seen if crosses were done using pure-breeding plants (wild type x mutant), then selfing the F_1 progeny. If **several genes** were involved there **would be several different combinations of the petal color, markings and stem position phenotypes** in the F_2 generation. If all 3 traits were due to an allele present at **one gene**, the **three phenotypes would always be inherited together** and the F_2 plants would be either yellow, dark brown, erect OR white, no markings and prostrate.

3-29.

a. blood types: **I-1 AB; I-2 A; I-3 B; I-4 AB; II-1 O; II-2 O; II-3 AB; III-1 A; III-2 O.**

b. genotypes: **I-1 $Hh\ I^AI^B$; I-2 $Hh\ I^Ai$ (or I^AI^A); I-3 $H\text{-}\ I^BI^B$ (or I^Bi); I-4 $H\text{-}\ I^AI^B$; II-1 $H\text{-}\ ii$; II-2 $hh\ I^AI^A$ (or I^Ai or I^AI^B); II-3 $Hh\ I^AI^B$; III-1 $Hh\ I^Ai$; III-2 $hh\ I^AI^A$ (or I^AI^B or I^Ai or I^bi or I^BI^B)**

At first glance, you find inconsistencies between expectations and what could be inherited from a parent. For example, I-1 (AB) $\times$ I-2 (A) could not have an O child (II-2). The epistatic *h* allele (which causes the Bombay phenotype) could explain these inconsistencies. If II-2 has an O phenotype because she is *hh*, her parents must both have been *Hh*. The Bombay phenotype would also explain the second seeming inconsistency of two O individuals (II-1 and II-2) having an A child. II-2 could have received an *A* allele from one of her parents and passed this on to III-1 together with one *h* allele. Parent II-1 would have to contribute the *H* allele so that the *A* allele would be expressed; the presence of *H* means that II-1 must also be *ii* in order to be type O. A third inconsistency is that individuals II-2 and II-3 could not have an *ii* child since II-3 has the I^AI^B genotype, but III-2 has the O phenotype. This could also be explained if II-3 is *Hh* and III-2 is *hh*.

3-30.

a. Diagram one of the crosses:

white-1 x white-2 $\rightarrow$ red F_1 $\rightarrow$ 9 red : 7 white

Even though there are only 2 phenotypes in the F_2, this is not controlled by one gene - the 9:7 ratio shows that this is an epistatic variation of 9:3:3:1, so there are 2 genes controlling these phenotypes. Individuals must have at least one dominant allele of both genes in order to get the red color. Thus the genotypes of the two white parents in this cross are *aa BB* x *AA bb*. The same conclusions hold for the other 2 crosses. If white-1 is mutant in gene *A* and white-2 is mutant in gene *B*, then white-3 must be mutant in gene *C*. Therefore, **three genes are involved**.

b. **White-1 is *aa BB CC*; white-2 is *AA bb CC* and white-3 is *AA BB cc*.**

c. ***aa BB CC* (white-1) x *AA bb CC* (white-2) $\rightarrow$ *Aa Bb CC* (red) $\rightarrow$ 9/16 *A- B- CC* (red) : 3/16 *A- bb CC* (white) : 3/16 *aa B- CC* (white) : 1/6 *aa bb CC* (white).**

3-31. Diagram the cross. Figure out an expected monohybrid ratio for each gene separately, then apply the product rule to generate the expected dihybrid ratio. Also recall that the albino phenotype is epistatic to all other coat colors.

A^yA Cc x A^yA cc → monohybrid ratio for the A gene alone: 1/4 A^yA^y (dead) : 1/2 A^yA (yellow) : 1/4 AA (agouti) = 2/3 A^yA (yellow) : 1/3 AA (agouti); monohybrid ratio for the C gene: 1/2 Cc (non-albino) : 1/2 cc (albino).

Overall there will be 2/6 A^yA Cc (yellow) : 2/6 A^yA cc (albino) : 1/6 AA Cc (agouti) : 1/6 AA cc (albino) = **2/6 A^yA Cc (yellow) : 3/6 -- cc (albino) : 1/6 AA Cc (agouti)**.

3-32.

a. **No,** a single gene cannot account for this result. While the 1:1 ratio seems like a testcross, the fact that **the phenotype of one class of offspring (linear) is not the same as either of the parents** argues against this being a testcross.

b. The appearance of four phenotypes means **two genes** are controlling the phenotypes.

c. The 3:1 ratio suggests that **two alleles of one gene** determine the difference between the wild-type and scattered patterns.

d. The true-breeding wild-type fish are homozygous by definition, and the scattered fish have to be homozygous recessive according to the ratio seen in part c, so the cross is: ***bb* (scattered) x *BB* (wild type)** → **F$_1$ *Bb* (wild type)** → **3/4 *B-* (wild type) : 1/4 *bb* (scattered)**.

e. The inability to obtain a true-breeding nude stock suggests that the nude fish are heterozygous (Aa) and that the **AA genotype dies**. Thus ***Aa* (nude) x *Aa* (nude)** → **2/3 *Aa* (nude) : 1/3 *aa* (scattered)**.

f. Going back to the linear cross from part b, the fact that there are four phenotypes led us to propose two genes were involved. The 6:3:2:1 ratio looks like an altered 9:3:3:1 ratio in which some genotypes may be missing, as predicted from the result in part e that AA animals do not survive. The 9:3:3:1 ratio results from crossing double heterozygotes, so **the linear parents are doubly heterozygous *Aa Bb*. The lethal phenotype associated with the AA genotype produces the 6:3:2:1 ratio. The phenotypes and corresponding genotypes of the progeny of the linear x linear cross are: 6 linear, *Aa B-* : 3 wild-type, *aa B-* : 2 nude, *Aa bb* : 1 scattered, *aa bb*.** Note that the *AA BB, AA Bb, AA Bb,* and *AA bb* genotypes are missing due to lethality.

3-33.

a. Based upon the Comprehensive Example in the textbook at the end of Chapter 3, we can deduce some information about the genotypes of the parents from their phenotypes. The rest we have to deduce based on the phenotypes of the progeny. The yellow parent must have an A^y allele, but we don't know the second allele of the A gene (A^y-). A^y is epistatic to the B gene so we don't know what alleles this yellow mouse has at the B gene (we'll leave these alleles as *??*). Since this mouse does show color we know it is not *cc* (albino), so it must have at least one *C* allele (*C-*). The brown agouti parent has at least one A allele (*A -*); it must be *bb* at the B gene; and since there is color it must also be *C-*. The mating between these two can be represented as *A^y- ?? C-* x *A- bb C-*. Now consider the progeny. Because one pup was albino (*cc*), the parents must both be *Cc*. A brown pup (*bb*) indicates that both parents had to be able to contribute a *b* allele, so we now know the first mouse must have had at least one *b* allele. The fact that this brown pup was non-agouti means both parents carried an *a* allele. The black agouti progeny tells us that the first mouse must have also had a B allele (but it was yellow because *A^y* is epistatic to B). The complete genotypes of the mice are therefore: **A^{y}a Bb Cc x Aa bb Cc.**

b.   Think about each gene individually, then the effect of the other genes in combination with that phenotype. *C-* leads to a phenotype with color; *cc* gives albino (which is epistatic to all colors determined by the other genes). The possible genotypes of the progeny of this cross for the A gene are *A^{y}A, A^{y}a, Aa* and *aa*, giving yellow, yellow, agouti and non-agouti phenotypes, respectively. Since yellow is epistatic to B, non-albino mice with *A^y* will be yellow regardless of the genotype of the B gene. *Aa* is agouti; with the *aa* genotype there is no yellow on the hair (non-agouti). The type of coloration depends on the B gene. For B the offspring could be *Bb* (black) or *bb* (brown). In total, **six different coat color phenotypes are possible: albino (-- -- cc), yellow (A^y(A or a) -- C), brown agouti (A- bb C-), black agouti (A- B- C-), brown (aa bb C-), black (A- B- C-).**

3-34. In **Figure 3.22** the A^0 and B^0 alleles are non-functional. The A^1 and B^1 alleles each have the same effect on the phenotype (plant height in this example). Thus, the shortest plants are $A^0A^0B^0B^0$, and the tallest plants are $A^1A^1B^1B^1$. **The phenotypes are determined by the total number or A^1 and B^1 alleles in the genotype.** Thus, $A^1A^0B^0B^0$ plants are the same phenotype as $A^0A^0B^0B^1$. In total there will be 5 different phenotypes: 4 '0' alleles (total = 0); 1 '1' allele + 3 '0' alleles (total = 1); 2 '1' alleles + 2 '0' alleles (total = 2); 3 '1' alleles + 1 '0' allele (total = 3); and 4 '1' alleles (total = 4).

In **Figure 3.17** the *a* allele = *b* allele = no function (in this case no color = white). If the A allele has the same level of function as a B allele then you would see 5 phenotypes as was the case for **Figure 3.22b**. But since there are a total of 9 phenotypes, this cannot be true so $A \neq B$. Notice that *aa Bb* is lighter than *Aa bb* even though both genotypes have the same number of dominant alleles. Thus **an A allele has more effect on coloration than a B allele.** If you assume, for example, that $B = 1$ unit of color and $A = 1.5$ unit of color, then 16 genotypes lead to 9 phenotypes.

3-35.

a. *Aa Bb Cc* x *Aa Bb Cc* → 9/16 *A- B-* x 3/4 *C-* : 9/16 *A- B-* x 1/4 *cc* :3/16 *A- bb* x 3/4 *C-* : 3/16 *A- bb* x 1/4 *cc*: 3/16 *aa B-* x 3/4 *C-* :3/16 *aa B-* x 1/4 *cc* : 1/16 *aa bb* x 3/4 *C-* : 1/16 *aa bb* x 1/4 *cc* = 27/64 *A- B- C-* (wild type) :9/64 *A- B- cc* : 9/64 *A- bb C-* : 3/64 *A- bb cc* : 9/64 *aa B-C-* : 3/64 *aa B-cc* : 3/64 *aa bb C-* : 1/64 *aa bb cc* = **27/64 wildtype : 37/64 mutant**.

b. Diagram the crosses:

   1. unknown male x *AA bb cc* → 1/4 wild type (*A- B- C-*) : 3/4 mutant

   2. unknown male x *aa BB cc* → 1/2 wild type (*A- B- C-*) : 1/2 mutant

   3. unknown male x *aa bb CC* → 1/2 wild type (*A- B- C-*) : 1/2 mutant

   The 1:1 ratio in test crosses 2 and 3 is expected if the unknown male is heterozygous for one of the genes that are recessive in the test cross parent. The 1 wild type : 3 mutant ratio arises when the male is heterozygous for two of the genes that are homozygous recessive in the test cross parent. (If you apply the product rule to 1/2 *B-* : 1/2 *bb* and 1/2 *C-* : 1/2 *cc* in the first cross, then you find 1/4 *B- C-*, 1/4 *B- cc*, 1/4 *bb C-*, and 1/4 *bb cc*. Only *B- C-* will be wild type, the other 3 classes will be mutant.) Thus the unknown male must be *Bb Cc*. In test cross 1 the male could be either *AA* or *aa*. Crosses 2 and 3 show that the male is only heterozygous for one of the recessive genes in each case - gene *C* in test cross 2 and gene *B* in test cross 3. In order to get wild type progeny in both crosses, the male must be *AA*. Therefore the genotype of the unknown male is **AA Bb Cc**.

3-36.

a. Answer 3EQ#1 and #2 for all 5 crosses. Cross 1 - 1 gene, red>blue. Cross 2 - 1 gene, lavender>blue. Cross 3 - 1 gene, codominance/incomplete dominance (1:2:1), bronze is the phenotype of the heterozygote. Cross 4 - 2 genes with epistasis (9 red : 4 yellow : 3 blue). Cross 5 - 2 genes with epistasis (9 lavender : 4 yellow : 3 blue). **In total there are 2 genes. One gene controls blue (c^b), red (c^r) and lavender (c^l) where $c^r = c^l > c^b$. The second gene controls the yellow phenotype: *Y* seems to be colorless (or has no effect on color), so the phenotype is determined by the alleles of the *c* gene. The *y* allele makes the flower yellow, and is epistatic to the *c* gene.**

b. **cross 1 – $c^r c^r$ *YY* (red) x $c^b c^b$ *YY* (blue) → $c^r c^b$ *YY* (red) → 3/4 c^r- *YY* (red) : 1/4 $c^b c^b$ *YY* (blue)**

   cross 2 – $c^l c^l$ *YY* (lavender) x $c^b c^b$ *YY* (blue) → $c^l c^b$ *YY* (lavender) → 3/4 c^l- *YY* (lavender) : 1/4 $c^b c^b$ *YY* (blue)

   cross 3 – $c^l c^l$ *YY* (lavender) x $c^r c^r$ *YY* (red) → $c^l c^r$ *YY* (bronze) → 1/4 $c^l c^l$ *YY* (lavender) : 1/2 $c^l c^r$ *YY* (bronze) : 1/4 $c^r c^r$ *YY* (red)

cross 4 – $c^r c^r YY$ x $c^b c^b\ yy$ (yellow) → $c^r c^b\ Yy$ (red) → 9/16 c^r- Y- (red) : 3/16 c^r- yy (yellow) : 3/16 $c^b c^b$ Y- (blue): 1/16 $c^b c^b\ yy$ (yellow)

cross 5- $c^l c^l\ yy$ (yellow) x $c^b c^b\ YY$ (blue) → $c^l c^b\ Yy$ (lavender) → 9/16 c^l- Y- (lavender) : 3/16 c^l- yy (yellow) : 3/16 $c^b c^b$ Y- (blue) : 1/16 $c^b c^b\ yy$ (yellow)

c. $c^r c^r\ yy$ (yellow) x $c^l c^l\ YY$ (lavender) → $c^r c^l\ Yy$ (bronze) → monohybrid ratio for the *c* gene is 1/4 $c^r c^r$: 1/2 $c^r c^l$: 1/4 $c^l c^l$ and monohybrid ratio for the *Y* gene is 3/4 Y- : 1/4y. Using the product rule, these generate a dihybrid ratio of **3/16 $c^r c^r$ Y- (red) : 3/8 $c^r c^l$ Y- (bronze) : 3/16 $c^l c^l$ Y- (lavender) : 1/16 $c^r c^r\ yy$ (yellow) : 1/8 $c^r c^l\ yy$ (novel genotype) : 1/16 $c^l c^l\ yy$ (yellow).** You expect the $c^r c^l\ yy$ genotype to be yellow as *y* is normally epistatic to the *c* gene. However, you have no direct evidence from the data in any of these crosses that this will be the case, so it is possible that this genotype could cause a different and perhaps completely new phenotype.

3-37.

a. Analyze each cross by answering 3EQ#1 and #2. In cross 1 – there are 2 genes because there are 3 classes in the F2 showing an epistatic 12:1:3 ratio, and LR is the doubly homozygous recessive class. In cross 2 – only 1 gene is involved because there are 2 phenotypes in a 3:1 ratio; WR>DR. In cross 3 – again, there is only 1 gene (2 phenotypes in a 1:3 ratio); DR>LR. In cross 4 - 1 gene (2 phenotypes, with a 3:1 ratio); WR>LR. In cross 5 - 2 genes (as in cross 1, there is a 12:1:3 ratio of three classes); LR is the double homozygous recessive. In total, **there are 2 genes controlling these phenotypes in foxgloves**.

b. Remember that all four starting strains are true-breeding. In cross 1 the parents can be assigned the following genotypes: **AA BB (WR-1)** x **aa bb (LR)** → *Aa Bb* (WR) → 9 *A- B-* (WR) : 3 *A- bb* (WR; this class displays the epistatic interaction) : 3 *aa B-* (DR) :1 *aa bb* (LR). The results of cross 2 suggested that DR differs from WR-1 by one gene, so **DR is *aa BB***; cross 3 confirms these genotypes for DR and LR. Cross 4 introduces WR-2, which differs from LR by one gene and differs from DR by 2 genes, so **WR-2 is *AA bb***. Cross 5 would then be *AA bb* (WR-2) x *aa BB* (DR) → *Aa Bb* (WR) → 9 *A- B-* (WR) : 3 *A- bb* (WR) : 3 *aa B-* (DR) : 1 *aa bb* (LR) = 12 WR : 3 DR : 1 LR.

c. WR from the F$_2$ of cross 1 x LR → 253 WR : 124 DR : 123 LR. Remember from part b that LR is *aa bb* and DR is *aa B-* while WR can be either *A- B-* or *A- bb = A- ??*. The experiment is essentially a test cross for the WR parent. The observed monohybrid ratio for the *A* gene is 1/2 *Aa* : 1/2 *aa* (253 *Aa* : 124 + 123 *aa*), so the WR parent must be *Aa*. The DR and LR classes of progeny show that the WR parent is also heterozygous for the *B* gene (DR is *Bb* and LR is *bb* in these progeny). Thus, the cross is *Aa Bb* (WR) x *aa bb* (LR).

3-38. The hairy x hairy → 2/3 hairy : 1/3 normal cross tells us that the hairy flies are heterozygous, that the hairy phenotype is dominant to normal, and that the homozygous hairy progeny are lethal (that is, hairy is a recessive lethal). Thus, hairy is *Hh*, normal is *hh,* and the lethal genotype is *HH*. Normal flies therefore should be ***hh* (normal-1)** and a cross with hairy (*Hh*) would be expected to always give 1/2 *Hh* (hairy) : 1/2 *hh* (normal) as seen in cross 1. In cross 2, the progeny MUST for the same reasons be 1/2 *Hh* : 1/2 *hh*, yet they ALL appear normal. This suggests the normal-2 stock has another mutation that suppresses the hairy wing phenotype in the *Hh* progeny. The hairy parent must have the recessive alleles of this suppressor gene (*ss*), while the normal-2 stock must be homozygous for the dominant allele (*SS*) that suppresses the hairy phenotype. Thus cross 2 is ***hh SS* (normal-2)** x *Hh ss* (hairy) → 1/2 *Hh Ss* (normal because hairy is suppressed) : 1/2 *hh Ss* (normal). In cross 3, the normal-3 parent is heterozygous for the suppressor gene: ***hh Ss* (normal-3)** x *Hh ss* (hairy) → the expected ratios for each gene alone are 1/2 *Hh* : 1/2 *hh* and 1/2 *Ss* : 1/2 *ss*, so the expected ratio for the two genes together is 1/4 *Hh Ss* (normal) : 1/4 *Hh ss* (hairy) : 1/4 *hh Ss* (normal) : 1/4 *hh ss* (normal) = 3/4 normal : 1/4 hairy. In cross 4 you see a 2/3 : 1/3 ratio again, as if you were crossing hairy x hairy. After a bit of trial-and-error examining the remaining possibilities for these two genes, you will be able to demonstrate that this cross was ***Hh Ss* (normal-4)** x *Hh ss* (hairy) → expected ratio for the individual genes are 2/3 *Hh* : 1/3 *hh* and 1/2 *Ss* : 1/2 *ss*, so the expected ratio for the two genes together from the product rule is 2/6 *Hh Ss* (normal) : 2/6 *Hh ss* (hairy) : 1/6 *hh Ss* (normal) : 1/6 *hh ss* (normal) = 2/3 normal : 1/3 hairy.

3-39. Diagram the cross:

D1d1 D2d2 d3d3 x *d1d1 D2d2 D3d3* → calculate the expected monohybrid ratios for each gene (here we consider only the phenotype for each of the three genes for simplicity rather than the corresponding genotypes): 1/2 *D1* : 1/2 *d1*; 3/4 *D2* : 1/4 *d2*; 1/2 *D3* : 1/2 *d3*. Use the product rule to determine the expected dihybrid ratio for D1 and D2 = 3/8 *D1 D2* : 1/8 *D1 d2* : 3/8 *d1 D2* : 1/8 *d1 d2*. Then multiply in the third gene to obtain the expected trihybrid ratio considering all three genes simultaneously = 3/16 *D1 D2 D3* (normal) : 3/16 *D1 D2 d3* (deaf 1 gene) : 1/16 *D1 d2 D3* (deaf 1 gene) : 1/16 *D1 d2 d3* (deaf 2 genes) : 3/16 *d1 D2 D3* (deaf 1 gene) : 3/16 *d1 D2 d3* (deaf 2 genes) : 1/16 *d1 d2 D3* (deaf 2 genes) : 1/16 *d1 d2 d3* (deaf 3 genes). The totals are 3/16 normal : 7/16 deaf due to 1 gene : 5/16 deaf due to 2 genes : 1/16 deaf due to 3 genes. Now apply the product rule to account for the incompletely penetrant lethality for those mutant at 2 or 3 of the genes. For the double mutant individuals, 1/4 die and 3/4 are alive and deaf. Thus 5/16 double mutant x 3/4 alive and deaf = 15/64 live deaf children with double mutations. For the triple mutant individuals 3/4 die and 1/4 are alive and deaf. Thus 1/16 triple mutant x 1/4 alive and deaf = 1/64 live, deaf triple mutants. In total, there are 3/16 normal + 7/16 single mutant deaf + 15/64 live double mutant deaf + 1/64 live triple mutant deaf children = 12/64 normal + 28/64 single mutant deaf + 15/64 double mutant deaf + 1/64 triple mutant deaf = 56/64 live children and 8/64 dead fetuses. Of the live-born children, **44/56 would be deaf. This means that there is a 78.6% chance that any live-born child of these two parents would be deaf.**

Chapter 4   The Chromosome Theory of Inheritance

Synopsis:

Chapter 4 is extremely critical for understanding basic genetics because it connects Mendel's Laws with chromosome behavior during meiosis. While you may have learned mitosis and meiosis in your basic biology class, now is the time to make sure you understand these processes in the context of inheritance. The physical basis for inheritance is chromosome segregation during meiosis. You should have an increased understanding of the importance of meiosis for genetic diversity through both independent assortment and recombination.

Genes are located on chromosomes and travel with them during cell division and gamete formation.

In the first division of meiosis, homologous chromosomes in germ cells segregate from each other, so each gamete receives one member of each matched pair, as predicted by Mendel's first law and as seen in **Figure 4.13**. Also, during the first meiotic division the independent alignment of each pair of homologous chromosomes results in the independent assortment of genes carried on different chromosomes, as predicted by Mendel's second law. The second meiotic division generates gametes with a haploid number of chromosomes (n). Fertilization of an egg and a sperm restores the diploid number of chromosomes (2n) to the zygote.

The experiments that showed the correlation between chromosome behavior and inheritance using X-linked genes in *Drosophila* are described in this chapter. X-linked traits have characteristic inheritance patterns recognized in pedigrees or in results of reciprocal crosses (**Table 4.5**).

Significant Elements:

After reading the chapter and thinking about the concepts you should be able to:

♦ Understand homologs and alleles in meiosis as seen in **Figure 4.17**. Think of a good analogy. For example, think of the road or street that you live on as a copy of a chromosome. Nearby there is another copy of the same street (homologous chromosome). The 2 copies of the street are very similar, but not identical. For instance, any building (gene) found on one copy will be found in the same position on the other copy (alleles). The 2 copies of your residence are identical on both homologous streets (your gene is homozygous). Your next-door neighbor's residence has minor differences between the 2 copies - the front door is green on one and yellow on the other (heterozygous). What will happen to these 2 streets during meiosis? During mitosis?

♦ Draw chromosome alignments during metaphase of mitosis, meiosis I and meiosis II.

- Describe how chromosome behavior explains the laws of segregation and independent assortment.

- Identify sex-linked inheritance patterns - <u>see 3EQ#3 below</u>. Determine genotypes in sex-linked pedigrees and probabilities of specific genotypes and phenotypes.

- If you truly understand meiosis, you can explain how the results seen in the genotype of a child enable you to figure out if non-disjunction occurred in meiosis I or meiosis II and the parent in which it occurred. Your explanation will include sister and non-sister chromatids.

- Understand the differences between sex determination in humans and *Drosophila*, see **Table 4.1**.

Problem Solving Tips:

- Keep clear the distinction between sister chromatids (identical, replicated copies of a chromosome) and homologs (chromosomes carrying the same genes but different alleles).

- Compare and contrast mitosis and meiosis as in **Table 4.3**.

- Two features that lead you to consider X-linked inheritance are criss-cross inheritance (inheritance of a characteristic from mother to son and father to daughter) and a sex-dependent phenotypic difference in the progeny of a cross (see <u>Problem 4-33</u>).

Problem Solving - How to Begin:

THREE ESSENTIAL QUESTIONS (3EQ):

1. How many genes are involved in the cross?

2. For **each gene** involved in the cross: what are the phenotypes associated with the gene? Which phenotype is the dominant one and why? Which phenotype is the recessive one and why?

3. For **each** gene involved in the cross: is it X-linked or autosomal?

   From this point on, all 3 questions are valid.

   Hints:

   For 3EQ#1. look for the number of phenotypic classes in the F_2 progeny.

   For 3EQ#2. if the parents of a cross are true-breeding, look at the phenotype of the F_1

individuals. Also, look at the <u>monohybrid</u> ratios for each gene in the F_2 progeny. If there is a 3:1 ratio then the phenotype associated with the 3/4 portion is the dominant one.

   For 3EQ#3 the determination of whether or not a gene is X-linked is more subtle. In general X-linkage is seen as a clear phenotypic difference between the sexes in one generation's progeny of a cross. This is NOT a difference in the absolute numbers of males and females of a certain phenotype, but instead a phenotype that is present in one sex and totally absent in the other sex. This difference between the sexes will be seen in <u>either</u> the F_1 generation <u>or</u> the F_2 generation but **not** in both generations in the same cross. It is not possible to make a definitive conclusion about X-linkage based on just one generation of a cross - you **must** see the data from **both** the F_1 and F_2 progeny. If the sex difference is seen in the F_1 generation then the female parent had the X-linked phenotype. If the sex difference is seen in the F_2 generation then the male parent had the X-linked trait. X-linked genes usually show a 1:1 monohybrid ratio.

After you answer questions 3EQ#1, #2 and #3 to the best of your ability, use the answers to diagram the cross, i.e. assign genotypes to the parents of the cross. Then follow the cross through, figuring out the expected genotypes and phenotypes in the F_1 and F_2 generations. Remember to assign the expected phenotypes based on those initially assigned to the parents. Next, compare your predicted results to the observed data you were given. If the 2 sets of information match then your initial genotypes were correct! In many cases there may be two possible set of genotypes for the parents. If your predicted results do not match the data given, try the other possible set of genotypes for the parents.

Solutions to Problems:

Vocabulary

4-1. a. **13**; b. **7**; c. **11**; d. **10**; e. **12**; f. **8**; g. **9**; h. **1**; i. **6**; j. **15**; k. **3**; l. **2**; m. **16**; n. **4**; o. **14**; p. **5**.

Section 4.1 – Chromosomes: The Carriers of Genes

4-2. A diploid number of 46 means there are 23 homologous pairs of chromosomes.

a.   A child receives **23 chromosomes from the father**.

b.   Each somatic cell has **44 autosome**s (22 pairs) and **2 sex chromosomes** (1 pair).

c.   **A human ovum (female gamete) contains 23 chromosomes** - one of each homologous pair (22 autosomes and one X chromosome).

d.   **One sex chromosome (an X chromosome) is present in a human ovum.**

4-3.

a.   There are **7 centromeres**.

b.   There are **7 chromosomes** in the diagram. Note that by definition the number chromosomes is equal to the number of centromeres (page 92 bottom of right column).

c.   There are **14 chromatids**.

d.   There are **3 pairs** of homologous chromosomes and one chromosome without a partner.

e. There are **4 metacentric** chromosomes and **3 acrocentric** chromosomes.

f. There would be **4 colors** visible. Each of the 3 pairs would be a different color because they have different chromosomes, but chromosome painting cannot distinguish between homologous chromosomes. The unpaired chromosome would also have a different color; note that it has a different size and shape from any other chromosome.

g. This is an organism in which males are XO, so it is most likely that **females would be XX**. The karyotype of females would be the same as the one shown, except that there would be 2 copies of the left-most chromosome instead of one copy.

Section 4.2 – Mitosis: Cell Division That Preserves Chromosome Number

4-4. Mitosis produces 2 daughter cells each with 14 chromosomes (2n, diploid). Mitosis maintains the chromosome number.

4-5. a. **iii**; b. **i**; c. **iv**; d. **ii**; e. v.

4-6.

a. **G_1, S, G_2 and M (Figure 4.7).**

b. **G_1, S and G_2** are all part of interphase.

c. **G_1 is the time of major cell growth that precedes DNA synthesis and chromosome replication. Chromosome replication occurs during S phase. G2 is another phase of cell growth after chromosome replication during which the cell synthesizes many proteins needed for mitosis.**

4-7.

	G_1	S	G_2	Prophase	Metaphase	Anaphase	Telophase
a.	1	1 → 2	2	2	2	1	1
b.	yes	yes	yes	yes → no	no	no	no → yes
c.	no	no	no	no → yes	yes	yes	yes → no
d.	yes	yes	yes	yes → no	no	no	no → yes

4-8. No, mitosis can (and does) occur in haploid cells. This is because each chromosome aligns independently at the metaphase plate; there is no requirement for homologous chromosome pairing.

Section 4.3 – Meiosis: Cell Divisions That Halve Chromosome Number

4-9. Meiosis produces 4 cells (n, haploid), each with 7 chromosomes (one half the number of chromosomes of the starting cell).

4-10. In order to answer this question count the number of independent centromeres. If two sister chromatids are connected by one centromere this counts as one chromosome. **Meiosis I** is the only division that **reduces the chromosome number by half**, so it is a reductional division. A cell undergoing meiosis II or mitosis has the same number of centromeres (and thus chromosomes) as the each daughter cell. Therefore the **meiosis II and mitotic divisions are both equational**.

4-11.

a. **mitosis, meiosis I, meiosis II**

b. **mitosis, meiosis I**

c. **mitosis**

d. Mitosis is obviously excluded, but the rest of the answer depends on whether your definition of ploidy counts chromosomes or chromatids (review the answer to problem 4-20). Meiosis I in a diploid organism produces daughter cells with n chromosomes but $2n$ chromatids; meiosis II produces haploid daughter cells with n chromosomes or n chromatids. To avoid potential confusion, geneticists usually use the terms "n", "$2n$", and "ploidy" only to describe cells with unreplicated chromosomes. Thus there are **2 possible answers: meiosis II (and meiosis I, see explanation)**.

e. **meiosis I**

f. **none**

g. **meiosis I**

h. **meiosis II, mitosis**

i. **mitosis, meiosis I**

4-12. Remember that the problem states that all cells are from the same organism. This determines the designation of mitosis, meiosis I and II. The stage of the cell cycle can be inferred from the morphology of the spindle, the presence or absence of the nuclear membrane, and whether homologous chromosomes (or sister chromatids) are paired (or connected through a centromere) or separated. **The n number is 3 chromosomes.**

a. **anaphase of meiosis I**

b. **metaphase of mitosis** (<u>not</u> meiosis II! because there are 6 chromosomes or 2n in this cell)

c. **telophase of meiosis II**

d. **anaphase of mitosis**

e. **metaphase of meiosis II**

4-13.

a. The cell is in **metaphase/early anaphase of meiosis I in a male**, assuming the heterogametic sex (that is, the sex with two different sex chromosomes) in *Tenebrio* is male. Note the heteromorphic chromosome pair in the center of the cell.

b. It is not possible to distinguish **centromeres, telomeres or sister chromatids**, among other structures.

c. $n = 5$.

4-14.

a. metaphase of mitosis:

b. metaphase of meiosis I: (Note the pairing of homologous chromosomes and the two possible alignments of the 2 non-homologous chromosome pairs.)

<table>
<tr><td>A
A</td><td>HD⁺
HD⁺</td><td rowspan="2">OR</td><td>A
A</td><td>HD
HD</td></tr>
<tr><td>a
a</td><td>HD
HD</td><td>a
a</td><td>HD⁺
HD⁺</td></tr>
</table>

c. metaphase of meiosis II: Shown below is only one of the two products of meiosis I from the cell diagrammed at the left in part b; the other product of this meiosis I would have chromosomes bearing a and HD^+. The other alignment in meiosis I (shown at the right in part b) will give A HD^+ and a HD daughter cells.

$$A \quad HD$$
$$A \quad HD$$

4-15. It is **very realistic to assume that homologous chromosomes carry different alleles of some genes**. As we will see later in the book there are about 3 million differences between the DNA sequences in the two haploid human genomes in any one human being, or on average about 130,000 differences between any two homologous chromosomes. In contrast, **recombination almost always occurs between homologous chromosomes in any meiosis**; you will remember that crossing-over is needed to allow the homologous chromosomes to segregate properly during meiosis I. Thus **the second assumption is much less realistic**. Using these assumptions, each person can produce 2^{23} genetically different gametes. Thus, the couple could potentially produce 2^{23} x $2^{23} = 2^{46}$ or **70,368,744,177,664 different zygotic combinations**. That is 70 trillion, 368 billion, 744 million, 177 thousand, 664 genetically different children. When someone says they have never known anyone genetically like you, she is exactly right (unless you are an identical twin or triplet)!

4-16. The diploid sporophyte contains 7 homologous pairs of chromosomes. One chromosome in each pair came from the male gamete and the other from the female gamete. When the diploid cell undergoes meiosis one homolog of each pair ends up in the gamete. For each homologous pair, the probability that the gamete contains the homolog inherited from the father = 1/2. **The probability that a gamete contains only the homologs inherited from the father is $(1/2)^7 = 0.78\%$.**

4-17. Absolutely. **Meiosis requires the pairing of homologous chromosomes during meiosis I**, so meiosis can not occur in a haploid organism.

4-18.

a. **The diagram at left is correct. In this diagram the connection between sister chromatids is maintained in the region between the chiasma and the telomeres**; in the diagram on the right this connection is incorrectly shown as it occurs between the chromatids of homologous chromosomes. If you trace out the chromatids you will notice that the sister chromatid cohesion in the region between the chiasma and the telomeres provides a physical connection that joins homologous chromosomes, allowing them to stay together at the metaphase plate during meiosis I. This shows why recombination is so important for proper meiotic chromosome behavior.

b. **The cohesin complexes at the centromere must be different than those along the chromosome arms, at least during meiosis I. At anaphase I the cohesin complexes along the arms must be destroyed in order to allow the homologous chromosomes to separate. However, the cohesin at the centromeres must remain because sister chromatids stay together until anaphase of meiosis II.** Recent research indicates that this difference between the cohesin complexes occurs because centromeric cohesin is specifically protected from destruction at anaphase I by a protein called *shugoshin* (Japanese for *guardian spirit*).

Section 4.4 - Gametogensis

4-19.

a. **400 sperm** are produced from 100 primary spermatocytes;

b. **200 sperm** are produced from 100 secondary spermatocytes;

c. **100 sperm** are produced from 100 spermatids;

d. **100 ova** are formed from 100 primary oocytes. Remember that although each primary oocyte will produce three or four meiotic products (depending on whether the first polar body undergoes meiosis II), only one will become an egg (ovum);

e. **100 ova** develop from 100 secondary oocytes - the other 100 haploid products are polar bodies;

f. **No ova** are produced from polar bodies.

4-20. Remember that this textbook uses the convention that the number of chromosomes = number of separate centromeres (**page 75 left column**). Thus, after DNA synthesis each chromosome has replicated and so has 2 chromatids held together at the replicated by attached centromere. This structure is one chromosome with two chromatids. The centromeres separate at anaphase in mitosis or at anaphase II in meiosis; at this point, the two chromatids now become two chromosomes. For an overview of egg formation in humans **see** **Figure 4.17**; for an overview of sperm formation in humans see **Figure 4.18**.

a. **96 chromosomes** with 1 chromatid each = **96 chromatids**;

b. **48 chromosomes** with 2 chromatids each = **96 chromatids**;

c. **24 chromosomes** with 1 chromatid each = **24 chromatids**;

d. **48 chromosomes** unreplicated in G_1 = **48 chromatids**;

e. **48 chromosomes** with 2 chromatids each (replicated in G2) = **96 chromatids**;

f. **48 chromosomes** unreplicated in G_1 preceding meiosis = **48 chromatids**;

g. **48 chromosomes** with 2 chromatids each = **96 chromatids**;

h. **48 chromosomes** unreplicated before S = **48 chromatids**;

i. **24 chromosomes** with 1 chromatid each = **24 chromatids**;

j. **48 chromosomes** with 2 chromatids each = **96 chromatids**;

k. **24 chromosomes** with 2 chromatids each = **48 chromatids**;

l. **24 chromosomes** with 1 chromatid each = **24 chromatids**;

m. **24 chromosomes** with 1 chromatid each = **24 chromatids**.

4-21. Remember that ZW is ♀, ZZ is ♂ and WW is lethal.

a. The ZW eggs would give rise to **only ZW females**.

b. Cells resulting from meiosis will be 1/2 Z : 1/2 W. Upon chromosomal duplication they would become ZZ or WW. ZZ cells develop into males and WW is lethal, so **only males** are produced by this mechanism.

c. After eggs have gone through meiosis I, they will contain either a replicated Z or a replicated W chromosome. If the sister chromatids separate to become the chromosomes you will have 1/2 ZZ:1/2 WW. Again **only ZZ males** are produced since WW cells are inviable.

d. Meiosis of a ZW cell produces 4 haploid products, 2 Z : 2 W, one of which is the egg and the other 3 are the polar bodies. If the egg is Z then it has 1/3 chance of fusing with a Z polar body and 2/3 chance of fusing with a W polar body = 1/3 ZZ (♂) : 2/3 ZW (♀). If the egg is W then the fusion products will be 1/3 WW (lethal) : 2/3 ZW (♀). Because these 2 types of eggs are mutually exclusive, you add these 2 probabilities: 1/2 (probability of Z egg) x (1/3 ZZ : 2/3 ZW) + 1/2 (probability of W egg) x (1/3 WW : 2/3 WZ) = 1/6 ZZ : 2/6 ZW + 1/6 WW : 2/6 WZ = 1/6 ZZ (♂) : 4/6 ZW (♀) : 1/6 WW (lethal) = **1/5 ZZ (♂) : 4/5 ZW (♀)**.

4-22. The primary oocyte is arrested in prophase I, so it contains a duplicated set of the diploid number of chromosomes (46 chromosomes and 92 chromatids). During meiosis I, the homologous chromosomes segregate into two separate cells, so the chromosome carrying the *A* alleles will segregate into one cell while the chromosome carrying the *a* alleles will segregate into the other cell. One of these cells becomes the secondary oocyte (containing 23 chromosomes each with 2 chromatids and more of the cytoplasm) and the other becomes the polar body. **The genotype of the dermoid cyst that develops from a secondary oocyte could be either *AA* or *aa*.** Note that the attached sister chromatids in the secondary oocyte must separate from each other before the first mitosis leading to cyst formation.

Section 4.5 – Validation of Chromosome Theory

4-23.

a. Ivory eyes is the recessive phenotype and brown is the dominant phenotype; females have 2 alleles of every gene and males have only one. Diagram the cross: ivory ♀ (*bb*) x brown ♂ (*B*) → fertilized eggs are *Bb* ♀ (**brown**); unfertilized eggs are *b* ♂ (**ivory**).

b. The cross is *Bb* F1 ♀ x *B* ♂ → fertilized eggs are **1/2 *Bb* ♀ (brown) : 1/2 *BB* ♀ (brown) = all brown ♀ progeny**; unfertilized eggs are **1/2 *B* ♂ (brown) : 1/2 *b* ♂ (ivory)**.

4-24. In birds males are ZZ, females are ZW. Diagram the crosses between true-breeding birds:

yellow ♀ x brown ♂ → brown ♀ and ♂; brown ♀ x yellow ♂ → yellow ♀ and brown ♂

Answer 3EQ#1: there are 2 phenotypes in the second cross, so there is 1 gene controlling color in canaries. For 3EQ#2: because the parents are true-breeding, the first cross shows that brown > yellow. For 3EQ#3: these two crosses are reciprocal crosses and the progeny show a sex-associated difference in phenotypes - both sexes are the same phenotype (brown) in the first cross but different phenotypes in the second cross (brown ♀ and yellow ♂). This indicates the **trait is sex-linked**. Also, the second cross shows crisscross inheritance - brown females x yellow males → brown sons and yellow daughters. This is also characteristic of a sex-linked trait. **Therefore, the alleles of the gene are Z^B (brown allele on Z chromosome) and Z^b (yellow allele on Z chromosome)**; the W chromosome does not carry the feather color gene. The first cross was Z^bW x Z^BZ^B → Z^BW (brown females) and Z^BZ^b (brown males). The second cross was Z^BW x Z^bZ^b → Z^BZ^b (brown males) and Z^bW (yellow females).

4-25. In birds females are the heterogametic sex, (ZW); Z^B represents the Z chromosome with the barred allele, Z^b is non-barred.

a. Z^BW (barred hen) x Z^bZ^b(non-barred rooster) → **Z^bW (non-barred females) and Z^BZ^b (barred males)**.

b. Z^BZ^b F_1 x Z^bW → **Z^BW (barred) and Z^bW (non-barred) females and Z^BZ^b and Z^bZ^b (barred and non-barred) males**.

4-26. Pedigrees 1-4 show examples of each of the four modes of inheritance.

a. **Pedigree 1** represents a recessive trait because two unaffected individuals have affected children. If the trait were X-linked, then I-1 would have to be affected in order to have an affected daughter. The trait is **autosomal recessive. Pedigree 2** represents recessive inheritance (see part a). This is **X-linked recessive** inheritance as autosomal recessive inheritance was already accounted for in part a. This conclusion is supported by the data showing that the father (I-1) is unaffected and only sons show the trait in generation II, implying that the mother must be a carrier. **Pedigree 3** shows the inheritance of a dominant trait because affected children always have an affected parent (remember that all four diseases are rare). The trait must be **autosomal dominant** because the affected father transmits it to a son. **Pedigree 4** represents an **X-linked dominant** trait as characterized by the transmission from affected father to all of his daughters but none of his sons.

b.   **Pedigree 1** - both parents are carriers for this autosomal recessive trait so there is a **1/4 chance** that the child will be affected (*aa*). **Pedigree 2** – individual I-2 is a carrier for this X-linked trait. The probability is 1/2 that she will pass on X^a to her daughter II-5. The unaffected father (I-1) contributes a normal X^A chromosome, so the probability that is II-5 is a carrier = 1/2. The **probability of an affected son** = 1/2 (probability II-5 is a carrier) x 1/2 (probability II-5 contributes X^a) x 1/2 (probability of Y from father II-6) = **1/8**. The **probability of an affected daughter = 0** because II-6 must contribute a normal X^A. **Pedigree 3** - for an autosomal dominant trait there is a **1/2 chance** that the heterozygous mother (II-5) will pass on the mutant allele to a child of either sex. **Pedigree 4** - the father (I-1) passes on the mutant X chromosome to all his daughters and none of his sons. Therefore II-6 does not carry the mutation, as shown by his normal phenotype. The **probability of an affected child = 0**.

4-27. It is most likely that **the bag-winged females have one mutation on the X chromosome that has a dominant effect on wing structure and that also causes lethality in homozygous females or hemizygous males**. This is analogous to the A^y allele in mice discussed in Chapter 3, except here the bag gene is on the X chromosome. Thus all of the bag winged flies are heterozygous for this mutation. Diagram the cross: $Bg\ Bg^+$ (bag-winged female) x Bg^+ Y (normal male) $\rightarrow$ 1/4 $Bg\ Bg^+$ (bag-winged females) : 1/4 $Bg^+\ Bg^+$ (normal females) : 1/4 Bg Y (dead) : 1/4 Bg^+ Y (normal males). Thus the ratio of the surviving progeny is 1/3 bag-winged females : 1/3 normal females : 1/3 normal males.

4-28.

a.   Draw the pedigree:

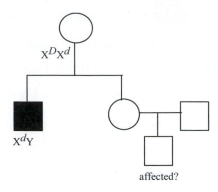

I-1 must have been heterozygous for the *d* allele. The probability that II-2 will have an affected

son = 1/2 (the probability that II-1 inherited the X^d chromosome from I-1) x 1/2 (the probability

that II-1's son receives the X^d chromosome from her if she is $X^D X^d$) = **1/4**.

b.

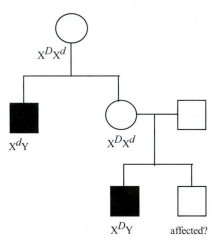

We now know that II-2 is in fact a carrier since she had an affected son. Therefore the probability

is **1/2** that she will pass on the X^d chromosome to III-2.

c.

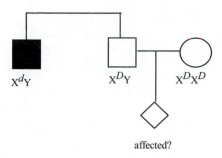

affected?

The mother of these two men was a carrier. She passed on the X^d chromosome to the affected son and she passed on the X^D chromosome to the unaffected son. There is **no chance** that the unaffected man will pass the disease allele to his children.

d.

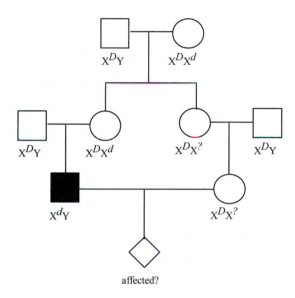

affected?

The probability of IV-1 being an affected son = 1/2 (probability that II-3 is a carrier) x 1/2 (probability that III-2 inherits X^d) x 1/2 (probability that IV-1 inherits X^d) x 1/2 (probability IV-1 inherits the Y from III-1) = **1/16**. **The probability that IV-1 is an affected girl** = 1/2 (probability that II-3 is a carrier) x 1/2 (probability that III-2 inherits X^d) x 1/2 (probability that IV-1 inherits X^d) x 1/2 (probability IV-1 inherits X^d from III-1) = **1/16**. **The chance that IV-1 is unaffected** = 1 - (1/16 probability of an affected male + 1/16 probability of an affected female) = 1 - 2/16 = **7/8**.

e.

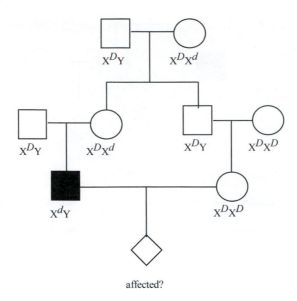

affected?

II-2 must be a carrier in order for her son to inherit the disease. II-3 is the brother and he is normal, so his genotype must be $X^D Y$. Therefore III-2, his daughter, must be homozygous normal. The probability that the fetus (diamond) will be an **affected boy = 0**; the probability of an **affected daughter = 0**; the probability of an **unaffected child = 100%.**

4-29. Color-blindness in this family is an X-linked recessive condition, because two unaffected parents have affected children, and because two unrelated individuals would have to carry rare alleles if the trait were autosomal. Since the males are hemizygous with only one allele for this trait, their phenotype directly represents their genotype. **II-2 and III-3 are affected, so they must be $X^{cb}Y$.** Now consider the parents of II-2 - **I-2 is normal and therefore $X^{CB}Y$. I-1 must be a carrier so her genotype is $X^{CB} X^{cb}$. II-1 can be either $X^{CB} X^{CB}$ or $X^{CB} X^{cb}$** (but she is more likely to be a normal homozygote if the trait is rare). **II-3 must be a carrier ($X^{CB} X^{cb}$) since she had an affected son. II-4 is $X^{CB}Y$; III-1 must be $X^{CB} X^{cb}$** (because she has normal color vision yet she must have received X^{cb} from her father); **III-2 is either $X^{CB}X^{CB}$ or $X^{CB}X^{cb}$; III-4 is an unaffected male and therefore must be $X^{CB}Y$.**

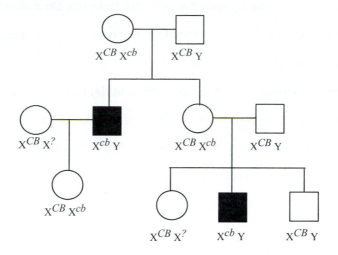

4-30.

a. If the trait is an autosomal dominant, then the fathers must be heterozygous *Rr* because they have unaffected daughters. The mothers would then be homozygous normal, *rr*. The probability of an affected male child from such matings = 1/2 (probability of inheriting *R*) x 1/2 (probability the child will be male) = 1/4. The probability of an unaffected daughter = 1/4. The probability of having 6 affected sons and 5 unaffected daughters = $(1/4)^6$ x $(1/4)^5$ = **$(1/4)^{11}$ = 2 x 10^{-7}** or extremely unlikely!

b. This trait could be an example of **Y-linked inheritance or it could represent sex-limited expression of the mutant allele** (that is, the allele is dominant in males and recessive or otherwise unexpressed in females, like male-pattern baldness). There is no direct evidence to support one hypothesis over the other. However there are no known examples of Y-linked inheritance other than maleness, while there are several examples of sex-influenced traits that are male dominant and female recessive.

4-31.

a. Unaffected individuals have affected children, so albinism is **recessive**.

b. If albinism were X-linked, then I-9 would have to be an affected hemizygote in order to have an affected daughter. As this is not the case, albinism is **autosomal**.

c. ***aa;*** d. ***Aa;*** e. ***Aa;*** f. ***Aa;*** g. ***Aa;*** h. ***Aa.***

4-32. The answer to this question depends upon when the nondisjunction occurs. Nondisjunction in meiosis I in the male results in one secondary spermatocyte that contains both sex chromosomes (the X and the Y) and the other secondary spermatocyte that lacks sex chromosomes. If the red-eyed males are $X^{w+}Y$, then the sperm that are formed after nondisjunction in meiosis I are $X^{w+}Y$ and nullo (O) sex chromosome. The $X^{w}X^{w}$ female makes X^{w} eggs, so fertilization will produce **$X^{w}O$ (white-eyed, sterile male) and $X^{w+}X^{w}Y$ (red-eyed female)** progeny. If nondisjunction occurred in meiosis II in the male, the sperm would be $X^{w+}X^{w+}$ or YY and nullo (O) sex chromosome. After fertilization of the X^{w} eggs, the zygotes would be $X^{w+}X^{w+}X^{w}$ (lethal) or **$X^{w}YY$ (fertile white-eyed males) and $X^{w}O$ (sterile white-eyed males)**. Notice that the normal progeny of this cross will be $X^{w+}X^{w}$ (red-eyed females) and $X^{w}Y$ (white-eyed males), which are indistinguishable from the nondisjunction progeny in terms of their eye colors.

4-33. wild type ♀ x yellow vestigial ♂ → F$_1$ wild type ♀ and ♂ → F$_2$ 16 yellow vestigial ♂ : 48 yellow ♂ : 15 vestigial ♂ : 49 wild type ♂ : 31 vestigial ♀: 97 wild type ♀

3EQ#1 - there are 4 different phenotypes in the F$_2$, so 2 genes involved. 3EQ#2 - one gene determines body color and the F$_1$ shows that wild type is dominant to yellow; the other gene determines wing length and the F$_1$ shows that wild type is dominant to vestigial. This conclusion is reinforced by the F$_2$ progeny where there are 48 + 49 + 97 normal wing individuals : 16 + 15 + 31 vestigial winged flies = 194 normal : 61 vestigial = 3:1. 3EQ#3 - in the F$_2$ there are wild type and vestigial males and females in a 3:1 ratio, so wing shape is an autosomal trait; in the F$_2$ there are yellow males but no yellow females, so body color is an X-linked trait. Thus **vestigial is the recessive allele of an autosomal gene and yellow is the recessive allele of an X-linked gene**. The original cross was thus $X^{y+}X^{y+}vg^{+}vg^{+}$ ♀ x $X^{y}Y$ $vgvg$ ♂.

4-34. When nondisjunction occurs in meiosis I (MI), there are two types of affected gametes. Half of the gametes receive no sex chromosomes (nullo) while the rest of the gametes receive one copy of each of the sex chromosomes in the parent. Thus MI nondisjunction in a male will give XY sperm and nullo sperm. If the nondisjunction occurs in MII, then half of the affected gametes are nullo while the rest receive 2 copies of one of the original sex chromosomes. MII nondisjunction in a male will therefore give either XX sperm or YY sperm and nullo sperm with no sex chromosomes.

a.   The boy received both an X^{B} and a Y from his father. Because both the X and Y segregated into the same gamete, the nondisjunction occurred in **meiosis I in the father**.

b. The son could have received both the X^A and X^B from the mother and the Y from the father, or alternatively the X^B from the mother and the X^A and Y from the father. You can determine that the nondisjunction occurred in **meiosis I, but you cannot determine in which parent**.

c. The son has two X^A chromosomes and a Y. Therefore the nondisjunction occurred in the **mother, but because she is homozygous $X^A X^A$ you cannot determine during which meiotic division**.

4-35.

a. Hypertrichosis is almost certainly an **X-linked dominant trait**.

b. The trait is **dominant because every affected child has an affected parent**. If the trait were recessive then several of the people marrying into the family would have to be carriers, yet this trait is rare. Hypertrichosis is highly likely to be **X-linked because two affected males (II-4 and IV-3) pass the trait on to all of their daughters but none of their sons**. Using the same logic as in <u>problem 4-30a</u>, there is only a $(1/4)^{13}$ chance that the trait is autosomal dominant.

c. **III-2 had 4 husbands and III-9 had 6 husbands!**

4-36. Brown eye color (b) and scarlet (s) are autosomal recessive mutations while vermilion (v) is an X-linked recessive trait. The genes interact such that both brown, vermilion double mutants and brown, scarlet double mutants are white-eyed. When diagramming a cross involving more than one gene you <u>must</u> start with a genotype for each parent that includes information on <u>both genes</u>. Then figure out the genotype of the F_1 progeny. In order to predict the F_2 results, find the expected ratio for each gene separately, and then cross multiply the ratios to generate the F_2. Diagram the following crosses [<u>Note</u> that X-linked genes will no longer be written as a superscript of the X symbol. Instead only the gene symbol will be used, as is done for the autosomal genes. You can determine that the gene in question is X-linked because it will be paired with the Y chromosome in the male genotypes.]

a. vermilion ♀ ($vv\ b^+b^+$) x brown ♂ ($v^+Y\ bb$) → **F_1 vv^+ b^+b (wild type females)** x **$vY\ b^+b$ (vermilion males)** → F_2 ratio for brown alone = 3/4 b^+- : 1/4 bb; the ratio for vermilion alone in both F_2 females and males = 1/2 v^+ : 1/2 v (the other sex chromosome will be either v or Y); **the ratio for both genes in both F_2 females and males = 3/8 v^+ b^+- (wild type) : 3/8 v b^+ (vermilion) : 1/8 v^+ bb (brown) : 1/8 v bb (white).**

b. brown ♀ ($v^+v^+\ bb$) x vermilion ♂ ($vY\ b^+b^+$) → **F1 vv^+ b^+b (wild type females)** x **$v^+Y\ b^+b$ (wild type males)** → F_2 ratio for brown alone = 3/4 b^+- : 1/4 bb; ratio for vermilion alone in the F_2 females = 1 v^+- and in the males = 1/2 v^+: 1/2 v. The ratio for both genes in the F_2 females is **3/4 v^+- b^+- (wild type) : 1/4 v^+- bb (brown)** and the dihybrid ratio in the **F2 males is 3/8 v^+Y b^+- (wild type) : 3/8 vY b^+- (vermilion) : 1/8 v^+Y bb (brown) : 1/8 vY bb (white).**

c.   scarlet ♀ (b^+b^+ ss) x brown ♂ (bb s^+s^+) → **F₁ b^+b s^+s (wild type**, males and females are the same because both genes are autosomal) → F₂ monohybrid ratio for scarlet = 3/4 s^+- : 1/4 ss and for brown = 3/4 b^+-:1/4 bb. **The F_2 dihybrid ratio** (which hold for both sexes) = **9/16 s^+- b^+- (wild type) : 3/16 s^+- bb (brown) : 3/16 ss b^+- (scarlet) : 1/6 ss bb (white).**

d.   brown ♀ (bb s^+s^+) x scarlet ♂ (b^+b^+ ss) → **F₁ b^+b s^+s (wild type) → F₂ as in part c above.**

4-37.

a.   The father contributed the i or I^A, Rh^+, M or N, and $Xg^{(a-)}$ alleles. Only **alleged father #3 fits** these criteria.

b.   Alleged father #1 was ruled out based on his X chromosome genotype. He is $Xg^{(a+)}$ and can not have an XX daughter who is $Xg^{(a-)}$. If the daughter was XO (Turner), she could have inherited the one X from either mother or father. If the X came from her mother (who must have been heterozygous for Xg), then her father did not contribute an X chromosome. Therefore **alleged fathers #1 and #3** both fit the criteria for paternity.

4-38.

a.   Eosin is an X-linked recessive mutation. The cream colored variants were found occasionally in the eosin true-breeding stock. This suggests that the cream flies have the eosin genotype and <u>also</u> have an second mutation that further lightens the eosin phenotype. Diagram the cross:

wild type ♀ x cream ♂ → F₁ wild type → F2 104 wild type ♀ : 52 wild type ♂ : 44 eosin ♂ :

   14 cream ♂ (8:4:3:1 ratio)

3EQ#1 - there are 3 phenotypes seen in the F₂ males. This could be due to one gene with codominance/incomplete dominance, but the phenotypes are not present in a 1:2:1 ratio. Instead the F₂ phenotypic ratio seems to be an epistatic modification of 9:3:3:1. Therefore the eye colors must be controlled by 2 genes whose mutant alleles are recessive. The fact that the cream eye color initially arose in the eosin mutant stock suggests that cream is a modifier of eosin. This sort of modifier only alters the mutant allele of the gene it is modifying and has no effect on the wild type allele. 3EQ#2 - the F₁ shows you that the wild type phenotype is dominant. Question #3 - as expected, the difference in the phenotypes between the F_2 males and females shows that eosin (e) is X-linked, but the cream modifier (cr) is probably not on the X chromosome (that is, it is autosomal) because some F_2 animals are eosin but not cream.

e^+e^+ cr^+cr^+ ♀ x eY $crcr$ ♂ → $\mathbf{F_1}$ e^+e cr^+c ♀ x e^+Y cr^+cr ♂ → F_2 ratios for each gene

alone = 1/2 e^+- ♀ : 1/4 e^+Y : 1/4 eY and 3/4 cr^+- : 1/4 $crcr$. When these are cross multiplied, the

F_2 ratio for both genes = 3/8 e^+- cr^+- (wild type) ♀ : 1/8 e^+- $crcr$ (wild type) ♀ : 3/16 e^+Y

cr^+- (wild type) ♂ : 1/16 e^+Y $crcr$ (wild type) ♂ : 3/16 eY cr^+- (eosin) ♂ : 1/16 eY $crcr$

(cream) ♂ = 8 wild type ♀ : 4 wild type ♂ : 3 eosin ♂ : 1 cream ♂.

b. eY cr^+cr^+ (eosin ♂) x ee $crcr$ (cream ♀) → $\mathbf{F_1}$ ee cr^+cr ♀ x eY cr^+cr ♂ → F_2 ratios for

each gene alone = 1/2 ee ♀ : 1/2 eY ♂ and 3/4 cr^+- : 1/4 $crcr$. **F_2 ratios for both genes = 3/8 ee**

cr^+- (eosin ♀) : 3/8 eY cr^+- (eosin ♂) : 1/8 ee $crcr$ (cream ♀) : 1/8 eY $crcr$ (cream ♂).

c. ee cr^+cr^+ (eosin ♀) x eY $crcr$ (cream ♂) → $\mathbf{F_1}$ ee cr^+cr ♀ x eY cr^+cr ♂ → F_2 ratios for

each gene alone = 1/2 ee ♀ : 1/2 eY ♂ and 3/4 cr^+- : 1/4 $crcr$. **F_2 ratios for both genes = 3/8 ee**

cr^+- (eosin ♀) : 3/8 eY cr^+- (eosin ♂) : 1/8 ee $crcr$ (cream ♀) : 1/8 eY $crcr$ (cream ♂). This is

the same result as in part b.

4-39.

a. The white eye mutation is an X-linked, recessive mutation. Diagram the cross:

white ♂ x purple ♀ → F_1 wild type eye color → F_2 3/8 wild type ♀ : 1/8 purple ♀ : 3/16

wild type ♂ : 1/4 white ♂ : 1/16 purple ♂.

3EQ#1 – white and purple eye colors cannot be caused by two different alleles of the same gene

because then the F_1 males would be purple due to criss-cross inheritance. This is a also a

complementation test and wild type phenotype of the F_1 heterozygotes means there are 2

different genes controlling eye color (**Figure 3.16**). 3EQ#2 – the mutant allele of one of the

genes causes white eye color, the mutant allele of the other gene causes purple eye color. The

wild type alleles of both genes contribute to wild type eye color. **In both cases the mutant allele**

is recessive ($w < w^+$ and $p < p^+$) as demonstrated by the wild-type phenotype of the F_1 animals.

You also expect that white (the absence of pigment) should be epistatic to eyes with red or

purple pigmentation. Question #3 – **white is X-linked; purple is an autosomal trait** because it

shows no difference between the sexes in either the F_1 or the F_2 generation - there is a 3 p^+- : 1

pp ratio in both males and females in the F2. The original cross was thus: $wY\ p^+p^+$ (white ♂) x $w^+w^+\ pp$ (purple ♀).

b. $ww\ p^+p^+$ (white ♀) x $w^+Y\ pp$ (purple ♂) → **F₁** $w^+w\ p^+p$ (wild type eye color ♀) x $wY\ p^+p$ *(white ♂)* → F₂ ratio for white alone = 1/2 w^+w : 1/2 ww females and 1/2 w^+Y : 1/2 wY males; for the purple gene alone the ratio is 3/4 p^+- : 1/4 *pp*. **The F₂ ratio for both genes in females = 3/8 w^+- p^+- (wild type) : 1/8 w^+- pp (purple) : 3/8 ww p^+- (white) : 1/8 ww pp (white) = 3/8 wild type : 1/2 white : 1/8 purple. The ratio for the males is the same**. Note that in this reciprocal cross there is a phenotypic difference between the sexes in the F1 generation and no difference in the F2 generation.

4-40. Remember that white tigers are shown by <u>unshaded</u> symbols. Test each possibility by assigning genotypes based on the mode of inheritance.

a.   **No**. Y linked genes are only expressed in males. Since there are white females, the trait could not be Y-linked.

b.   **No**. White males would have to have white daughters but not white sons. Mohan's daughter is not white and Tony has a normally colored daughter, Kamala, and a white son, Bim.

c.   **Yes**. The information in the pedigree is consistent with dominant autosomal inheritance.

d.   **Yes**. The information in the pedigree is consistent with a recessive X-linked trait.

e.   **Yes**. There is no data in the cross scheme to allow you to rule out recessive autosomal inheritance in this highly inbred family.

4-41.

a.   **Individual III-5** was not related to anyone in the previous generation. Therefore the cancer with which he was afflicted is not likely to be caused by the rare mutation in BRCA2. Instead his cancer may have been due to a mutation in another gene that can predispose to cancer or exposure to environmental influences like carcinogens.

b.   There is a vertical pattern of inheritance of the cancers, so **the BRCA2 mutation has a dominant effect on causing cancer**. Most of the affected children in this pedigree have an affected parent, although there are a couple of cases where this is not seen (example V-1 and V-8). It is very unlikely that the mutant BRCA2 allele is recessive as then many unrelated people would have to be carriers of the rare mutant allele.

c.   The BRCA2 mutation is not Y-linked as affected males (for example III-9 and IV-6) pass the trait to their daughters. **The data do not clearly distinguish between X-linked and autosomal**

inheritance. The X-linked hypothesis is supported by the fact that all of the daughters of men who could transmit the mutant allele are affected (for example the five daughters of IV-6). BRCA2 is actually on autosome 13.

d. **The penetrance of the cancer phenotype is incomplete**. If the mutant allele is dominant then women II-4 and III-8 should have had cancer if the mutant disease-causing allele is dominant. It is critical to note that the presence of the BRCA2 mutation does not always cause cancer. Instead the mutation increases the predisposition to cancer. The reasons for this will be explained in Chapter 17.

e. **The expressivity is variable**. Some of the women who have inherited the mutant cancer causing allele do not have breast cancer but instead have ovarian or other cancers. This suggests that the BRCA2 mutation can predispose to cancers other than breast cancer, even though breast tumors are the most likely outcome.

f. **Ovarian cancer is sex-limited** since males don't have ovaries. Although the data are fragmentary, **the penetrance of breast cancer may be sex-influenced**. Of the five males who should be carrying the BRCA2 mutant allele only one has breast cancer (III-9), giving a rate of 20% affected men. Of the 21 women who must be carrying the mutant BRCA2 allele 16 have breast cancer - a rate of 80% affected women. Perhaps the hormonal environment or the amount of breast tissue helps determine whether a person with the BRCA mutation will develop breast cancer.

g. The absence of cancer in the first two generations **could be explained by the low penetrance of the cancer phenotype, particularly among men**. Three individuals in generations I and II must have carried the mutant allele of BRCA2 but only one of these was a woman. Other explanations are also possible. One interesting hypothesis has to do with the fact that cancer is a disease whose frequency increases with age and life expectancies have improved only in recent times. Thus the individuals in the first two generations of this pedigree may have died early of other causes before they had the chance to develop cancer.

Chapter 5   Linkage, Recombination, and the Mapping of Genes on Chromosomes

Synopsis:

This chapter is devoted to a very important topic: linkage of genes. The concept of linkage is the basis for genetic mapping. Genes on the same chromosome are physically connected or linked. Gene pairs that are close together on the same chromosome are genetically linked because the alleles on the same homolog are transmitted together (parental types) into gametes more often than not during meiosis (**Figures 5.2 and 5.3**).

Gene pairs that assort independently exhibit a recombination frequency of 50% because the number of parental types = the number of recombinants (**Table 5.2**). Genes may assort independently either because they are on different chromosomes or because they are far apart on the same chromosome. The recombination frequencies of pairs of genes indicate how often 2 genes are transmitted together. For linked genes, the frequency is less than 50%. The greater the distance between linked genes, the higher the recombination frequency (rf). Recombination frequencies become more inaccurate as the distance between genes increases.

The relationship between relative recombination frequency and distance is used to create genetic maps. The greater the density of genes on the map, and the smaller the distances between the genes, the more accurate the map.

Statistical analysis can help determine whether or not 2 genes assort independently. The Chi square test can also be used to determine how well the outcomes of crosses fit other genetic hypotheses (**Figure 5.5**).

Tetrad analysis is done in certain species of yeast in which the meiotic products are kept together in a sac (ascus) so the results of a single meiosis are displayed in each tetrad. This array of 4 spores reveals the relation between genetic recombination and the segregation of chromosomes during the 2 meiotic divisions. Solving tetrad analysis problems will increase your understanding of chromosome segregation during meiosis.

In diploid organisms heterozygous for 2 alleles of a gene rare <u>mitotic</u> recombination between the gene and its centromere can produce genetic mosaics in which some cells are homozygous for one allele or the other (**Figure 5.24** and <u>Problem 5-43</u>).

Significant Elements:

After reading the chapter and thinking about the concepts you should be able to:

♦ Determine if genes are linked or not based on the frequency of different types of gametes or progeny.

- Determine whether observed results are statistically consistent with expected results using the chi square test.

- When given the genotypes of parents and information on linkage of genes, list the gametes that can be produced with and without recombination.

- Beginning with 3 genes, map the genes. Determine the order of the genes and the distance between linked genes based on the percentage of recombination that occurred during meiosis.

- Determine if there is crossover interference in a three point cross.

- Identify different types of tetrads [parental ditype (PD), non-parental ditype (NPD), tetratype (T)] and understand what events during meiosis gives rise to these different types of asci.

- Determine linkage between genes given either random spore results or numbers of different types of asci.

- Identify centromere linked genes and determine distance from gene to centromere in *Neurospora* tetrads.

Problem Solving Tips:

This chapter discusses the analysis of linked genes. One way to analyze recombination frequency (rf) between linked genes is using 2 point crosses, as is done in Tetrad Analysis. A more accurate map of linked genes can be derived from analyzing the data from 3 linked genes (3 point crosses). Both of these types of analysis of linked genes lead to maps, but each type of analysis has its own peculiarities.

- The minimum requirement for detecting recombination is heterozygosity for 2 genes. The recombination events that can be detected are the ones that occur between the 2 genes, giving recombinant gametes instead of parental gametes.

- It is easiest to detect the parental vs recombinant gametes if you do a test cross.

- In a test cross of $aa^+ bb^+$ x $aa\ bb$ the expected phenotypic frequencies and classes of progeny are 1 $a^+- b^+-$: 1 $aa\ bb$: 1 $a^+- bb$: 1$aa\ b^+-$ if the genes are assorting independently. The genes are genetically linked if you see more parental than recombinant progeny.

- Recombination frequency = # recombinant progeny / total # progeny = rf x 100 = mu or cM. Depending on the vagaries of the particular system or problem we may derive some basic variations of this formula.

- When discussing recombination, you must write the genotypes in a way that represents linkage. Remember that there is one allele per homolog, so *aa* becomes *a / a*. When discussing more than one gene on a chromosome you must initially assume an order. While solving a problem, always write the genes in the same arbitrary, assumed order. Thus, a^+ b^+ c / a b c^+ is more accurately written as:

$$\frac{a^+ \qquad b^+ \qquad c}{a \qquad b \qquad c^+}$$

Genes on the X chromosome can be mapped without a test cross. Just use the hemizygous male progeny as in Problem 5-8.

♦   The Chi Square calculation is $(o - e)^2/e$.

Problem Solving - How to Begin:

THREE ESSENTIAL QUESTIONS (3EQ):

1. How many genes are involved in the cross?

2. For **each gene** involved in the cross: what are the phenotypes associated with the gene? Which phenotype is the dominant one and why? Which phenotype is the recessive one and why?

3. For **each** gene involved in the cross: is it X-linked or autosomal?

The 3EQ are still useful, as you often have to diagram the cross and assign genotypes! For a more thorough list of Hints, review **Solving Problems - How to Begin** in Chapter 4 of this Study Guide.

THREE POINT CROSSES:

♦   In a three point cross first rewrite the classes of progeny (data) assigning genotypes to each trait. These genotypes are based on your answers to the 3EQ!

♦   The offspring are generated by a heterozygous individual. Therefore, all classes (parental, etc.) will occur as reciprocal pairs of progeny. These reciprocal pairs will be both genetic reciprocals and numerically equivalent.

♦   Designate the different gametes or offspring as noncrossover (parental), single crossover or double crossover. The noncrossover classes are those classes of progeny who have one of the intact, non-recombinant homologs from the parent. The noncrossover classes will be represented by the greatest numbers of offspring. The double crossover classes will be represented by the smallest numbers of offspring (**Figure 5.10**). Sometimes one or both double crossover classes are missing because they are rare.

♦   By examining the pattern of data seen in a problem, you can often start solving the problem with a basic understanding of the linkage relationships of the genes. Some of the more common patterns of data are:

- 3 linked genes give 8 classes of data that occur as 4 reciprocal pairs genetically and numerically;

- 3 unlinked genes gives 8 classes of data that occur as 4 genetically reciprocal pairs, but all classes are seen in a 1:1:1:1:1:1:1:1 ratio;

- 4 linked genes give 16 classes of progeny that occur in 8 reciprocal pairs;

- 4 unlinked genes give 16 classes of progeny in a 1:1:1:1:1:1:1:1:1:1:1:1:1:1:1:1 ratio;

- 3 linked genes and 1 gene assorting independently gives 16 classes of data occurring as 8 reciprocal pairs genetically and 4 groups of 4 numerically.

♦ Begin the process of mapping the genes by ordering the genes. To figure out which gene is in the middle of a group of three genes, choose one of the double crossover classes. Compare it to the most similar parental class of progeny where two of the three genes will have the same combination of alleles. The gene that differs is the gene in the middle. See problem 5-24c for further explanation.

♦ The last step is to determine the distance between the genes on each end and the gene in the middle. Use the formula rf = # recombinants between the 2 genes / total # of progeny.

♦ See problem 5-24 for an example of using the data to make a map.

♦ See problem 5-29 for an example of how to generate the data when you start with the map.

♦ See problem 5-27 for some hints on how to solve more complicated arrangements of genes.

TETRAD ANALYSIS:

♦ Some of the things that can be done here when analyzing recombination frequency (rf) are based on the biology of the organisms. Thus you must understand some of the basic biology of *Neurospora* and yeast (**Figure 5.14**). There is no three point cross analysis in tetrads. Instead you map two genes at a time. Study **Figure 5.15** (two genes on different chromosomes) and **Figure 5.17** (two linked genes) to understand the derivation of the Three Easy Rules for Tetrad Analysis.

♦ Rule #1 - If the number of PD tetrads is about equal to the number of NPD tetrads then the genes are unlinked. If the # PD tetrads >>> # NPD tetrads, the genes are linked.

♦ Independent assortment dictates that these two classes of asci will be about equal when genes are unlinked, see **Figure 5.15**.

♦ Rule #2 - Distance between linked genes is determined by calculating recombination frequency using the equation: rf = (NPD + 1)/2T/ total # tetrads × 100.

♦ Rule #3 - Crossovers between a gene and the centromere mean that the two alleles are separated at the second meiotic division (second division segregation) instead of during the first meiotic division. Centromere distance can be measured in yeasts with ordered asci, such as *Neurospora* and *Ascobolus*.

♦ These rules are based on the effects of meiosis on two unlinked genes (**Figure 5.15**) versus the effects on two linked genes (**Figure 5.17**). Further intricacies of single and double crossovers are discussed in problem 5-38.

Solutions to Problems:

Vocabulary

5-1. a. **8**; b. **4**; c. **1**; d. **11**; e. **2**; f. **5**; g. **6**; h. **3**; i. **10**; j. **12**; k. **9**; l. **7**.

Section 5.1 – Gene Linkage and Recombination

5-2.

a. Diagram the cross.

scabrous ♂ ($scsc\,j^+j^+$) x javelin ♀ ($sc^+sc^+\,jj$) → F_1 WT ♀ ($scsc^+\,j^+j$) x scabrous javelin ♂ ($scsc\,jj$) → 1/4 scabrous (P) : 1/4 javelin (P) : 1/4 WT (R) : 1/4 scabrous javelin (R).

This F_1 female will make four different types of gametes in equal frequency – 1/4 $sc^+\,j$: 1/4 $sc\,j^+$: 1/4 $sc^+\,j^+$: 1/4 $sc\,j$. Because this is a test cross the male parent will always give the recessive alleles of the genes. Thus the phenotypes of the progeny will be determined by the gamete they receive from the heterozygous F_1 female.

Of course these genes may be linked. In this case the cross would be diagrammed as follows:

scabrous ♂ ($sc\,j^+$ / $sc\,j^+$) x javelin ♀ ($sc^+\,j$ / $sc^+\,j$) → F_1 WT ♀ ($sc\,j^+$ / $sc^+\,j$) x scabrous javelin ♂ ($sc\,j$ / $sc\,j$) → scabrous ($sc\,j^+$) (P) : javelin ($s^+\,j$) (P) : WT ($sc^+\,j^+$) (R) : scabrous javelin ($sc\,j$) (R). Again, this F_1 female will make 4 different types of gametes. **There will be the two Parental (P) types and two Recombinant (R) types - $s^+\,j$ (P) : $sc\,j^+$ (P) : $sc^+\,j^+$ (R) : $sc\,j$ (R).** The two Parental types must be equal frequency and the two Recombinant types must be equal to each other. The relative proportion of Parental to Recombinant will depend on the genetic distance (recombination frequency = rf) between the genes on the chromosome. The recombination frequency could be anything from 0 mu (the genes are very closely linked) to 50 mu (the genes are far apart on the same chromosome. In the latter case P = R and the proportions of the four types of gametes will also be 1 : 1 : 1 : 1.

b. In these results the parental (77 scabrous + 74 javelin) are equal frequency to the recombinants (76 wild type + 73 scabrous javelin). **Thus the genes assort independently (are 50 mu apart).**

c. F_1 WT ♀ ($scsc^+ j^+j$) x wild type ♂ ($sc^+sc^+ j^+j^+$) → all wild type progeny. The F_1 female will still make four types of gametes in equal frequency. However the male parent in this cross can only contribute $sc^+ j^+$ gametes so **all the progeny will be phenotypically wild type. Thus you could not determine the frequency of parental and recombinant gametes from the F_1 female**.

d. Diagram the cross.

    javelin ♀ ($sc^+sc^+ jj$) x scabrous javelin ♂ ($scsc\, jj$) → F_1 javelin ♀ ($scsc^+ jj$). This F_1 female will make $sc^+ j$ and $sc\, j$) parental gametes. **Her recombinant gametes will be the same genotypes as her parental gametes thus making it impossible to detect crossing over**. In order to detect crossing over (or independent assortment) the parent must be heterozygous for two genes or markers. The cross overs detected will occur between the two genes.

5-3.

a. Diagram this cross.

    $B_1B_2\, D_1D_4$ ♂ x $B_3B_3\, D_2D_3$ ♀ → $B_1B_3\, D_1D_3$ ♂ (John). Thus John received a $B_1\, D_1$ gamete from his father and a $B_3\, D_3$ gamete from his mother. John will produce **parental gametes of these types - $B_1\, D_1$ and $B_3\, D_3$**.

b. The **recombinant gametes produced by John will be $B_1\, D_3$ and $B_3\, D_1$**.

c. All 100 sperm have the parental genotypes. There are no recombinant sperm. Therefore **the B and D DNA loci are linked**. Based on this data they are 0 mu apart.

Section 5.2 – Chi-Square Test

5-4. The null hypothesis is that there is independent assortment of 2 genes yielding a dihybrid phenotypic ratio of 9/16 R-Y- : 3/16 R- yy : 3/16 rr Y- : 1/16 rr yy. Use the Chi square (X^2) test to compare Mendel's observed data with the 9:3:3:1 ratio expected for 2 genes that assort independently.

Genotypes	Observed #	Expected #	X^2 square equation	Sum of X^2
R-Y-	315	9/16 (556) = 313	$(315 - 313)^2/313$	0.01
R- yy	108	3/16 (556) = 104	$(108 - 104)^2/104$	0.15
rr Y-	101	3/16 (556) = 104	$(101 - 104)^2/104$	0.09
rr yy	32	1/16 (556) = 35	$(32 - 35)^2/35$	0.26
				0.51

The number of classes is 4, so the degrees of freedom is 4–1 or 3. Using **Table 5-1** in the text, the probability of having obtained this level of deviation by chance alone is between 0.9 and 0.99 (90 - 99%). Thus we can NOT reject the null hypothesis. In other words, **the data are consistent with independent assortment and we therefore conclude that Mendel's data could indeed result from the independent assortment of the 2 genes.**

5-5.

a. Diagram the cross.

orange (O- bb) x black (oo B-) $\rightarrow$ F_1 brown (O- B-) $\rightarrow$ 100 brown (O- B-) : 25 orange (O- bb) : 22 black (oo B-) : 13 albino (oo bb)

Because the F_1 snakes were all brown, we know that the orange snake could not have contributed an o allele, or there would have been some black snakes. The orange snake must be OO bb. The black snake could not have contributed a b allele or there would have been some orange snakes, so the black parent must be oo BB. Therefore **the F_1 snakes must be Oo Bb**.

b. The F_1 snakes are heterozygous for both genes (Oo Bb). If the two loci assort independently, we expect the F_2 snakes to show a 9 brown : 3 orange : 3 black : 1 albino ratio. The total number of F_2 progeny is 160. **We expect 90 (160 x 9/16) of these progeny to be brown, 30 (160 x 3/16) to be orange, 30 to be black and 10 (160 x 1/16) to be albino**.

c.

Genotypes	Expected #	Observed #	X^2 square equation	Sum of X^2
O- B-	90	100	$(100–90)^2/90$	1.11
O- bb	30	25	$(25–30)^2/30$	0.83
oo B-	30	22	$(22–30)^2/30$	2.13
oo bb	10	13	$(13–10)^2/10$	0.9
				4.97

There are three degrees of freedom (4 classes − 1) and the p value is between 0.5 and 0.1. **The observed values do not differ significantly from the expected.**

d. There is a 10% - 50% probability that these results would have been obtained by chance if the null hypothesis were true; this is simply another way of writing the meaning of the p value.

5-6.

a. The cross is:

normal (DD) x dancer (dd) $\rightarrow$ F_1 normal (Dd) F_2 3/4 D- (normal) : 1/4 dd (dancer)

1/4 of the F$_2$ mice will be dancers if the trait is determined by a single gene with complete dominance.

b. Diagram the cross:

normal (*AA BB*) x dancer (*aa bb*) → F$_1$ normal (*Aa Bb*)   F$_2$ 15/16 normal (*A- B- + A- bb + aa B-*) : 1/16 dancer (*aa bb*)

1/16 of the mice would be expected to be dancers given the second hypothesis that dancing mice must be homozygous for the recessive alleles of two genes.

c. Calculate the chi square values for each situation. Null hypothesis #1: Dancing is caused by the homozygous recessive allele of one gene, so 1/4 of the F$_2$ mice should be dancers. Calculating the expected numbers, $1/4 \times 50$ mice or 13 should have been dancers, 37 should have been nondancers.

Genotypes	Observed #	Expected #	X^2 square equation	Sum of X^2
nondancers	42	(3/4) x 50 = 37	$(42–37)^2/37$	0.68
dancers	8	(1/4) x 50 = 13	$(8–13)^2/13$	1.92
				2.60

With one degree of freedom the *p* value is between 0.5 and 0.1 and we cannot reject the null hypothesis. The hypothesis that dancing is caused by the homozygous recessive allele of one gene is therefore a good fit with the data. Null hypothesis #2: Dancing is caused by being homozygous for the recessive alleles of two genes (*aa bb*), so 1/16 of the F$_2$ mice should be dancers.

Genotypes	Observed #	Expected #	X^2 square equation	Sum of X^2
nondancers	42	47	$(42–47)^2/47$	0.53
dancers	8	3	$(8–3)^2/3$	8.33
				8.86

With one degree of freedom, the *p* value is < 0.005, so the null hypothesis that two genes control the dancer phenotype is not a good fit; in fact, the hypothesis can be rejected by the criteria employed by most geneticists. **The one gene hypothesis is a better fit with the data.**

5-7. This question asks you to weigh the advantages and disadvantages of using Chi square analysis to test for linkage between two genes using two different assumptions.

a. In assumption #1, as seen in Figure 5.5, the expectation of independent assortment is that the frequency of parentals is the same as the frequency of recombinants. In assumption #2, the expectation of independent assortment is the parental and recombinant progeny should be present in a ratio of 1:1:1:1. **Notice that the null hypothesis is the same in both cases: that the genes are assorting independently.**

b.   Assumption #1, 50 progeny:

Genotypes	Observed #	Expected #	X^2 square equation	Sum of X^2
parental	31	25	$(31 - 25)^2/25$	1.44
recombinant	19	25	$(9 - 25)^2/25$	1.44
				2.88

1 dof (degree of freedom), $0.05 < p < 0.1$ and the null hypothesis cannot be rejected - the data is compatible with the hypothesis that the genes are assorting independently.

Assumption #2, 50 progeny:

Genotypes	Observed #	Expected #	X^2 square equation	Sum of X^2
AB	17	12.5	$(17 - 12.5)2/12.5$	1.62
ab	14	12.5	$(14 - 12.5)2/12.5$	0.18
Ab	8	12.5	$(8 - 12.5)/12.5$	1.62
aB	11	12.5	$(11 - 12.5)2/12.5$	0.18
				3.6

3 dof, $0.10 < p < 0.5$ and the null hypothesis cannot be rejected - the data is compatible with the hypothesis that the genes are assorting independently.

Assumption #1, 100 progeny:

Genotypes	Observed #	Expected #	X^2 square equation	Sum of X^2
parental	62	50	$(62 - 50)2/50$	2.88
recombinant	38	50	$(38 - 50)2/50$	2.88
				5.76

1 dof, $0.01 < p < 0.05$ and the null hypothesis is rejected - the genes are not assorting independently, so they must be linked.

Assumption #2, 100 progeny:

Genotypes	Observed #	Expected #	X^2 square equation	Sum of X^2
AB	34	25	$(34 - 25)2/25$	3.24
ab	28	25	$(28 - 25)2/25$	0.36
Ab	16	25	$(16 - 25)2/25$	3.24
aB	22	25	$(22 - 25)2/25$	0.36
				7.2

3 dof, $0.05 < p < 0.1$ and the null hypothesis cannot be rejected - the data is compatible with the hypothesis that the genes are assorting independently.

   In the first experiment with only 50 progeny, $p < 0.05$ using both the 2 class and 4 class assumptions, so neither assumption supports the idea that the genes are linked (although the test with 2 classes is closer to significance). When more progeny are scored, the null hypothesis (that the genes are assorting independently) is rejected when you look at the data as 2 classes but not as

4 classes. This suggests that **using 2 classes is a more sensitive test for linkage than using 4 classes**.

c. There is actually a subtle difference in the null hypotheses: in the 2 class situation the null hypothesis is linkage; in the 4 class situation it is 1:1:1:1, which not only means linkage but also equal viability of the 4 classes. You could imagine a situation in which certain classes are sub-viable, for example any class with the *a* allele. In such a case, you might see linkage with the 2 class test, but you would miss the even more important point that one allele causes reduced viability. **This ability to see the relative viability of the alleles is an advantage to the 4 class method**.

Section 5.3 – Recombination Results from Crossing-Over During Meiosis

5-8. The parents are from true breeding stocks. Diagram the cross:

raspberry eye color ♂ x sable body color ♀ → F_1 wild type eye and body color ♀ x sable body color ♂ (if no mention is made of the eye color, then it is assumed to be wild type) → F_2 216 wild type ♀ : 223 sable ♀ : 191 sable ♂ : 188 raspberry ♂ : 23 wild type ♂ : 27 raspberry sable ♂ 3EQ #1 - the phenotypes seem to be controlled by 2 genes, one for eye color and the other for body color. 3EQ #2 - the F_1 female progeny show that the wild type allele is dominant for body color ($s^+ >$ s) and the wild type allele is also dominant for eye color ($r^+ > r$). Question #3 - sable body color is seen in the F_1 males but not the F_1 females, so the s gene is X-linked. In the F_2 generation the raspberry eye color is seen in the males but not in the females, so *r* is also an X-linked gene. We can now assign genotypes to the true-breeding parents in this cross:

$r s^+$ / Y (raspberry ♂) x $r^+ s$ / $r^+ s$ ♀ → F_1 $r s^+$ / $r^+ s$ ♀ (wild type) x $r^+ s$ / Y (sable ♂) →

[the heterozygous F_1 female can make the following gametes: (parentals) $r s^+$, $r^+ s$ and (recombinants) $r s$, $r^+ s^+$; the F_1 male can make Y and $r^+ s$ gametes] → F_2 will be $r s^+$ / $r^+ s$ (wild type females), $r^+ s$ / $r^+ s$ (sable females), $r s$ / $r^+ s$ (sable females), $r^+ s^+$ / $r^+ s$ (wild type females), $r^+ s$ / Y (sable males), $r s^+$ / Y (raspberry males), $r s$ / Y (raspberry sable males), $r^+ s^+$ / Y (wild type males)

The F_1 female is heterozygous for both genes and will therefore make parental and recombinant gametes. The F_1 male is not a true test cross parent, because he does not carry the recessive alleles for both of the X-linked genes. However, this sort of cross can be used for mapping, because F_2 sons receive only the Y chromosome from the F_1 male and are hemizygous for the X chromosome from the F_1 female. Thus the phenotypes in the F_2 males represent the array of parental and recombinant gametes generated by the F_1 female as well as the frequencies of these gametes. Using the F_2 males, the **rf between sable and raspberry = 23 (wild type males) + 27 (raspberry sable males) / 429 (total males) = 0.117 x 100 = 11.7 cM.**

5-9.

a. The parental females are heterozygous for the X-linked dominant Greasy fur (*Gs*) and Broadhead (*Bhd*) genes. They are crossed to homozygous recessive males. The male progeny of this cross will show four phenotypes, which will fall into two genetic reciprocal pairs. One pair will be Gs Bhd and wild type (*Gs Bhd* and *Gs$^+$ Bhd$^+$*) while the other pair will be Gs and Bhd (*Gs Bhd$^+$* and *Gs$^+$ Bhd*). If the genes are far apart on the X chromosome then they will assort independently and all four classes will show equal frequency. If the genes are more closely linked then there will be a more frequent reciprocal pair (the parental pair) and a less frequent pair (the recombinant pair). There are 51 *Gs Bhd$^+$* and 48 *Gs$^+$ Bhd* male progeny. Thus, this is the parental pair and the genotype of the female is *Gs Bhd$^+$ / Gs$^+$ Bhd*. **The cross can be diagrammed:**

$\quad$ **Gs Bhd$^+$ / Gs$^+$ Bhd ♀ x Gs$^+$ Bhd$^+$ / Y ♂ → 51 Gs Bhd$^+$ ♂ : 48 Gs$^+$ Bhd ♂ : 2 Gs Bhd ♂**

$\quad$ **: 1 Gs$^+$ Bhd$^+$ ♂.**

The distance between these two genes: rf = 2 + 1 / 100 = 0.03 = 3mu.

b. The daughters in this cross must inherit the *Gs$^+$ Bhd$^+$* X chromosome from their fathers. This chromosome carries the recessive alleles of both genes, so this is a true test cross. Thus the **genotypes, phenotypes and frequencies of the female progeny would be the same as their brothers.**

5-10.

a. Diagram the cross:

$\quad$ *CC DD* x *cc dd* → F_1 *C D / c d* x *cc dd* → 903 *Cc Dd*, 897 *cc dd*, 98 *Cc dd*, 102 *cc Dd*.

Because the gamete from the homozygous recessive parent is always *c d*, we can ignore one *c* and one *d* allele (the *c d* homolog) in each class of the F_2 progeny. The remaining homolog in each class of F_2 is the one contributed by the doubly heterozygous F_1, the parent of interest when considering recombination. In the F_2 the two classes of individuals with the greatest numbers represent parental gametes (*C D* or *c d* from the heterozygous F_1 parent combining with the *c d* gamete from the homozygous recessive parent). The other two types of progeny result from a recombinant gametes (*C d* or *c D* combining with the *c d* gamete from the homozygous recessive parent). The number of recombinants divided by the total number of offspring x 100 gives the map distance: $(98 + 102)/(903 + 897 + 98 + 102) = 200/2000 = 0.01 \times 100 = 10\%$ rf or **10 map units (mu) or 10 cM**.

b. $CC\,dd \times cc\,DD \rightarrow F_1\ C\,d\,/\,c\,D \times c\,d\,/\,c\,d \rightarrow$ as determined in part a, c and d are 10 cM apart. Thus the gametes produced by the heterozygous F_1 will be 45% $C\,d$, 45% $c\,D$, 5% $C\,D$, 5% $c\,d$. After fertilization with $c\,d$ gametes, there would be **45% $Cc\,dd$, 45% $cc\,Dd$, 5% $Cc\,Dd$, 5% $cc\,dd$**. Because this is a test cross, the gametes from the doubly heterozygous F1 parent determine the phenotypes of the progeny.

5-11. To determine the probability that a child will have a particular genotype, look at the gametes that can be produced by the parents. In this example, A and B are 20 mu apart, so in a doubly heterozygous individual 20% of the gametes will be recombinant and the remaining 80% will be parental. The *aa bb* homozygous man can only produce *a b* gametes. The doubly heterozygous woman, with a genotype of $A\,B\,/\,a\,b$, can produce 40% $A\,B$, 40% $a\,b$, 10% $A\,b$ and 10% $a\,B$ gametes. (Total of recombinant classes = 20%.) **The probability that a child receives the $A\,b$ gamete from the female (and is therefore $A\,b\,/\,a\,b$) = 10%.**

5-12.

a.   Designate the alleles: H = Huntington allele, h = normal allele; B = brachydactyly, b = normal fingers. **John's father is *bb Hh*; his mother is *Bb hh***.

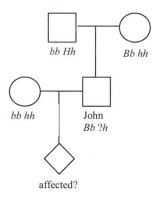

b.   We know the John is *Bb* because he has brachydactyly. **His complete genotype could be *Bb Hh* or *Bb hh***. The probability that John's genotype is Hh = 1/2 (chance that he inherited the H allele from his father) x 1/3 (probability of >50 years of age for onset of symptoms) = 1/6.

c.   The probability that the child will have and express brachydactyly = 1/2 (probability of inheriting B from John) x 9/10 (probability of expressing the phenotype) = 0.45. The probability that the child will express Huntington's disease by age 50 = 1/6 (probability that John is Hh) x 1/2 (probability that the child will inherit the H allele) x 2/3 (probability the child will show symptoms of Huntington's by age 50) = 0.056. **If brachydactyly and Huntington's assort independently the probability that the child will express both phenotypes by age 50 = 0.45 (probability of expressing brachydactyly) x 0.056 (probability of expressing Huntington's by age 50) = 0.025.**

d.   If the two loci are linked, the alleles on each of John's homologs will be either *B h / b H* or *B h / b h*. If it is the former (and John has Huntington's), the *B H* gamete could only be produced by recombination. The probability of that specific recombinant gamete = 10% (*b h* constitutes the other 10% of the recombinant gametes). As shown in part a, the probability that John's genotype is *B h / b H* = 1/6. **The probability that John's child will inherit both the Huntington and brachydactyly alleles = 1/6 (probability that John is *B h / b H*) x 1/10 (probability of child inheriting *B H* recombinant gamete from John) x 9/10 (probability of expressing brachydactyly) x 2/3 (probability of expressing Huntington's by age 50) = 0.01.**

5-13. Designate the alleles of the genes: A = normal pigmentation and a = albino allele; $Hb\beta^A$ = normal globin and $Hb\beta^S$ = sickle allele.

a.   Because both traits are rare in the population, we assume that the parents are homozygous for the wild type allele of the gene dictating their normal traits (that is, they are not carriers). Diagram the cross: *a Hbβ^A / a Hbβ^A* (father) x *A Hbβ^S / A Hbβ^S* (mother) → *a Hbβ^A / A Hbβ^S* (son). Given that the genes are separated by 1 map unit, parental gametes = 99% and recombinant gametes = 1% of the gametes. **The son's gametes will consist of: 49.5% *a Hbβ^A*, 49.5% *A Hbβ^S*, 0.5% *a Hbβ^S* and 0.5% *A Hbβ^A*.**

b.   In this family, the cross is *A Hbβ^A / A Hbβ^A* (father) x *a Hbβ^S / a Hbβ^S* (mother) → *a Hbβ^S / A Hbβ^A* (daughter). **The daughter's gametes will be: 49.5% *a Hbβ^S*, 49.5% *A Hbβ^A*, 0.5% *a Hbβ^A* and 0.5% *A Hbβ^S*.**

c.   The cross is: *a Hbβ^A / A Hbβ^S* (son) x *a Hbβ^S / A Hbβ^A* (daughter). **The probability of an *a Hbβ^S / a Hbβ^S* child (sickle cell and anemic) = 0.005 (probability of *a Hbβ^S* from son) x 0.495 (probability of *a Hbβ^S* from daughter) = 0.0025.**

5-14. Diagram the cross:

blue smooth x yellow wrinkled → 1447 blue smooth, 169 blue wrinkled, 186 yellow smooth, 1510 yellow wrinkled.

a.   To determine if genes are linked, first predict the results of the cross if the genes are unlinked. In this case, a plant with blue, smooth kernels (*A- W-*) is crossed to a plant with yellow, wrinkled kernels (*aa ww*). Since there are four classes of progeny, the parent with blue, smooth kernels must be heterozygous for both genes (*Aa Ww*). From this cross, **we would predict equal numbers of all four phenotypes in the progeny if the genes were unlinked. Since the numbers are very skewed, with the smaller classes representing recombinant offspring, the genes are linked. rf = (169 + 186) / (1447 + 169 + 186 + 1510) = 355/3312 = 10.7% = 10.7 cM**.

b. The genotype of the blue smooth parent was *Aa Ww*. The arrangement of alleles in the parent is determined by looking at the phenotypes of the largest classes of progeny (the parental reciprocal pair). Since blue, smooth and yellow wrinkled are found in the highest proportion, *A W* must be on one homolog and *a w* on the other = *A W / a w*.

c. The genotype of the blue, wrinkled progeny is *A w /a w*. This genotype can produce *A w* or *a w* gametes in equal proportions. Recombination can NOT be detected here, as this genotype is only heterozygous for one gene. Notice that recombination between these homologs yields the same two combinations of alleles (*A w* and *a w*) as the parental, so each type of gamete is expected 50% of the time. The yellow smooth progeny plant has a genotype of *a W / a w*. Again since recombination cannot be detected, the frequency of each type is 50%. Thus the cross is *A w / a w* (blue wrinkled) x *a W / a w* (yellow smooth). **Four types of offspring are expected in equal proportions: 1/4 *A w / a W* (blue smooth) : 1/4 *A w / a w* (blue wrinkled) : 1/4 *a w / a W* (yellow smooth) : 1/4 *a w / a w* (yellow wrinkled).**

5-15. Diagram the cross: *CC bb* (brown rabbits) x *cc BB* (albinos) → F_1 *Cc Bb* x *cc bb* → 34 black : 66 brown : 100 albino.

a. If the genes are unlinked, the F_1 will produce *C B, c b, C b* and *c B* gametes in equal proportions. A mating to animals that produce only *c b* gametes will produce four genotypic classes of offspring: 1/4 *Cc Bb* (black) : 1/4 *cc Bb* (albino) : 1/4 *Cc bb* (brown) : 1/4 *cc BB* (albino). The ratio would be **1/4 black : 1/2 albino : 1/4 brown**.

b. If *c* and *b* are linked, then the genotype of the F_1 class is *C b / c B* and you would expect the parental type *C b* and *c B* gametes to predominate; *c b* and *C B* are the recombinant gametes and are therefore present at lower levels. The parental gametes are represented in the F_2 by the *Cc bb* (brown) and *cc Bb* (albino) classes. Since we cannot distinguish between the albinos resulting from fertilization of recombinant gametes (*c b*), and those resulting from parental gametes (*c B*), we have to use the proportion of the *C B* recombinant class and assume that the other class of recombinants (*c b*) is present in equal frequency. Since the crossing-over is a reciprocal exchange, this assumption is reasonable. There were 34 black progeny; assuming that 34 of the 100 albino progeny were the result of recombinant gametes, **the genes are (34 + 34) recombinant / 200 total progeny = 34% rf = 34 cM apart.**

5-16. Notice that you are asked for the <u>number of different kinds of phenotypes</u>, not the number of individuals with each of the different phenotypes.

a.   **2** (*A*- and *aa*);

b.   **3** (*AA*, *Aa* and *aa*);

c.   **3** (*AA*, *Aa* and *aa*);

d.   **4** (*A- B-*, *A- bb*, *aa B-*, *aa bb*);

e.   **4** (*A- B-*, *A- bb*, *aa B-*, *aa bb*; because the genes are linked, the frequency of the four classes will be different than that seen in part d);

f.   **Nine** phenotypes in total. There are three phenotypes possible for each gene. The total number of combinations of phenotypes is $(3)^2 = 9$ ((*AA*, *Aa* and *aa*) x (*BB*, *Bb* and *bb*)).

g.   Normally there are four phenotypic classes, as in parts d and e (*A-B-*, *A-bb*, *aaB-* and *aabb*). In this case, one of the genes is epistatic so two of the classes have the same phenotype, **giving 3 phenotypic classes**.

h.   Two genes means four phenotypic classes (*A- B-*, *aa B-*, *A- bb* and *aa bb*). Because gene function is duplicated the first three classes are all phenotypically equivalent in that they have function, and only the *aa bb* class will have a different phenotype, being without function. Thus there are only **2 phenotypic classes**.

i.   There is 100% linkage between the two genes. The number of phenotypic classes will depend on the arrangement of alleles in the parents. **If the parents are *A B / a b* x *A B / a b*, the progeny will be 3/4 *A- B-* : 1/4 *aa bb* and there will be two phenotypic classes. If the parents are *A b / a B* x *A b / a B*, the progeny will be 1/4 A- b : 1/2 A- B- : 1/4 a B- and there will be three phenotypic classes in the offspring.**

5-17.

a.   *AA BB* x *aa bb* → F$_1$ *A B / a b* (*A B* on one homolog and *a b* on the other homolog). The F1 progeny will produce gametes of the parental types *A B* and *a b;* and of the recombinant types *A b* and *a B*. Because the genes are 40 cM apart, **the recombinants will make up 40% of the gametes (20% *Ab* and 20% *aB*). The parental gametes make up the remaining 60% of the gametes, 30% *A B* and 30% *a b*.** Set up a Punnett square to calculate the frequency of the 4 phenotypes in the F$_2$ progeny. In the Punnett square the phenotypes are shown in parentheses (*A-bb*). **The F$_2$ phenotypic ratio is: 0.59 *A- B-* : 0.16 *A- bb* : 0.16 *aa B-* : 0.09 *aa bb*.**

	0.3 *A B*	0.3 *a b*	0.2 *A b*	0.2 *a B*
0.3 *A B*	*A B / A B* (0.09 *A- B-*)	*A B / a b* (0.09 *A- B-*)	*A B / A b* (0.06 *A- B-*)	*A B / a B* (0.06 *A- B-*)
0.3 *a b*	*a b / A B* (0.09 *A- B-*)	*a b / a b* (0.09 *aa bb*)	*a b / A b* (0.06 *A- bb*)	*a b / a B* (0.06 *aa B-*)
0.2 *A b*	*A b / A B* (0.06 *A- B-*)	*A b / a b* (0.06 *A- bb*)	*A b / A b* (0.04 *A- bb*)	*A b / a B* (0.04 *A- B-*)
0.2 *a B*	*a B / A B* (0.06 *A- B-*)	*a B / a b* (0.06 *aa B-*)	*a B / A b* (0.04 *A- B-*)	*a B / a B* (0.04 *aa B-*)

b.   If the original cross was *AA bb* x *aa BB*, the allele combinations in the F_1 would be *A b / a B*. Parental gametes in this case are 30% *A b* and 30% *a B* and the recombinant gametes are 20% *A B* and 20% *a b*. Set up a Punnett square, as in part a. **The F_2 phenotypic ratio is: 0.54 *A- B-* : 0.21 *A- bb* : 0.21 *aa B-* : 0.04 *aa bb*.**

5-18. v^1, v^2, and v^3 are codominant alleles of the DNA variant marker locus, while *D* and *d* are alleles of the disease gene. The marker and the disease locus are linked. Diagram the cross:

$v^1 D / v^2 d$ (father) x $v^3 d / v^3 d$ (mother) → $v^2 ? / v^3 d$.

The fetus <u>must</u> get a $v^3 d$ homolog from the mother. The fetus also received the v^2 allele of the marker. The father could have given a $v^2 d$ (non-recombinant) or a $v^2 D$ (recombinant) gamete.

a.   If the *D* locus and the v^1 allele of the marker are 0 mu apart (there is no recombination between them; they appear to be the same gene), **the probability that the child, who received the v^2 allele of the marker, has the *D* allele = 0**.

b.   If the distance between the disease locus and marker is 1 mu, 1% of the father's gametes are recombinant between *D* and the marker locus. Half of the recombinant gametes (0.5%) will be v^2 *D;* the other half will be $v^1 d$. Because we are only considering cases where the child has the v^2 marker, **the probability that the child inherited *D* = 0.05 (probability of inheriting v^2 *D* / 0.5 (probability of inheriting v^2) = 0.01 = 1%.**

c.   **5%**.

d.   **10%**.

e.   **50%**.

Section 5.4 – Mapping: Locating Genes Along a Chromosome

5-19.

a. There are four equally frequent phenotypic classes in the progeny must include a reciprocal pair of parentals and a reciprocal pair of recombinants. The parental reciprocal pair could either be wild type wings, wild type eyes and dumpy wings, brown eyes or wild type wings, brown eyes and dumpy wings, wild type eyes. The first possibility means the genotype of the heterozygous female was $dp^+ bw^+ / dp\ bw$ while the second pair means the genotype of the female was $dp\ bw^+ / dp^+ bw$. In either case the recombination frequency is 50% (178 + 181 / 716 or 185 + 172 / 716) so **the two genes are assorting independently**.

b. In this cross the heterozygous parent is the male and there is no recombination in *Drosophila* males! If dumpy and brown were on separate chromosomes then they must assort independently and there would be four classes of progeny as in part a. If **the two genes are on the same chromosome** then the lack of recombination in the male parent means that only the two parental classes of progeny are seen. Also, **the genotype of the heterozygous male must have been $dp^+ bw^+ / dp\ bw$**.

c. If the genes are far enough apart on the same chromosome they will assort independently in the first cross. This is because (1) **recombination occurs at the four strand stage of meiosis**, and (2) **so many crossovers occur between genes when they are far apart on the same chromosome that the linkage between alleles of these genes will be randomized**. Independent assortment does not occur in the second cross because there is no recombination in *Drosophila* males. In *Drosophila*, therefore, **it is a simple matter to decide if genes are syntenic** (linked to the same chromosome). A cross between a male heterozygous for the two genes of interest and a recessive female will clearly distinguish between genes on separate chromosomes and syntenic genes. In the case of genes on separate chromosomes there will be four equally frequent classes of progeny. In the case of genes on the same chromosome (even genes that are very far apart on the same chromosome) there will only be two classes of progeny – the parental classes.

d. This distance can not be measured in any two-point cross. The measured recombination frequency can be no higher than 50% when looking at data involving a single pair of genes. Large genetic distances can be measured accurately only **by summing up the values obtained for smaller distances separating other genes in between those at the ends**.

5-20. Diagram the cross:

wild type ♀ x reduced cinnabar ♂ → F₁ ♀ x F₁ ♂ → 292 wild type, 9 cinnabar, 7 reduced, 92 reduced cinnabar.

Two genes are involved in this cross, but the frequencies of the phenotypes in the second generation offspring do not look like frequencies expected for a cross between double heterozygotes for two independently assorting genes (9:3:3:1). The genes must be linked. Designate the alleles: cn^+ = wild type, cn = cinnabar; rd^+ = wild type, rd = reduced. The cross is:

$cn^+ rd^+ / cn^+ rd^+$ ♀ x $cn\,rd / cn\,rd$ ♂ → F₁ $cn^+ rd^+ / cn\,rd$.

Recombination occurs in *Drosophila* females but <u>not</u> in males. Thus males can only produce the parental $cn^+ rd^+$ or $cn\,rd$ gametes. The females produce both the parental gametes and the recombinant gametes $cn^+ rd$ and $cn\,rd^+$.

female gametes	male gamete $cn^+ rd^+$	male gamete $cn\,rd$
$cn^+ rd^+$ (parental)	$cn^+ rd^+ / cn^+ rd^+$ (wild type)	$cn^+ rd^+ / cn\,rd$ (wild type)
$cn\,rd$ (parental)	$cn\,rd / cn^+ rd^+$ (wild type)	$cn\,rd / cn\,rd$ (cinnabar reduced)
$cn^+ rd$ (recombinant)	$cn^+ rd / cn^+ rd^+$ (wild type)	$cn^+ rd / cn\,rd$ (reduced)
$cn\,rd^+$ (recombinant)	$cn\,rd^+ / cn^+ rd^+$ (wild type)	$cn\,rd^+ / cn\,rd$ (cinnabar)

The reduced flies and the cinnabar flies are recombinant classes. However, there should be an equal number of recombinant types that have a wild-type phenotype because they got a $cn^+ rd^+$ gamete from the male parent. If we assume that these recombinants are present in the same proportions, then rf = 2 x (7+9)/400 = 8% recombination. **The genes are separated by 8 cM.**

5-21.

The shortest distances are the most accurate. Therefore begin assembling a map using the genes that are closest together. MAT-LEU2 are 16 cM apart and MAT-THR4 are 16 cM apart. This gives two possible maps: MAT - (16 cM) - LEU2/THR4 or LEU2 - (16 cM) - MAT - (16 cM) - THR4. Because THR4 is 35 cM from LEU2 the second map must be the correct one. Note that the two smaller distances to not sum to the longer distance because these recombination frequencies are based on two point crosses. The HIS4 gene is 23 cM from LEU2. This initially leads to two possible maps: HIS4 - (23 cM) - LEU2 - MAT - THR4 or LEU2 - (23 cM) - HIS4/MAT - THR4. The first map is the most likely because HIS4 is 37 cM from MAT instead of being very close to MAT. So **the order of the genes is HIS4 - LEU2 - MAT - THR4** (or the inverse).

5-22.

a. A test cross is the best way to find the map distance between two genes. One parent must the heterozygous for both genes and the other parent must the homozygous recessive. Thus the cross would be **Bb Cc x bb cc**. Because these genes are syntenic the heterozygous parent may be either B C / b c or B c / b C.

b. There are two possible orders for these genes:

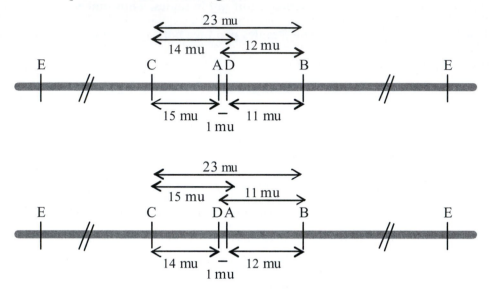

c. Because the map positions of A and D are so close, **the relative order of A and D is not clear. Another uncertainty is the location of gene E**. This gene must be on this chromosome because we are told all of these genes are syntenic. However the E gene is genetically unlinked to any of the other genes (A, B, C and D).

d. To figure out the actual order of the genes you must **do a three point cross with either B A D or A D C and order the genes** by finding the double cross over class, comparing this to the Parental class and ascertaining the order of the genes. This is described in the 'How to Begin - Three Point Crosses' section at the beginning of this chapter (page 66). Also see problem 5-24c for a further explanation of this method of determining the gene order.

   The location of the E gene can be determined by finding new genes that are genetically linked to the left of C and to the right of B in the maps seen in part b. This **extension of the linkage group** will eventually allow the placement of gene E when shows genetic linkage to one of these new genes.

5-23.

Although this problem considers three different genes each cross only considers two at a time. You are thus asked to construct a map of the three genes based on 3 two point crosses. Note the unusual usage of a comma to separate the genes in the genotypes written for the parents of each cross. This is meant to show that we don't know if any of the genes are on the same chromosome. If the two genes in each cross are assorting independently then the four phenotypic classes in the test cross progeny must occur in equal frequencies (see problem 5-17). This is clearly not the case in any of the three crosses. Therefore the *mb*, *e* and *k* genes are all linked. The genotypes of the parents in the first cross may be correctly written as $mb^+ e^+ / mb^+ e^+$ x *mb e* / *mb e*.

In the first cross, the $mb^+ e^+$ and *mb e* classes are the parentals. This is based both on the known genotypes of the true breeding parental generation and on the fact that this reciprocal pair is the most frequent. The recombinants are the $mb^+ e$ and the $mb e^+$ flies, so the rf = 11 + 15 / 250 = 0.104 = 10.4mu. In the second cross the recombinant reciprocal pair is $k^+ e^+$ and *k e* so the recombination frequency is 11 + 7 / 312 = 0.058 = 5.8mu. The $k mb^+$ and $k^+ mb$ classes are the recombinant reciprocal pair for the third cross. In this case the rf = 11 + 15 / 422 = 0.062 = 6.2mu. The *mb* and *e* genes in the first cross are the furthest apart, so the *k* gene must be in between them. Thus the map of this data is:

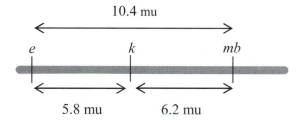

Note that the distance calculated between *e* and *mb* in cross one (10.4mu) does NOT equal the sum of the two shorter distances (12.0mu). The distance calculated from cross one is less accurate because it is based on single crossovers between e and mb and does not include any of the double crossovers that occurred in the e – k and k – mb regions simultaneously. Thus, **the best map of these genes is:**

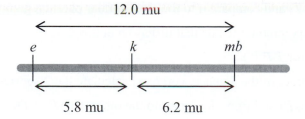

5-24. In foxgloves wild type flower color is red and the mutant color is white; the mutation peloria causes the flowers at the apex of the stem to be very large; normal foxgloves are very tall, and dwarf affects the plant height. When you describe the phenotype of an individual you usually only refer to the non-wild type traits. The cross is:

white flowered x dwarf peloria → F_1 white flowered x dwarf peloria → 172 dwarf peloria, 162 white, 56 dwarf peloria white, 48 wild type, 51 dwarf white, 43 peloria, 6 dwarf, 5 peloria white.

a.   Because there is only 1 phenotype of F_1 plant, the parents must have been homozygous for all three genes. The phenotype of the F_1 heterozygote indicates the **dominant alleles: white flowers , tall stems, and normal-sized flowers** (3EQ #2).

b.   Designate the alleles for the 3 genes: W = white, w = red; P = normal-sized flowers, p = peloria; T = tall, t = dwarf. The cross is: ***WW PP TT* (white flowered) x *ww dd pp* (dwarf peloria)**.

c.   Note that all 3 of these genes are genetically linked. There are only 2 classes of parental progeny as defined both phenotypically and numerically. In order to draw a map of these genes, organize the test cross data, figure out which of the 3 genes is in the middle and calculate the recombination frequencies in regions 1 and 2. By definition, the parentals are the class with the same phenotype as the original parents of the cross, and the DCO class is the least frequent reciprocal pair of progeny. At this point, arbitrarily one of the remaining 2 reciprocal pairs of progeny is SCO 1 and the last remaining pair is SCO 2.

Classes of gametes	Genotype	Numbers
Parental (P)	$W\,P\,T$	162
	$w\,p\,t$	172
SCO 1	$W\,p\,t$	56
	$w\,P\,T$	48
SCO 2	$W\,P\,t$	51
	$w\,p\,T$	43
DCO	$w\,P\,t$	6
	$W\,p\,T$	5

Compare the DCO class with the parental class. Remember that a DCO comes from a meiosis with a simultaneous crossover in regions 1 and 2. As a result, the allele of the gene in the middle switches with respect to the alleles of the genes on the ends. In this data set, take one of the DCO classes (for example $W\,p\,T$) and compare it to the most similar parental gamete ($W\,P\,T$). Two alleles out of three are in common; the one that differs (p in this case) is the gene in the middle. Thus, the order is $W\,P\,T$ (or $T\,P\,W$).

Once you know the order of the 3 genes, you must calculate the 2 shortest distances - from the end to the center (W to P) and from the center to the other end (P to T). The total number of progeny in this test cross = 543. rf W-P = (56 + 48 + 6 + 5)/543 = 115/543 = 21.2 cM; rf P-T = (51 + 43 + 6 + 5)/543 = 19.3 cM.

The map is:

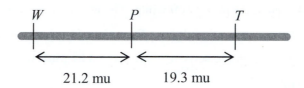

21.2 mu   19.3 mu

d. Interference (I) = 1 - coefficient of coincidence (coc)

coc = frequency of observed DCO / frequency of expected DCO

As discussed in <u>problem 5-25</u>, the expected percentage of double crossovers is the product of the frequency of recombination in each interval: (0.193) x (0.212) = 0.0409. The observed DCO frequency = 11/543 = 0.0203; **coc = 0.0203/0.0409 = 0.496; I = 1 - 0.496 = 0.504.**

5-25.

a. See <u>problem 5-29</u> for a detailed explanation of the methodology. Diagram the cross:

$a^+ b^+ c^+ / a\,b\,c\,♀$ x $a\,b\,c / a\,b\,c\,♂$ → ??. Recombination occurs in *Drosophila* females, so the female parent will make the following classes of gametes. Because this is a test cross, these gametes will determine the phenotypes in the progeny.

Classes of gametes	Genotype	Frequency of reciprocal pair	Numbers	Frequency of each class	# of progeny freq x 1,000
Parental (P)	$a^+ b^+ c^+$ $a\ \ b\ \ c$	1 - all else	1 - (0.18 + 0.08 + 0.02) = 0.72	0.36 0.36	**360** **360**
SCO 1	$a^+ b\ \ c$ $a\ \ b^+ c^+$	rf in region 1 = SCO 1 + DCO	SCO 1 = 0.2 - 0.02 = 0.18	0.09 0.09	**90** **90**
SCO 2	$a^+ b^+ c$ $a\ \ b\ \ c^+$	rf in region 2 = SCO 2 + DCO	SCO 2 = 0.1 - 0.02 = 0.08	0.04 0.04	**40** **40**
DCO	$a^+ b\ \ c^+$ $a\ \ b^+ c$	(rf in region 1) x (rf in region 2)	(0.2) x (0.1) = 0.02	0.01 0.01	**10** **10**

b. Diagram the cross: $a^+ b^+ c^+ / a\,b\,c\,♂$ x $a\,b\,c / a\,b\,c\,♀$ → ??. Here, the heterozygous parent is the male. Recombination DOES NOT OCCUR in male *Drosophila*. Thus, the heterozygous parent will only give 2 types of gametes, the parental types. If you score 1,000 progeny of this cross, you will find **500 $a^+ b^+ c^+$ and 500 $a\,b\,c$ progeny.**

5-26. Diagram the cross:

pink petals, black anthers, long stems x pink petals, black anthers, long stems →

a. The cross is pink x pink → (78 + 6 + 44 + 15) red : (39 + 13 + 204 + 68) pink : (2 + 2 + 117 +
    39) white = 143 red : 324 pink : 160 white. The appearance of two new phenotypes (red and
    white) suggest that flower color shows incomplete dominance. This is confirmed by the 1:2:1
    monohybrid ratio seen in the self-cross progeny. **The pink flowered plants are *Pp*, red are *PP***
    and white are *pp*.

b. The expected ratio of red: pink : white would be 1:2:1. Calculating for the 650 plants, this equals
    162.5 red, 325 pink, and 162.5 white.

c. The monohybrid ratio of the black and with tan anthers =. (78+26+39+13+5+2) tan : (44 + 15 +
    204 + 68 + 117 + 39) black = 163 tan : 487 black = ~ **1 tan : 3 black. Therefore black (B) is**
    dominant to tan (b). The monohybrid ratio for the stem length = 487 long stems : 163 short
    stems. = ~ **3 long :1 short, so long (L) is dominant to short (s)**.

d. Designate the alleles: *P* = red, *Pp* = pink, *p* = white; *B* =black, *b* =tan, *L* =long, *l* =short. Because
    all 3 monohybrid phenotypic ratios are characteristic of heterozygous crosses, **the genotype of**
    the original plant is *Pp Bb Ll*.

e. If the stem length and anther color genes assort independently, the 9:3:3:1 phenotypic ratio
    should be seen in the progeny. Totaling all the progeny in each of the classes: 365 long black :
    122 short black : 122 long tan : 41 short tan. The observed dihybrid ratio is close to a 9:3:3:1
    ratio, so **the genes for anther color (B) and stem length (L) are unlinked**.

    The expected monohybrid ratio for flower color is 1 red : 2 pink :1 white, while that for stem
    length is 3 long :1 short. If the two genes are unlinked, the expected dihybrid ratio can be
    calculated using the product rule to give 3/8 long pink : 3/16 long red : 3/16 long white : 1/8 short
    pink : 1/16 short red : 1/16 short white, or a 6 L- Pp :3 L- PP :3 L- pp :2 ll Pp :1 ll PP :1 ll pp
    ratio. The observed numbers are: 243 long pink : 122 long red : 122 long white : 81 short pink :
    41 short red : 41 short white. This observed ratio is close to the predicted ratio, so **the genes for**
    petal color (P) and stem length (L) are unlinked.

    The same analysis is done for flower color and anther color. The expected dihybrid ratio here
    is also 6:3:3:2:2:1. The observed numbers are: 272 black pink : 59 black red : 156 black white :
    52 tan pink : 104 tan red : 7 tan white. Because these numbers do not fit a 6:3:3:2:1:1 ratio, we
    can conclude that **flower color (P) and anther color (B) are linked genes**.

f. **It is clear that the original snapdragon was heterozygous for *Pp* and *Bb*. However the**
    genotype of this heterozygous plant could have been either *P B / p b* or *P b / p B*. If it is the
    former and the genes are closely linked, then the *pp bb* phenotype will be very frequent
    (almost 1/4 of the progeny because it is non-recombinant). Instead the *pp bb* class is very

infrequent, accounting for only about 1% (7/650) of the progeny. Thus, the parental genotype must have been *P b* / *p B*. In this case, the infrequent *pp bb* class received a *p b* recombinant gamete from both parents (see problem 5-30 parts a & b). **The frequency of the *pp bb* genotype = (probability of the *p b* recombinant gamete)^2. The recombination frequency (rf) between the P gene and the B gene = frequency of *P B* + *p b* gametes = 2($\sqrt{}$(#*pp bb*/# total progeny)) = 2($\sqrt{}$(7/650)) = ~20 map units.**

5-27.

a.   wild type ♀ x scute echinus crossveinless black ♂ → 16 classes of data! Among the 16 classes there are wild type and sc, ec, cv and bl. This tells you that the parental female was heterozygous for all 4 traits. The fact the parental female is wild type also tells you that the wild-type alleles of all 4 genes are dominant. Remember that if you do a test cross with a female that is heterozygous for <u>3</u> linked genes, the data shows a very specific pattern. Because each type of meiosis (no recombination, DCO, etc.) gives a pair of gametes, you will see 8 classes of data that will occur in 4 pairs, both genetically and numerically. Thus, if you begin a cross with a female that is heterozygous for 4 linked genes, you should see 16 classes of progeny in a pattern of 8 genetic and numeric pairs. Although we have 16 classes of data, they do NOT occur in numeric pairs - instead we see numeric groups of 4. If one (or more) of the genes instead assorts independently of the others, then in a test cross you must see numeric groups of 4. For example, the most frequent classes will be parental, and there will be a 1:1:1:1 ratio of the 4 parental types.

 Which gene is assorting independently relative to the other 3 genes? To answer this question, list the genotypes of the gametes that came from the heterozygous parent in the largest group of 4. This group should include the parental classes for the 3 linked genes; there are 4 genotypes here to account for the independent assortment of the unlinked gene. Then choose one of the 4 genes, and remove the allele of that gene from all 4 groups. When you do this with the gene that is assorting independently of the rest, you will find there are only 2 reciprocal classes of data left, which are the parental classes for the linked genes. If you choose one of the linked genes to remove, you will still have 4 different phenotypic classes left. Try removing the *b* gene first. You see that when the *b* allele is removed there are only 2 classes left, *s e c* and + + +; when this analysis is repeated removing the allele of the *s* gene there are still 4 different phenotypes left.

Genotype	Numbers	remove b	remove s
$b s e c$	653	$s e c$	$b e c$
$+ s e c$	670	$s e c$	$+ e c$
$+ + + +$	675	$+ + +$	$+ + +$
$b + + +$	655	$+ + +$	$b + +$

The b gene is therefore assorting independently of the other 3 genes. When the b gene is removed from all 16 classes of data, you can see that the data reduces to 8 classes that form 4 genetic and numeric reciprocal pairs, just as in any three-point cross (see answer to part b below). Thus, the genotype of the parental female is:

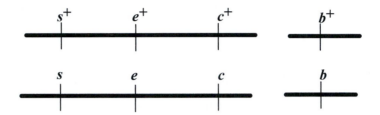

b. Write the classes out as reciprocal pairs.

Classes of gametes	Genotype	Numbers
Parental (P)	$+ + +$	1323
	$s e c$	1330
SCO 1	$s + +$	144
	$+ e c$	147
SCO 2	$s e +$	171
	$+ + c$	169
DCO	$s + c$	2
	$+ e +$	2

Compare the DCO $s + c$ to the parental $s e c$; this shows that e is in the middle. Calculate rf $s - e = (144 + 147 + 2 + 2)/3288 = 9.0$ cM; rf $e - c = (171 + 169 + 2 + 2)/3288 = 10.5$ cM.

```
         s                    e                    c
         |                    |                    |
         +--------------------+--------------------+
         <------------------->|<------------------->
              9 mu                   10.5 mu
```

c. Interference = 1 - coefficient of coincidence

coc = observed DCO frequency / expected DCO frequency = $(4/3288) / (0.09 \times 0.105) = 0.001 / 0.009 = 0.11$

I = 1 − 0.11 = **0.89; yes**, there is interference.

5-28.

First re-write the data as genotypes grouping the genetic reciprocal pairs.

Classes of gametes	Genotype	Numbers
Parental (P)	*th h st*	432
	+ + +	429
recombinant	*th h +*	37
	+ + *st*	33
recombinant	*th + st*	35
	+ *h* +	34

The th h st and wild type classes are defined as parental because they are the most frequent reciprocal pair.

a.   Each class of the parental reciprocal pair corresponds to one homolog of the heterozygous

Drosophila female that was test crossed to the homozygous recessive male. Thus the genotype of

this female is ***th h st / th⁺ h⁺ st⁺***.

b.   This is a test cross with three linked genes, but there are only 6 classes of data (3 reciprocal pairs)

instead of 8 classes. The missing reciprocal pair is *th* + + and + *h st*. This pair must be the least

frequent DCO class. Comparing the *th* + + DCO class to the + + + parental class (the most

similar) shows that *th* is the gene that differs so *th* must be the gene in the middle. The *h* to *th*

distance is 35 + 34 / 1000 = 0.069 = 6.9mu. The *th* to *st* distance is 37 + 33 / 1000 = 0.07 = 7mu.

The best map is:

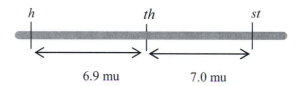

6.9 mu   7.0 mu

c.   **I = 1 – coefficient of coincidence; coefficient of coincidence = observed DCO / expected**

DCO. Thus coefficient of coincidence = 0 / (0.069)(0.07) = 0 and I = 1 – 0 = 1. The

interference is complete.

5-29. Diagram the cross:

MCS/MCS (Virginia strain) x mcs/mcs (Carolina strain) → F$_1$ MCS/mcs x mcs/mcs

→ ?

Assume no interference, and remember the map of these 3 genes:

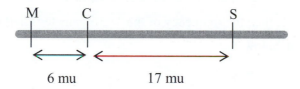

6 mu   17 mu

You are being asked to calculate the proportion of the test cross progeny that will have the Virginia parental phenotype. In Chapter 3 we could answer this question by calculating the monohybrid ratios each pair of alleles: $M : m$, $C : c$ and $S : s$ in the heterozygous parent. The test cross parent can only provide the recessive alleles for each gene, so the probabilities of the various phenotypes can be determined by applying the product rule. Unfortunately, this method of arriving at the probability of the progeny phenotypes only works when the genes under discussion are assorting independently! When the 3 genes are linked, as in this problem, to calculate the frequency of a parental class we must calculate first calculate the frequencies of ALL of the phenotypes expected in the test cross progeny!

A parent that is heterozygous for 3 genes will give 8 classes of gametes. In a test cross, the gamete from the heterozygous parent determines the phenotype of the progeny. When the 3 genes are genetically linked, these 8 classes will be found as 4 reciprocal pairs. In other words, each meiotic event in the heterozygous parent must give 2 reciprocal products that occur in equal frequency. For instance, a meiosis with no recombination will produce the parental gametes, $M\,C\,S$ and $m\,c\,s$ in equal frequency. This particular reciprocal pair will also be the most likely event and so the most frequent pair of products. The least probable meiotic event is a double crossover which is a recombination event in the region between M and C (region 1) and simultaneously in the region between C and S (region 2). The remaining gametes are produced by a single crossover in region 1 (the reciprocal pair known as single crossovers in region 1) and a single crossover in region 2 (the reciprocal pair known as single crossovers in region 2).

The numbers shown on the map of this region of the chromosome represent the recombination frequencies in the gene-gene intervals. There are 6 mu between the M and C genes, so 6% (rf = 0.06) of the progeny of this cross will have had a recombination event in region 1. This recombination frequency includes all detectable recombination events between these 2 genes - both SCO in region 1 AND DCO. Likewise, 17% of all the progeny will be the result of a recombination event in region 2. The DCO class is the result of a simultaneous crossover in region 1 and region 2. Recombination in two separate regions of the chromosome should be independent of each other, so we can apply the product rule to calculate the expected frequency of DCOs. Thus frequency of DCO = (0.06) x (0.17) = 0.01. Remember that the recombination frequency between M and C (region 1) includes both SCO 1 and DCO. Thus, 0.06 = SCO 1 + 0.01; solve for SCO 1 = 0.06 - 0.01 = 0.05. The same calculation for region 2 shows that SCO 2 = 0.16. The parental class = 1 - (SCO 1 + SCO 2 + DCO). Also remember that each class of gametes (parental, etc.) is made up of a reciprocal pair. If the frequency of the DCO class is 0.01, then the frequency of the $M\,c\,S$ gamete is half of that = 0.005. This is summarized in the table below.

Classes of gametes	Genotype	Frequency of reciprocal pair	Numbers	Frequency of each class
Parental (P)	M C S m c s	1 - all else	1 - (0.05 + 0.16 + 0.01) = 0.78	0.39 0.39
SCO 1	M c s m C S	rf in region 1 = SCO 1 + DCO	SCO 1 = 0.06 - 0.01 = 0.05	0.025 0.025
SCO 2	M C s m c S	rf in region 2 = SCO 2 + DCO	SCO 2 = 0.17 - 0.01 = 0.16	0.08 0.08
DCO	M c S m C s	(rf in region 1) x (rf in region 2)	(0.06) x (0.17) = 0.01	0.005 0.005

a. **The proportion of backcross progeny resembling Virginia (parental, *M C S*) = 0.39.**

b. **Progeny resembling *m c s* (P) = 0.39.**

c. **Progeny with *M c S* (DCO) = 0.005.**

d. **Progeny with *M C s* (SCO 2) = 0.8.**

5-30.

a. Recombination does NOT occur in male *Drosophila*. Therefore, in the cross *A b / a B* ♀ x *A b / a B* ♂ → F$_1$ the females will make 4 types of gametes (parental *A b* and *a B*; recombinant *A B* and *a b*). The frequencies of the 2 parental gametes will be equal to each other, and the frequency of the *A B* recombinant will be equal to the frequency of its reciprocal recombinant, *a b*.

female gametes	male gamete *A b*	male gamete *a B*
A b [parental]	*A b / A b* (*A- bb*)	*A b / a B* (*A- B-*)
a B [parental]	*a B / A b* (*A- B-*)	*a B / a B* (*aa B-*)
A B [recombinant]	*A B / A b* (*A- B-*)	*A B / a B* (*A- B-*)
a b [recombinant]	*a b / A b* (*A- bb*)	*a b / a B* (*aa B-*)

There is a ratio of 1/4 *A- bb* : 1/2 *A- B-* : 1/4 *aa B-* for the progeny receiving the parental gametes AND the same ratio among the progeny receiving the recombinant gametes. Thus, **the overall phenotypic dihybrid ratio will <u>always</u> be 1/4 *A- bb* : 1/2 *A- B-* : 1/4 *aa B-*, independent of the recombination frequency between the A and B genes.**

   This will <u>not</u> be true of the cross *A B / a b* ♀ x *A B / a b* ♂. The male will make the parental gametes, *A B* and *a b*, while the female will make 4 types of gametes: parental *A B* and *a* *b*; recombinant *A b* and *a B*. In this case, as in problem 5-17, half of the progeny will look *A- B-* because they received the *A B* gamete from the male parent irrespective of the gamete from the female parent. When the male parent donates the *a b* gamete, then it is the gamete from the female parent that determines the phenotype of the offspring. The progeny that are *A- bb* and *aa B-* have received a recombinant gamete from the female parent (and the *a b* gamete from the male). These classes of progeny can then be used **to estimate the recombination frequency between the *A* and *B* genes: rf = 2(# of *A- bb* + # of *aa B-*)/total progeny.**

b.   In mice, recombination occurs in both females and males. Therefore, in the cross *A b / a B* ♀ x *A b / a B* ♂ both sexes will make the same array of gametes (parental *A b* and *a B*; recombinant *A B* and *a b*). In this case, the only phenotype of progeny with a singular genotype will be *aa bb*. In the cross under consideration here, the *aa bb* phenotype can only arise if both parents donate the *a b* recombinant gamete. Of course, the other recombinant gamete, *A B*, occurs with equal frequency. The probability of the *aa bb* genotype = (probability of an *a b* recombinant gamete)^2. If you know the frequency of the *aa bb* phenotypic class in the progeny, **recombination frequency (frequency of recombinant products) between the *A* and *B* genes = 2($\sqrt{\#aa\ bb}$/# total progeny).**

In the case of the *A B / a b* ♀ x *A B / a b* ♂ cross in mice, both sexes are making the same parental gametes (*A B* and *a b*) and recombinant gametes (*A b* and *a B*) gametes. Again, the only phenotype of progeny with a singular genotype will be *aa bb*. In this example, this phenotype is the result of the fusion of the *a b* parental gamete from each parent. The frequency of the *aa bb* phenotype = (the frequency of the *a b* gamete) x (the frequency of the *a b* gamete). If you know the frequency of the *aa bb* phenotypic class in the progeny, **the frequency of non-recombinant products between the *A* and *B* genes = 2($\sqrt{\#aa\ bb}$)/# total progeny. Recombination frequency = 1 - frequency of non-recombinant products between the *A* and *B* genes.**

5-31. In cross #1 the criss-cross inheritance of the recessive alleles for dwarp and rumpled from mother to son tells you that all these genes are X-linked. In cross #2, you see the same pattern of inheritance for pallid and raven, so these genes are X-linked as well. The fact that the F1 females in both crosses were wild type tells you that the wild type allele of all 4 genes is dominant to the mutant allele. Designate alleles: *dwp*$^+$ and *dwp* for the dwarp gene, *rmp*$^+$ and *rmp* for the rumpled gene, *pld*$^+$ and *pld* for the pallid gene, and *rv*$^+$ and *rv* for the raven gene. Assign the genes an arbitrary order to write the genotypes. If you keep the order the same throughout the problem, it is sufficient to represent the wild type allele of a gene with +.

Cross 1: *dwp rmp + +* / *dwp rmp + +* x *+ + pld rv* / Y → *dwp rmp + +* / *+ + pld rv* (wild-type
   females) and *dwp rmp + +* / Y (dwarp rumpled males).

Cross 2: *+ + pld rv* / *+ + pld rv* x *dwp rmp + +* / Y → *+ + pld rv* / *dwp rmp + +* (wild-type
   females) and *+ + pld rv* / Y (pallid raven males).

dwp rmp + + / *+ + pld rv* (cross 1 F$_1$ females) x *dwp rmp pld rv* / Y → 428 *+ + pa ra*, 427 *dw ru +
+*, 48 *+ ru pa ra*, 47 *dw + + +*, 23 *+ ru pa +*, 22 *dw + + ra*, 3 *+ + pa +*, 2 *dw ru + ra*.

All 4 genes <u>must</u> be linked, as they are all on the X chromosome. You expect 16 classes of progeny (2
x 2 x 2 x 2) in 8 genetic and numeric pairs. Notice that there are only 8 classes of progeny. The data
that is seen shows <u>exactly</u> the pattern you expect for a female that is heterozygous for 3 linked genes!
If 2 of the 4 genes NEVER recombine then you would expect the pattern of data that is seen. If 2
genes never recombine, they will always show the parental configuration of alleles. Thus, examine
the various pairs of genes for the presence or absence of recombinants. Note that you <u>never</u> see
recombinants between *pallid* and *dwarp*: all the progeny are either pallid or dwarp, but never pallid
and dwarp nor wild type for both traits. This suggests that the two genes are so close together that
there is essentially no recombination between the loci. If a much larger number of progeny were
examined, you might observe recombinants. Treat *dwp* and *pld* as 2 genes at the same location, so one
of them (*dwp*, for instance) can be ignored, and this problem becomes a three-point cross between
pld, rv and *rmp*.

Classes of gametes	Genotype	Numbers
Parental (P)	*dwp rmp + +*	427
	+ + pld rv	428
SCO 1	*+ rmp pld rv*	48
	dwp + + +	47
SCO 2	*dwp + + rv*	22
	+ rmp pld +	23
DCO	*+ + pld +*	3
	dwp rmp + rv	2

When the *rmp + rv* DCO class is compared to the *rmp + +* parental class, you can see that **rv is in
the middle. The *rmp - rv* rf = (48 + 47 + 3 + 2)/1000 = 10 cM; the *pld - rv* rf = (22 + 23 + 3 +
2)/1000 = 5 cM**. I = 1 - coc; coc = observed frequency of DCO/expected frequency of DCO. So coc =
(5/1000)/(0.05)(0.1) = 0.005/0.005 = 1; I = 1 - 1 = 0, so **there is no interference**.

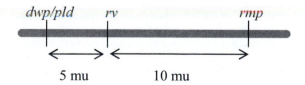

Section 5.5 – Tetrad Analysis

5-32.

a. This data is presented as phenotypes of individual spores. Because there is data for 3 genes, this may be analyzed as a 3 point cross. Organize the data into reciprocal pairs of spores.

Classes of gametes	Genotype	Numbers
Parental (P)	$\alpha + +$ afg	31 29
SCO 1	$a + g$ $af+$	6 6
SCO 2	$\alpha + g$ $af+$	13 14
DCO	$a + +$ αfg	1 1

The αfg DCO spore type is most similar to the afg parental spore type. Thus, the mating type (a/α) is the gene in the middle. The distances are: $f - a/\alpha = (6 + 6 + 1 + 1)/101 = 13.9$ cM and a/α to $g = (13 + 14 + 1 + 1)/101 = 28.7$ cM.

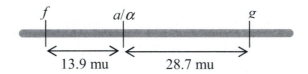

b. This problem says you have an ascus with an αfg spore. This spore is the result of a double crossover (see part a). The reciprocal product would be the $a + +$ spore, but this is not seen in the ascus. Remember that there are 3 different types of double crossovers - 2 strand DCOs, 3 strand DCOs and 4 strand DCOs. Each of these types of DCOs gives a different array of spores in the resulting ascus (**Figure 5.17**). Draw a meiotic figure of this chromosome and try some different types of DCOs. **A 3 strand DCO gives the desired result**.

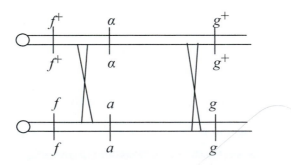

5-33.

a. The number of meioses represented here is the total of the number of asci = **334**. Each ascus contains the 4 products of one meiosis.

b. Diagram the cross: $a + c$ x $+ b +$

To map these genes, use the **Three Easy Rules for Tetrad Analysis**. First designate the type of asci represented. This has to be done for each pair of loci as PD (P), NPD (N) and T refer exclusively to the relationship between two genes. In the table below, the top row shows the designations for all three pairs: the *a-b* comparison is at the lower left, the *b-c* comparison is at the lower right and the *a-c* comparison is at the top of the pyramid.

P	P	T	T	N	P
P P	N N	T P	T N	N P	T T
$a + c$	$a b c$	$+ + c$	$+ b c$	$a b +$	$a + c$
$a + c$	$a b c$	$a + c$	$a b c$	$a b +$	$a b c$
$+ b +$	$+ + +$	$+ b +$	$+ + +$	$+ + c$	$+ + +$
$+ b +$	$+ + +$	$a b +$	$a + +$	$+ + c$	$+ b +$
137	141	26	25	2	3
I I I	I I I	II I I	II I I	I I I	I II I

Rule #1: For genes *a* and *b* PD = NPD, so these two genes are not linked. For genes *b* and *c* PD = NPD, so genes *b* and *c* are not linked. For genes *a* and *c*, PD>>NPD, so the genes are linked. Calculate rf between *a* and *c* = 2 + 1/2(26 + 25)/334 = 8.2 cM (**Rule #2**).

Gene-centromere distances can be calculated in *Neurospora* (**Rule #3**), so analyze the data for MI and MII segregation patterns for the alleles of each gene in each ascus type. This analysis is done <u>separately for each gene</u>, unlike the gene - gene analysis done above. The designation for each gene is presented under that gene at the bottom row of the table (I - MI, II = MII). **Rule #2** shows that the distance between *a* and its centromere = 1/2(26 + 25)/334 = 7.6 mu; the distance between *b* and its centromere = 1/2(3)/334 = 0.4 mu; the distance between *c* and its centromere = 1/2(0)/334 = 0.

Now compile all of these pieces of data into one map. **Rule #1** shows that gene *b* is 0.4 mu from its centromere and is on a different chromosome from genes *a* and *c*. Genes *a* and *c* are on the same chromosome, so the *a*-centromere distance and the *c*-centromere distance refer to the same centromere. Gene *c* is 0 mu from the centromere. As you can see from the map, there are 2 slightly different distances for the gene *a* to gene *c* region - the gene-gene distance is 8.2 mu and the a-centromere distance is 7.6 mu. In this case the longer gene-gene distance is more accurate as it includes the SCOs between *a* and *c* as well as some of the DCOs between *a* and *c* (the 4 strand DCO, **Figure 5.17**).

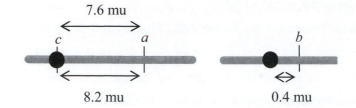

c. Carefully consider the information you have for the different chromosomes in the ascus type chosen (the group with 3 members). The *a* and *c* genes show PD segregation, which can mean either no crossing over between them or a 2 strand DCO. Both genes show MI segregation, which means there haven't been any single crossovers between either gene and the centromere. **Gene *b* shows MII segregation, which means there has been an SCO between the gene and its centromere.** This crossover also means the *a-b* and *c-b* comparisons in this class will show the tetratype pattern (T are due to crossovers between either gene and its centromere when the genes are on separate chromosomes).

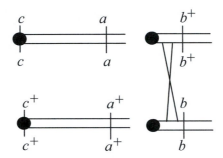

5-34. Use the Three Easy Rules for Tetrad Analysis to help you solve this problem.

a. In cross 1, the number of PD (parental ditypes) = NPD (nonparental ditypes) so the *ad* gene and the mating type locus assort independently. In cross 2 the number of PD >> NPD so we can conclude the *p* gene and the mating locus are linked; rf *p - mating type* = (NPD + 1/2T)/ total tetrads = (3) + (1/2)(27)/54 =16.5/54= .31 × 100 = 31 cM between the two genes.

b. **To calculate gene-centromere distances you need information on the order of ascospores in each ascus type.** Only with this information can you calculate gene - centromere distances based on 1/2(# of asci showing MII segregation for the gene)/total asci.

5-35. Diagram the cross and summarize the data: met^- lys^- x met^+ lys^+

P	T
met^+ lys^+	met^+ lys^+
met^+ lys^+	met^- lys^+
met^- lys^-	met^+ lys^-
met^- lys^-	met^- lys^-
89	11

a. **The two types of cells in the first group of 89 asci are met^+ lys^+ (could grow on all four types of media) and met^- lys^- (require the addition of met and lys to minimal medium).** These asci are parental ditypes (PD). The four types of cells in the second group of 11 asci are **met^+ lys^+; met^+ lys^- (grew on min + lys and on min + lys + met); met^- lys^+ (grew on min + met and on min + lys + met); and met^- lys^-.** These are tetratype (T) asci.

b. Because the number of PD>>> NPD, **the genes are linked.** The distance between them is:

(NPD + 1/2(T)) / total tetrads = (0 + 1/2(11))/100 × 100 = **5.5 m.u.**

c. NPD should be seen eventually and would result from four strand double crossovers. There would be two types of spores: met^- lys^+ (could grow on min + met and on min + met + lys) and met^+ lys^- (could grow on min + lys and on min + met + lys).

5-36. Genes *a, b,* and *c* are all on different chromosomes. Thus, all crosses between these genes are expected to give an equal number of PD and NPD asci. The fact that the cross involving genes *a* and *b* yields no T indicates that **genes *a* and *b* are <u>both</u> very close to their respective centromeres.** If two genes are on separate chromosomes, tetratypes arise when a crossover occurs between one of the genes and its centromere or between the other gene and its centromere. In the crosses involving genes *a* and *c* or genes *b* and *c,* many T asci are seen. Because genes *a* and *b* are tightly centromere-linked, **gene *c* must be very far from its centromere** in order to generate these T asci.

5-37.

a. The genotype of a true breeding wild type diploid strain of *Saccharomyces* can be written + / +. If this diploid undergoes meiosis, **all (100%) of the asci will have 4 viable spores.**

b. The genotype of this strain is + / *n*, where *n* = a null activity allele of an essential gene. The diploid cells will be viable, because they have functional enzyme for this essential gene from the + allele. If this diploid undergoes meiosis, then **all (100%) of the tetrads will have 2 + : 2 n spores (that is, only 2 spores in each tetrad will be viable).**

c. Gene *a* and gene *b* are different essential genes; *a* and *b* represent temperature-sensitive alleles of these genes. When you diagram a cross you must write complete genotypes for both haploid parents: $a\,b^+$ (strain a) x $a^+\,b$ (strain b) $\rightarrow$ $aa^+\,bb^+$. Each haploid parent strain will die when grown under restrictive conditions because they cannot produce a required product, while the diploid cells are viable because they have a wild type allele for both genes.

   If these genes are unlinked, then after meiosis PD asci = NPD asci ('3 Easy Rules for Tetrad Analysis' rule #1). When the genes are unlinked, also remember that T tetrads (asci) arise from SCOs between either gene and its centromere. Because both of these genes are 0 mu from their centromeres, you will not see T asci. Thus, after meiosis, you will have 50% PD (2 $a\,b^+$ spores : 2 $a^+\,b$ spores) : 50% NPD (2 $a^+\,b^+$: 2 $a\,b$). None of the spores in the PD tetrads are viable under restrictive conditions, while 2 of the 4 spores in the NPD asci are viable ($a^+\,b^+$). Thus, **50% of the asci will have 0 viable spores (PD) and 50% of the asci will have 2 viable spores (NPD).**

d. Again genes *a* and *b* are unlinked essential genes, but now gene *a*-centromere = 0 mu and gene *b*-centromere = 10 mu. A SCO between gene *b* and its centromere will happen in 20% of the asci (because only half of the ascospores of an SCO are recombinant) and will give T asci. A T ascus will have a spore ratio of 1 $a^+\,b^+$: 1 $a\,b$: 1 $a\,b^+$: 1 $a^+\,b$. There is 1 viable spore ($a^+\,b^+$) in a T ascus. The remaining 80% of the asci will be equally divided between PD and NPD because the genes are unlinked. Thus, **40% of the asci will have 0 viable spores (PD), 40% of the asci will have 2 viable spores (NPD) and 20% of the asci will have 1 viable spore (T).**

e. Because both genes are on the same chromosome, you can diagram this cross: $a\,b^+$ (strain a) x $a^+\,b$ (strain b) $\rightarrow$ $a\,b^+$ / $a^+\,b$. When the genes are linked, SCO and 3 strand DCO between the genes give T asci and 4 strand DCO between the genes gives NPD asci. The remainder of the asci are PD (no crossover and 2 strand DCO). If the recombination frequency between gene *a* and gene *b* = 0 mu, then the only possible result will be PD asci (2 $a\,b^+$: 2 $a^+\,b$) in which none of the spores are viable. Thus, **100% of the asci will have 0 viable spores.**

f. Now the genes are 10 mu apart and there are no DCO events. This means that 20% of the asci produced will be T (see part d) and the remaining 80% will be PD. Therefore, **80% of the asci will have 0 viable spores (PD) and 20% of the asci will have 1 viable spore (T).**

g. A 4 strand DCO gives an NPD ascus. The genotypes of the spores will be 2 $a^+\,b^+$: 2 $a\,b$. Thus, **2 of the 4 spores will be viable.**

5-38. Genes *c* and *d* are a total of 22 mu apart.

a. If interference = 1, then there are no double crossovers (DCO). Therefore, all the recombinants between the 2 genes are due to single crossovers (SCO). If two genes are linked, then SCOs between the genes give tetratype asci (T). Remember that the rf between *C* and *D* is 0.22, DCO between *C* and *D* is 0 (so **there can be no NPD tetrads**), and the formula for recombination frequency between 2 genes = NPD + 1/2(T)/total asci. Solve for T: 0.22 = 0 + 1/2(T) so T = 2(0.22) = 0.44. Therefore, **44% of the asci will be tetratypes and the remaining 56% will be parental ditypes.**

b. Interference = 0, so DCOs will occur in the expected frequency (see problems 5-29 and 5-25). The expected frequency of DCO = (recombination frequency in the *C-D* region) x (recombination frequency in the C-D region) = (0.22) x (0.22) = 0.0484. The rf between *C-D* = SCO + DCO; so 0.22 = SCO + 0.0484; SCO = 0.22 - 0.0484 = 0.1716. If SCO frequency between *C* and *D* = 0.1716 then T due to SCO = 0.3432 (see part a of this problem).

   Remember that when analyzing tetrads the three different types of DCOs can be distinguished: 2 strand DCOs, which give PD asci; 3 strand DCOs, which give T asci; and 4 strand DCOs, which give NPD asci (**Figure 5.17b-f**). These DCOs occur in the ratio of 1/4 (2 strand DCOs) : 1/2 (3 strand DCOs) : 1/4 (4 strand DCOs). If the total DCO frequency = 0.0484 then 1/4 (0.0484) = 0.0121 is the frequency of 4 strand DCOs (NPD), 1/2 (0.0484) = 0.0242 is the frequency of 3 strand DCOs (T) and the remaining 1/4 of the DCOs (0.0121) are 2 strand DCOs, which will be PD tetrads.

   In total, **NPD = 0.0121, T = 0.0242 (due to DCO) + 0.3674 (due to SCO) = 0.3795 and the remainder are PD = 0.6205**.

c. In *Neurospora* the recombination events that underlie the formation of PD, NPD and T are the same as in yeast. The difference is that the ordered tetrads allow you to distinguish whether a SCO event occurs between *C* and the centromere or between *D* and the centromere. If the SCO occurs between *C* and the centromere then you will see MII segregation for the alleles of the *C* gene and MI segregation for the alleles of the *D* gene. If the SCO occurs between *D* and the centromere then you will see the reverse - MI segregation for *C* and MII segregation for *D*.

   Here, I = 1, so there are no DCOs. Therefore, as in part a, T = 0.44 and PD = 0.56. However 7/22 of the SCO events occur between gene *C* and the centromere, while the remaining 15/22 of the SCOs occur between *D* and the centromere. **The expected results are summarized in the table below**.

crossover type and location:	no crossover	SCO *C*-cent.	SCO *D*-cent
ascus type:	PD	T	T
MI or MII gene *C*:	MI	MII	MI
MI or MII gene *D*:	MI	MI	MII
frequency:	0.56	7/22 x 0.44 = 0.14	15/22 x 0.44 = 0.30

d. Here, I = 0, so DCO events are occurring at the expected frequencies. The frequencies of each of the DCO events is as in part b. A DCO represents a single meiosis in which a crossover in the first region (*C* to centromere) happens simultaneously with a crossover in the second region (centromere to *D*). Thus any of the DCO events will show MII segregation for both genes. **The expected results are summarized in the table below**.

crossover type and location:	no crossover	SCO *C*-cent.	SCO *D*-cent	DCO 2 strand	DCO 3 strand	DCO 4 strand
ascus type:	PD	T	T	PD	T	NPD
MI or MII gene *C*:	MI	MII	MI	MII	MII	MII
MI or MII gene *D*:	MI	MI	MII	MII	MII	MII
frequency:	0.528	7/22 x 0.419 = 0.14	15/22 x 0.419 = 0.30	1/4 x 0.0105 = 0.003	1/2 x 0.0105 = 0.005	1/4 x 0.0105 = 0.003

5-39. Diagram the yeast cross: *his⁻ trp⁺* x *his⁺ trp⁻* → *his⁻his⁺ trp⁺trp⁻* → 233 PD, 11 NPD, 156 T.

a. The **PD asci must have 2 *his⁻ trp⁺* : 2 *his⁺ trp⁻* spores, the NPD asci have 2 *his⁺ trp⁺* : 2 *his⁻ trp⁻* spores and the T asci have 1 *his⁻ trp⁺* : 1 *his⁺ trp⁻* : 1 *his⁺ trp⁺* : 1 *his⁻ trp⁻* spores**.

b. **PD>>NPD so the genes are linked; rf between 2 genes = 11 + 1/2(156)/400 = 22.3 mu**.

c. Because the genes are linked, you know that NPD asci are the result of 4 strand DCOs. This sort of DCO is 1/4 of all the DCO events. The 3 strand DCOs (1/2 of all DCOs) give T while 2 strand DCOs (1/4 of all DCOs) give PD. In the data you see 11 NPD tetrads, or 11 meioses that underwent 4 strand DCOs. There are another 22 tetrads that are the result of 3 strand DCOs (and are Ts), and another 11 asci that underwent 2 strand DCOs (and are PDs), for **a total of (11 + 22 + 11 =) 44 asci that underwent 2 crossovers. The remaining (156 - 22 =) 142 T asci underwent a SCO, or 1 crossover, and the remaining (233 - 11=) 222 PD asci underwent 0 recombination events**.

d. There are 44 asci that underwent DCOs for a total of 88 crossover events. There were another 142 asci that underwent SCOs. Thus, there were 142 + 88 = 230 crossover events / 400 meioses = **0.575 crossovers/meiosis**.

e.   The equation for recombination frequency between 2 genes used in the textbook and in part b above <u>only</u> includes 4 strand DCO (NPD). **This formula ignores 3 strand and 2 strand DCOs.** This calculation for recombination frequency assumes that all T asci are due to SCOs. Though this is true for the majority of T asci, a small proportion of T asci are due to 3 strand DCO events. Also, it is an oversimplification to assume that all PD asci are non-recombinant, because a very few of them are due to 2 strand DCOs. Remember that the 3 types of DCO events occur in a ratio of 1/4 2 strand DCO : 1/2 3 strand DCO : 1/4 4 strand DCO. With this information we can derive a more accurate formula for recombination frequency between 2 genes = SCO + DCO/total asci: **rf = 1/2(T – 2NPD) + 4(NPD)/total asci** = 1/2(T) + 3(NPD)/total asci. Many yeast geneticists use this more accurate formula in preference to the one in your textbook.

f.   rf = 1/2(156 – 22) + 4(11)/400 = 1/2(134) + 44/400 = 111/400 = 0.278 = **27.8 mu**. As you can see, this is somewhat larger than the distance calculated in part b.

5-40.

a.   Refer to the three strains as trp1, trp2 and trp3. Remember that in *Neurospora* the 4 meiotic products undergo a subsequent mitosis to give 8 spores in the ascus. However, the first 2 spores are identical to each other (they are mitotic products of the same initial spore), the 3d and 4th spores are the same as each other, and so on. For the purposes of discussing the results, assume that the ascus is made up of the 4 original spores prior to this extra mitosis. You cross *trp1⁻* x wild type → diploid → 2 wild type : 2 *trp1⁻*. You are seeing a monohybrid ratio of 1:1, which means there is only one gene controlling the *trp⁻* phenotype in strain trp1. Each of the strains gives the same result, so **in each haploid strain a single mutant gene is responsible for the *trp⁻* phenotype**.

b.   First consider the cross between trp3 and wild type. The diploid is *trp3⁻ / trp3⁺*. After meiosis the first 2 spores were either both *trp3⁺* (could grow on minimal media) or they were both *trp3⁻* (could not grow on minimal media). Thus, all the asci were 2 *trp3⁻* : 2 *trp3⁺*, and the order of the alleles was either - - + + or + + - -. In other words, all the asci showed MI segregation for the *trp3* gene! In the case of the crosses with *trp1⁻* x wild type and *trp2⁻* x wild type, the resultant diploids gave some asci that gave a different result - only one spore of the top 2 was trp⁺ while the other one was trp⁻. In other words, these asci showed MII segregation for the *trp* gene in question. In summary, **the *trp3* gene is very closely linked to its centromere (no T = no crossovers between the gene and the centromere), while the *trp1* and *trp2* mutations are further from their centromere(s)**.

c. If 2 strains have mutations in the same gene, then the resulting diploid would be unable to grow on minimal media *trpx⁻ / trpx⁻* and would give asci where all 4 spores were *trpx⁻* (0 spores viable on minimial media). This result is never seen, so each mutant strain must have a mutation in a different gene, for a total of 3 different *trp⁻* genes in the 3 strains. Diagram the cross between **trp1 and trp2**: *trp1⁻ trp2⁺* x *trp1⁺ trp2⁻* → *trp1⁻ trp1⁺ trp2⁺ trp2⁻*. When this diploid is allowed to undergo meiosis you see **78 asci** with 0 viable spores (2 *trp1⁻ trp2⁺* : 2 *trp1⁺ trp2⁻* = **PD** - see problem 5-37f) **and 22 asci** with 2/8 or 1/4 viable spores (1 *trp1⁺trp2⁺* : 1 *trp1⁻trp2⁻* : 1 *trp1⁻ trp2⁺* : 1 *trp1⁺ trp2⁻* = **T).** In the **trp1 x trp3 cross** you see a new class of asci, those with 2 viable spores (2 *trp1⁺trp3⁺* : 2 *trp1⁻trp2⁻* = NPD). In this cross **there are 46 PD, 48 NPD, and 6 T asci.** In the last cross, **trp2 x trp3, there are 42 PD, 42 NPD and 16 T asci.**

d. In *trp1* x *trp2* PD>>NPD (= 0) so the genes are linked. The T asci are caused by SCOs between the 2 genes. The recombination frequency between *trp1* and *trp2* = (0 + 1/2(22))/100 = 0.11 = 11 mu. In *trp1* x *trp3* there are 46 PD = 48 NPD so the genes are unlinked; the 6 T asci arise from SCOs between *trp1* and its centromere. None of these T asci can arise from crossovers between *trp3* and its centromere because you found in part b that they were very tightly linked. Thus, *trp1* is 3 mu from its centromere. In the cross between *trp2* x *trp3*, PD = NPD so the genes are unlinked. There are 16 T asci, so gene 2 must be 8 mu from its centromere. **The map is shown below.**

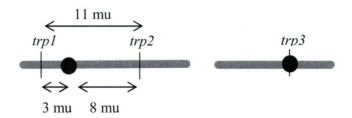

e. Ordered octads show the segregation pattern for each gene separately. In this example, **both mutant genes give the same phenotype (trp⁻). Thus, it is impossible to determine if a trp⁻ spore is + - or -- or - +.** If you cannot distinguish these different types of spores, then you cannot determine the segregation pattern (MI vs MII) for the individual genes.

f. **You can calculate gene-centromere distances because you discovered that one of the genes (trp3) is tightly linked to its own centromere.** Therefore, when *trp3⁻* is crossed with either of the other mutants, any T asci must be due to SCO between the other gene and its centromere, and these T asci can then be used to calculate a distance between the other gene and its centromere. This same sort of analysis can also be used in yeast to map gene-centromere distances if a strain is available with a mutation known to be right at a centromere.

Section 5.6 – Mitotic Recombination and Genetics Mosaics

5-41.

a. Red sectors arise when one cell in the growing colony becomes *ade2⁻* / *ade2⁻*. As all the cells continue to grow, the colony continues to expand and the *ade2⁻* / *ade2⁻* cells form a red sector within the white colony (**Figure 5.25**). **Red cells of the *ade2⁻* / *ade2⁻* genotype could arise by mitotic recombination (Figure 5.24), by loss of the entire chromosome containing the *ade2⁺* allele, by a deletion of the portion of the chromosome containing the *ade2⁺* allele, or by spontaneous mutation of the *ade2⁺* allele to an *ade2⁻* allele.**

b. The size of the red sectors depends on when in the formation of the colony the event occurred to form the initial *ade2⁻* / *ade2⁻* cell. If the event occurred early in the formation of the colony there will be a larger red sector than if the event occurred near the end of colony formation. All of the events mentioned in part a are rare, so in general they will occur later in colony formation when there are more cells in which they could occur. As a result, **most of the red sectors will be small**.

5-42. After replication in the heterozygous diploid, each chromosome would be composed of a pair of sister chromatids, as shown below. The centromeres on sister chromatids split and segregate from each other during mitosis (**Figure 5.24**). The centromeres are numbered in the figure below so you can follow them more easily. For example, cen1 and cen2 segregate from each other, so chromatids 1 or 2 and 3 or 4 will end up in a daughter cell. In a normal mitosis, this gives daughter cells that are genotypically *a b c leth d e* / *a⁺ b⁺ c⁺ leth⁺ d⁺ e⁺* and are phenotypically wild type, like the original cell. Mitotic recombination can rarely occur between non-sister chromatids, for example between chromatid 2 and 3. Several possible locations for such recombination events are indicated by X-I - X-V in the figure. Assume that no crossovers occur between gene *a* and the *cen* because they are so closely linked.

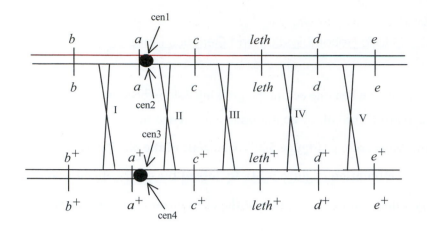

Consider the results of the X-I crossover. In one possible segregation, chromatids 2 and 4 segregate into one daughter cell; this cell will be homozygous for the b^+ allele and heterozygous for the rest of the genes on the chromosome. Thus, this cell will be phenotypically wild type and indistinguishable from the non-recombinant cells surrounding it. The reciprocal product will be the daughter cell with chromatids 1 and 3 whose genotype will be homozygous mutant for gene b and heterozygous wild type for everything else. This cell will continue to divide mitotically and give you a patch of *b* mutant tissue in a sea of wild type tissue. In effect, the genes that are further away from the centromere than the site of mitotic recombination become homozygous. One segregant (the 2 and 4 chromatids in this example) will be homozygous wild type, and thus indistinguishable from the non-recombinant cells; the other segregant (the 1 and 3 chromatids here) becomes homozygous for the mutant allele of the gene(s) that are further from the centromere than the crossover while those genes that are closer to the centromere than the recombination event (or on the other side of the centromere, i.e. those genes unaffected by the mitotic recombination event) remain heterozygous. Next consider crossover X-II. In this case the same segregations will give you a wild type cell (chromatids 2 and 4) while the reciprocal daughter cell will be $b^+ a^+ c$ *leth d e* (lethal, chromatids 1 and 3). Crossover X-III will also give a lethal recombinant product. Crossover X-IV will give a *d e* patch of mutant tissue, and crossover X-V will give a patch of *e* mutant tissue. Thus, **the only phenotypes that will be found in sectors as a result of mitotic recombination will be *b*, *e*, and *d e* (#2, 5, and 9).**

5-43. The genotype of the female fly is $y^+ \, sn^+ / y \, sn$:

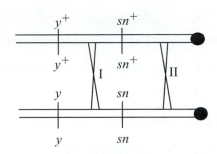

a. **The larger patch of yellow tissue arises from a mitotic crossover in region I. This mitotic recombination will generate a cell of the genotype $y \, sn^+ / y \, sn$ (<u>problem 5-42</u>). The smaller patch of tissue must have arisen from a second mitotic recombination between the centromere and sn, in region II**. This would produce cells that are $y \, sn / y \, sn$. Because the yellow patch is larger, the recombination in region I occurred earlier in development. The yellow, singed cells are a small patch within a larger patch of yellow cells, so the mitotic crossover in region II happened after (later in development than) the crossover in region I, and this second crossover happened in a recombinant daughter cell of the first recombination event. No wonder these patches within patches are rare!

b. If the genotype of the female was $y^+ \, sn / y \, sn^+$, then **a recombination event in the region I would give you a detectably recombinant cell with the genotype $y \, sn / y \, sn^+$. A subsequent second mitotic crossover in region II in one of these originally recombinant $y \, sn / y \, sn^+$ cells will give you a patch of $y \, sn$ tissue inside a patch of yellow tissue, as in part a**.

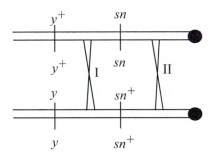

There is one mitotic recombination event which will give a different result in a y^+ sn^+ / y sn female (part a) than a y^+ sn / y sn^+ female (part b). After a recombination event in region II the female from part a will show a y sn patch of tissue. The other recombinant product will be homozygous y^+ sn^+, and thus indistinguishable from non-recombinant cells. The same sort of recombination event in region II in the female from part b will give one daughter cell of genotype y^+ sn / y^+ sn which will give a patch of sn tissue. The other product of the mitotic recombination will be y sn^+ / y sn^+ and will give an adjacent patch of yellow tissue. This phenomenon is called twin spots.

5-44. List the phenotypes seen in the normal tissue and the 20 tumors and their frequency of occurrence:

- normal tissue = $A^F A^S$ $B^F B^S$ $C^F C^S$ $D^F D^S$

- tumor type 1 = A^F $B^F B^S$ $C^F C^S$ D^F = 12 tumors

- tumor type 2 = A^F $B^F B^S$ $C^F C^S$ $D^F D^S$ = 6 tumors

- tumor type 3 = A^F B^S $C^F C^S$ D^F = 2 tumors

Remember that all of the tumor tissues were also homozygous for $NF1^-$ while the normal tissue is heterozygous $NF1^-NF1^+$.

a. **Mitotic recombination** could have caused all 3 types of tumors.

b. Remember that mitotic recombination causes genes further from the centromere to become homozygous, while genes between the recombination event and the centromere, or those on the other side of the centromere, remain heterozygous (review problem 5-42). Notice that all 3 types of tumor cells are homozygous for $NF1^-$, so by definition all 20 mitotic recombination events that gave rise to the 20 tumors occurred between the centromere and the $NF1$ gene. Notice also that all 20 tumors are homozygous for gene A, specifically the A^F allele. Because gene A is affected by all of the mitotic recombination events, it must be further from the centromere than the $NF1$ gene. Because these tumors become homozygous for the A^F allele, that allele must be on the same homolog with the $NF1^-$ allele. There are 6 tumors that are homozygous for these 2 genes (tumor type 2). As you work your way along the chromosome from $NF1$ toward the centromere the next gene to also become homozygous is the D gene (tumor type 1), specifically the D^F allele, then the B gene (tumor type 3), the B^S allele. Gene C never becomes homozygous, yet we are told it is on this chromosome as well. Thus it could either be on the other side of the centromere from $NF1$, A, D and B or it could be on the same side of the centromere as the rest of the genes but very closely linked to the centromere (like the a gene in problem 5-42). **The order of the genes and coupling of the alleles is shown below:**

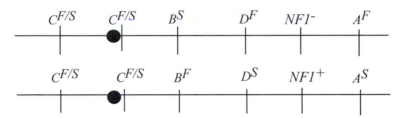

It is possible to use the **mitotic** recombination frequencies with which the various genotypes of tumors arose as a rough, relative approximation of the distances between the genes. These numbers are NOT to be confused with the meiotic recombination frequency we have been calculating in the other problems in this and other chapters! In tumor type 2 the mitotic recombination event happened between gene $NF1$ and gene D ($NF1^-$ is homozygous, and so further from the centromere than the recombination event while gene D is still heterozygous), and this occurred in 6/20 tumors = 0.3, so **the relative recombination frequency between $NF1$ gene D is 0.3**. In tumor type 1 the mitotic recombination event happened between gene B and gene D (B is still heterozygous in these tumors while D is homozygous D^F), and this happened in 12/20 tumors = 0.6, so **the relative recombination frequency between genes B and D is 0.6**. In tumor type 3 the mitotic recombination event happened between the centromere and gene B (or between gene C and gene B if you place gene C on the same side of the centromere as the NF1 gene). This event happened in 2/10 tumors = 0.1 so **the relative recombination frequency between the centromere and gene B is 0.1**.

c.   In this mechanism an entire homolog of one chromosome is lost. If the lost homolog was the one with the $NF1^+$ allele, then **the resulting cell would be hemizygous for the $NF1^-$ allele, and would develop into a tumor. This did NOT occur here, because ALL 3 genotypes of tumors are still heterozygous for at least the C gene.**

d. **Yes, deletions of portions of the *NF1*$^+$ homolog which cause loss of the *NF1*$^+$ allele** could cause tumors to develop in the resulting *NF1*$^-$ cells. If these deletions extended to the neighboring genes *A, D,* or *B,* the tumor cells could show the same protein variants as in the problem.

Likewise, **point mutations which inactivate the *NF1*$^+$ allele** could cause tumors to develop in the resulting *NF1*$^-$ cells.

Chapter 6   DNA Structure, Replication, and Recombination

Synopsis:

The statement "DNA's genetic functions flow directly from its molecular structure" is a good focus for reviewing DNA structure. By focusing on function, the beauty of the structure will become more evident. Make sure you really understand the structure and get a good mental image of the DNA molecule and its construction. Understanding where hydrogen and covalent bonds are found, the polarity of the strands of DNA, and why complementarity is important will provide a good basis for understanding many of the cellular processes (for example transcription, translation and recombination) and the manipulations of recombinant DNA technology, from cloning to genetic screening.

DNA is the nearly universal genetic material. Experiments showing that DNA causes bacterial transformation (**Figures 6.4** and **6.5**) and that DNA is the agent of virus production in phage-infected bacteria (**Figure 6.7**) demonstrated this fact.

According to the Watson-Crick model the DNA molecule is a double helix composed of two antiparallel strands of nucleotides; each nucleotide consists of one of four nitrogenous bases (A, T, G or C), a deoxyribose sugar and a phosphate. An A on one strand pairs with a T on the other, and a G pairs with a C. DNA carries information in the sequence of its bases, which may follow one another in any order.

The DNA molecule reproduces by semiconservative replication. In this type of replication the two DNA strands separate and the cellular machinery then synthesizes a complementary strain for each (**Feature Figure 6.20**). By producing exact copies of the base sequence information in DNA, semiconservative replication allows life to reproduce itself.

Recombination arises from a highly accurate cellular mechanism that includes the base pairing of homologous strands of nonsister chromatids (**Figure 6.22**, **Figure 6.23** and **Feature Figure 6.23**). Recombination generates new combinations of alleles.

Significant Elements:

After reading the chapter and thinking about the concepts, you should:

- Become familiar with semiconservative replication and understand the general processes of initiation and elongation.
- Think about why special enzymes (such as topoisomerases and telomerases) are needed for replication of the genome.

♦ In the process of recombination, understand how heteroduplex regions are formed and what effect they have on recombination outcomes.

♦ Work your way through the recombination process by drawing it out for yourself.

♦ Assign the 5' and 3' designations to backbone strands of DNA and RNA.

♦ Indicate complementary bases to an RNA or DNA template.

♦ Explain the importance of structural features of DNA for function in copying (replication) in recombining information.

♦ Think about experimental design.

Problem Solving Tips:

♦ The nature of the problems changes with this material. The problems still require that you have a basic understanding of the concepts and use that knowledge, but the problems are more based on experimental design. These problems can be viewed as more creative synthesis problems.

♦ The experimental type of question may cause you to go back and refresh your understanding of some techniques.

♦ A good way to approach these problems is to identify concepts relevant to the problem and review your knowledge of the topic and any relevant experimentation.

♦ One technique that comes up frequently is tagging a molecule with a label so the molecule can be followed. This is often done using radioactivity. Radioactive label can be incorporated into protein or DNA if a cell or organisms is grown in or fed a radioactive precursor that goes specifically into the type of molecule you want to follow. In designing experiments using radioactive labeling, be sure to consider how you can get a unique label into the molecule of interest.

Solutions to Problems:

<u>Vocabulary</u>

6-1. a. **6**; b. **11**; c. **9**; d. **2**; e. **4**; f. **8**; g. **10**; h. **12**; i. **3**; j. **13**; k. **5**; l. **1**; m. **7**.

Section 6.1 - DNA as the Genetic Material

6-2. (See **<u>Figure 6.5</u>**). The proof that DNA was the transforming principle was **the treatment of the transforming extract with an enzyme (DNase) that degrades DNA. After this treatment, the extract was no longer able to transform rough, nonvirulent strains of *Streptococcus pneumoniae* bacteria into smooth, virulent cells that could kill host mice.** Avery, MacCleod and McCarty also showed that treatments with RNase and proteinase did not abolish the transforming activity of their extracts, indicating that the transforming principle was not RNA or protein. These experiments using enzymatic treatments were important because it could be argued that the purified "transforming principle" these investigators isolated as DNA might have contained proteins or other molecules. Even with this extensive evidence, many scientists still remained unconvinced the DNA carries genes for almost 20 more years.

6-3. In DNA transformation, the DNA that enters the cell is in the form of randomly sized fragments, usually generated by mechanical forces that shear the DNA while it is prepared from the bacterial cell. Therefore, 2 genes that are closer together on the chromosome will end up on the same fragment more often than genes that are far apart. A high cotransformation frequency between 2 genes indicates that they are close together. **Gene *a* is closer to *c* than it is to gene *b* because there are many instances when *a* and *c* were cotransformed but only a few instances when *a* and *b* were cotransformed.**

6-4. Sulfur is found only in proteins, never in DNA, while phosphorus is a major constituent of the backbone of the DNA molecule and is found only very rarely in proteins (none of the amino acids in proteins contain phosphorous, though as we will see in Chapter 8 phosphorous can sometimes be added to certain proteins at certain times). Nitrogen and carbon, on the other hand, are found in both proteins and DNA. Hershey and Chase needed to differentiate between protein and DNA, so they needed to be able to specifically label the proteins and not the DNA and vice-versa. **If they had used labeled nitrogen or carbon there would be no way to differentiate protein and nucleic acid.**

Section 6.2 – DNA is a Double Helix

6-5. (See **Feature Figure 6.11b** for an overview of DNA structure) In **Tube #1** all the sugar phosphate bonds are broken. You would see **individual pairs of complementary nucleotides held together by hydrogen-bonds and attached to a sugar** with no the phosphate group and free phosphates (**<u>Figure 6.10</u>**). In reality, the hydrogen bonds that hold together individual complementary nucleotides (2 for A-T pairs and 3 for G-C pairs) are not very stable. These hydrogen bonds would be disrupted by the thermal forces working at room temperature. You usually need at least 4 nucleotide pairs in order to have DNA that is stably double stranded at room temperature. In **Tube #2** the bonds that attach the bases to the sugars are broken. You would see **base pairs** (similar to **Figure 6.10** without the 'sugar') **and sugar phosphate chains without the bases** (similar to **Feature Figure 6.9b** without the 'bases'). **Tube #3 would contain single strands of DNA** since the hydrogen bonds between bases were broken.

6-6. X-ray diffraction studies yielded a crosswise pattern of spots, indicating that **DNA is a helix** containing repeating units spaced every 3.4 Å. One complete turn of the helix occurs every 34 Å. The diameter of the molecule is 20 Å, indicating that DNA must be **composed of more than one polynucleotide chain** (**Feature Figure 6.9b-c**). The key X-ray diffraction pictures were taken by Rosalind Franklin and Maurice Wilkins in 1951-1952; James Watson and Francis Crick then built models based on the known chemistry of the nucleotide building blocks to fit the X-ray data.

6-7.

a.   Human DNA is double stranded. If 30% of the bases are A, and A pairs with an equal amount of T (30%), that leaves 40% to be C + G. Thus, there will be **20% C**.

b.   **30% T**.

c.   **20% G** (must equal the amount of C).

6-8. Remember that in double stranded DNA the amount of A (purine) = T (pyrimidine) and the amount of G (purine) = C (pyrimidine). Therefore one of the purines + one of the pyrimidines must equal the other purine + the other pyrimidine. The true statements are **a, b and e**. It is useful to keep in mind the fact that statement c is false; therefore A + T does not equal G + C. In fact, the DNA of different species can vary a great deal in the proportions of A-T base pairs relative to C-G base pairs. Moreover, this ratio can be very different in different regions of the same chromosome. In most organisms, the regions between genes have a higher proportion of A-T base pairs (they are "A-T rich") than the genes themselves.

6-9. In double stranded DNA adenine pairs with thymine so the percentage of A must equal the percentage of T (same for G and C). Here the amount of A is not equal to the amount of T (nor does C equal G), so **the chromosome of this virus must be single stranded**.

6-10.

a. **The A-T base pairs have only two hydrogen bonds, so it takes less heat energy to denature these base pairs.** G-C base pairs have three hydrogen bonds holding them together. It thus takes more energy to break the bonds between Cs and Gs. Remember that the DNA of different species can vary a great deal in the proportions of A-T base pairs relative to C-G base pairs. Moreover, this ratio can be very different in different regions of the same chromosome. In most organisms, the regions between genes have a higher proportion of A-T base pairs (they are "A-T rich") than the genes themselves. In the early stages research on genomes scientists sometimes tried to locate genes by looking for regions of DNA that were more resistant to heat denaturation, and thus had a higher G-C content.

b. **The denatured single stranded DNA must contain stretches of nucleotides that are complementary to a nearby sequence but in an inverted orientation.** These 'stem-and-loop' structures are regions where the single strand of DNA formed a double-stranded region. The loops and the strings holding the stems together are still single-stranded DNA.

6-11. 5'...CAGAATGGTCTCTGCTAT...3'. This could also be written backwards as 3'…..TATCGTGTCTGGTAAGAG…..5'. It is best to indicate polarity by showing the 5' and 3' directions on a strand of DNA or RNA. If no polarity is indicated then the 5' direction is at the right end of the sequence. **The dots show that this is a short region of a much longer nucleotide chain.**

Section 6.3 – DNA Stores Genetic Information

6-12. A note about nomenclature: depending on the context, the letters A, C, G and T can be used to represent either the nitrogenous bases or the nucleosides or the nucleotides containing those bases. Although it is not of great importance for most genetics courses, you might find it useful to know that organic chemists distinguish between bases, nucleosides and nucleotides with a complicated nomenclature. The name of the base is altered slightly to indicate a nucleoside (for example adenine becomes adenosine) and the name "deoxy", or more precisely "2-deoxy" is added to the nucleoside name if the sugar is deoxyribose. To name a nucleotide, the number of phosphate groups it contains and the position of their connection to the sugar are also specified. Thus 2'-deoxythymidine 5'-monophosphate signifies a nucleotide containing deoxyribose, the base thymine and a single phosphate group connected to the 5'-carbon atom of deoxyribose. See if you can find this nucleotide in Feature Figure 6.9.

a. **The information is contained in the order of those building blocks; there are 3 billion nucleotides in the complete haploid set of 23 human chromosomes.** This amount of sequence can potentially provide a huge amount of information. Although there are only 4 different building blocks they can be combined in a huge number of combinations. For instance, when considering a short 10 nucleotide long piece of DNA there are 4^{10} or 1,048,576 different possible sequence permutations. **The information may be recognized by proteins that bind directly to DNA (see <u>Chapter 16</u>), or, as you will see in <u>Chapter 8</u>, it can be copied into RNA to direct synthesis of proteins.**

b.   Each of the four building blocks is a nucleotide. **Each nucleotide is made of the sugar deoxyribose and one of the four nitrogenous bases (A, C, G or T) and a phosphate group (see Feature Figure 6.9).** Deoxyribose plus a base makes a nucleoside. When phosphate groups are added these become nucleotides. In the DNA strand the **adjacent nucleotides are connected by phosphodiester bonds that link a phosphate group to both the 3'-carbon atom of the deoxyribose of one nucleotide and the 5' carbon atom of the deoxyribose of the next nucleotide in the chain**.

c.   There are four major differences between DNA and RNA. **(i) In RNA the sugar is ribose instead of deoxyribose. (ii) RNA contains the base U instead of T. (iii) Most DNA molecules found in nature are double stranded and most RNA molecules are single stranded, but there are exceptions to both of these cases. (iv) DNA strands can be very long** – more than 100,000,000 nucleotides in a human chromosome, for example. The longest naturally occurring RNA molecules are much shorter – about 20,000 nucleotides at most.

6-13. The complementary DNA would have the complementary sequence with the opposite polarity. Note also the presence of T in DNA in contrast with U in RNA.

      3' GGGAACCTTGATGTTTCGGCTCTAATT 5'

6-14. RNA from virus type 1 was mixed with protein from virus type 2 to reconstitute a "hybrid" virus. In a parallel experiment, RNA from virus type 2 was mixed with protein from virus type 1. When these reconstituted hybrid viruses were used to infect cells, **the progeny viruses in each case had the protein that corresponded to the type of RNA in the parent hybrid virus**. The protein in the progeny did <u>not</u> correspond to the protein in the parent hybrid virus.

6-15.

a.   The possibility of finding any particular nucleotide at any position along such a DNA molecule is 1/4. The possibility that a sequence of 6 nucleotides is 5'…GAATTC…3' would be 1/4 x 1/4 x 1/4 x 1/4 x 1/4 x 1/4 = 1/4,096. **In other words, this particular sequence of 6 nucleotides would appear on average once every 4,096 nucleotides in the DNA molecule.**

b. If the proportion of the 4 different nucleotides is equal then the 6 nucleotide long *Bam*H1 site would also occur **once every 4,096 nucleotides** in the DNA molecule. Note that if there were more As and Ts than Gs and Cs the spacing between the *Eco*R1 sites would be less because the 6 nucleotides contain 4 As and Ts but only 2 Cs and Gs, while the spacing between the *Bam*H1 sites would be greater.

c. For *Hae*III, the recognition site is only 4 nucleotides long. In a DNA molecule with equal proportions of A, C, G and T in a random sequence the likelihood that any 4 base sequence would be 5'…GGCC…3' is $(1/4)^4 = 256$. **On average the *Hae*III-recognized sequences would be 256 nucleotides apart.**

Section 6.4 – DNA Replication

6-16.

a. **1/4 of the DNA is in an H/H band near the bottom of the gradient; 3/4 of the DNA is in an L/L band near the top.**

b. **1/4 of the DNA is in an L/H band in the middle of the gradient; 3/4 of the DNA is in an L/L band near the top.**

c. After one round, a single L/H band near the middle of the gradient would be expected. In subsequent rounds this band would be expected to spread gradually toward the top of the gradient as the heavy DNA is diluted across all of the new molecules in a random fashion. As a result **you would not expect discrete bands. Instead you would see a broad "smear."**

d. **1/8 of the DNA is in an H/H band near the bottom of the gradient; 7/8 of the DNA is in an L/L band near the top.**

6-17. It is best to consider the individual strands of DNA when calculating the amount of DNA of different densities. Meselson and Stahl started with H:H double stranded DNA (with nucleotides only containing ^{15}N). After one generation in 14N media (with L or light ^{14}N nitrogen), this original molecule becomes two molecules - H:L and H:L, both of which have an intermediate density. After a second generation these 2 molecules will become 4 - H:L, L:L and H:L, L:L. Thus there will be equal amounts of the intermediate band and the light band, as stated in the problem. After another round of DNA replication in the light media (round 3) these 4 molecules will become 8 - H:L, L:L, L:L, L:L and H:L, L:L, L:L, L:L. Thus **after 3 rounds 1/4 of the total DNA will be intermediate density and 3/4 will be light density**. After the next round of replication (round 4) the 8 molecules will become 16 - H:L, L:L, L:L, L:L, L:L, L:L, L:L, L:L and H:L, L:L, L:L, L:L, L:L, L:L, L:L, L:L. **At this point**

(after round 4), 1/8 of the double-stranded DNA molecules would be of the intermediate density and the remaining 7/8 would be light density.

6-18. After one S phase, the label would be in one strand of each DNA double helix (call this the round 1 helix), so each chromatid (a double-stranded molecule) contains label on one of its two strands. The labeled ^{3}H-thymidine was removed before the next S phase, so the next set of new strands are not labeled. When the unlabeled strand of each round 1 chromatid is used as a template in round 2, the resulting double stranded chromatid will be unlabeled. When the labeled strand in the round 1 helix is replicated, the resultant chromatid contains this labeled strand and an unlabelled, newly synthesized complementary strand. Thus, after this second round of replication is complete every chromosome would have one labeled chromatid (but the label is only on one strand) and one unlabeled chromatid. Therefore, for each homologous chromosome pair, **one chromatid of each of the two chromosomes would contain label. This is option c**.

6-19.

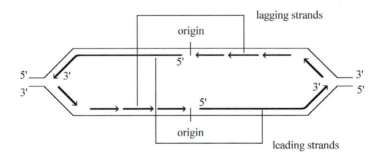

6-20.

a. **DNA polymerase needs to interact with the high energy phosphate groups at the 5'-end of the incoming nucleotide triphosphate and the 3'-end of the previously incorporated nucleotide**. This action of the DNA polymerase automatically dictates the 5'-to-3' growth of the DNA chain during replication.

b. DNA ligase also catalyzes the formation of a phosphodiester bond between the nucleotides located at the 3' end of one Okazaki fragment and the 5' end of the next Okazaki fragment. **Because only a single phosphate group is attached to the 5'-carbon atom of the nucleotide at the 5'-end of an Okazaki fragment after the primer is removed, DNA ligase needs an external source of energy to catalyze the formation of a phosphodiester bond. Ligase obtains this energy by hydrolyzing ATP. This lack of a triphosphate group at the 5' end of an Okazaki fragment also explains why DNA polymerase cannot be used to join Okazaki** fragments – this enzyme would have no energy source to catalyze the formation of a phosphodiester bond to a nucleotide without a high energy bond.

6-21. Primers for DNA synthesis are RNA molecules that are made by the "Primase" enzyme. The primer is complementary to the DNA sequence shown and has the opposite polarity:

5' UAUACGAAUU 3'

6-22. A bubble is formed by 2 replication forks proceeding in opposite directions form a single origin of replication.

a.   There are 3 bubbles in this figure, so there must be **3 origins of replication** in this DNA molecule.

b.   There are **6 replication forks,** one at each of the two ends of each of the 3 replication bubbles.

c.   If all replication forks move at the same rate, then the largest bubble was the first one activated. The **smallest bubble (the one in the middle) was the last origin of replication to be activated**.

6-23.

a.   **Topoisomerase relieves the stress of the over wound DNA ahead of the replication fork**.

b.   **Helicase unwinds the DNA by breaking hydrogen bonds between base pairs to expose the single strand templates for replication**.

c.   **Primase synthesizes a short RNA oligonucleotide, which DNA polymerase requires as a primer in order to copy the template**.

d.   **DNA ligase joins the sugar phosphate backbones of adjacent Okazaki fragments in order to construct a continuous strand of newly synthesized DNA**.

6-24.

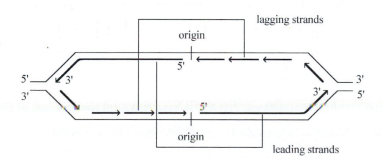

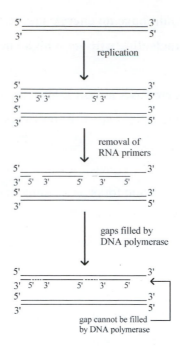

Normally, after an RNA primer is removed from the 5' end of an Okazaki fragment, the "lost" information in the strand being synthesized can be replaced. This occurs when DNA polymerase extends the 3' end of the preceding Okazki fragment by copying the template strand exposed by primer removal. However, this cannot occur at the 5' ends of the new DNA strands, because there is no preceding Okazaki fragment that could be extended. In other words, **when the primer is removed from the very 5' end of a newly synthesized strand in a linear chromosome, bases are exposed at the 3' end of the (old) template strand. There is no way to synthesize complementary DNA for those exposed nucleotides. As a result, you would expect that information equal to the length of the removed primer will be lost from the 5' end of the new strand at each end of the chromosome** if there is not some sort of alternate replication methodology at the chromosome ends. This leads to a successive shortening of the chromosomes in each generation of cells. It turns out that linear chromosomes in eukaryotic organisms have special structures called "telomeres" at their ends that allow them to overcome this obstacle. Chapter 12 discusses the nature of these telomeres in detail. The figure above shows the situation at the right end of a linear chromosome. The same loss of DNA sequence happens at the left end of the chromosome as well. You can picture this by rotating this diagram 180° while remembering that the RNA primer on the left end of the chromosome will be found on the pair of strands that is continuous at the right end of the chromosome.

6-25. The figure shows both strands of DNA being replicated in the same direction relative to the replication fork. We now know that each replication fork has a lagging strand and a leading strand because DNA polymerase only catalyzes the addition of nucleotides to the 3' end of the growing chain. Watson and Crick's figure also makes it seem as if DNA replication happens spontaneously without the participation of any of the many proteins that are now known to be involved in the process (**Feature Figure 6.20**). In spite of these simplifications this drawing was one of the most influential in the history of modern biology. Watson and Crick's simplification made it intuitively obvious how double helical DNA could be passed from one generation to the next. This was very important in convincing scientists of the time that their model was valid.

6-26. In order to unlock the intertwined helices, **you must break both strands of one of the intertwined molecules. The unbroken strand can then pass through the break in the other DNA molecule**. The broken helix must then be rejoined. This breakage/rejoining is mediated by topoisomerase.

6-27. Remember that DNA polymerase requires a primer supplying a free 3' end so that nucleotides can be added to the growing new strand, a single stranded template DNA to copy and a source of dNTPs. The dNTPs have been already added to the reaction tubes.

a.   The reaction tube with DNA #1 has only a single type of DNA. The DNA polymerase cannot function to produce a new strand, so **no new DNAs will be formed.**

b.   Again the reaction tube with DNA #2 has only a single type of DNA so **no new DNAs will be formed**.

c.   The reaction tube with DNA #1 + DNA #2 has two DNA strands. In order for DNA polymerase to work these DNAs must form a region of double helical DNA - the two strands must have some complementary base sequences. They can pair with each other as follows:

> 5' CTACTACGGATCGGG 3'
> 3' TGCCTAGCCCTGACC 5'

In this configuration both strands can act as a template and as a primer. Each strand supplies a free 3' end to which DNA polymerase can attach new nucleotides whose sequence will be determined by complementarity with the template supplied by the other strand. **Two new DNA molecules will be formed**, each 5 nucleotides longer than the two original DNAs:

> 5' CTACTACGGATCGGGACTGG 3' and 3' GATGATGCCTAGCCCTGACC 5'

DNA sequences are usually written with the 5' end at the left, so the second of these DNA molecules could also be written as: 5' CCAGTCCCGATCCGTAGTAG 3'.

d. The reaction tube with DNA #3 is a special case. **This single strand of DNA has two regions that have complementary base sequence so the DNA can form a so-called** *hairpin loop* (problem 6-10). The two regions can base pair with each other as follows:

$$5'\ \ AGTAGCCAGTGGGG\ A\ A$$
$$GGTCACCCC\ A\ A\ A$$

Note that there is now a free 3' end to which DNA polymerase can add nucleotides as well as a template to guide the addition. **The product will therefore be 5 nucleotides longer than the original:**

5' AGTAGCCAGTGGGGAAAAACCCCACTGGCTACT 3'

It is also possible for two of these DNA molecules to pair as shown:

5' AGTAGCCAGTGGGG<u>AAAAA</u>CCCCACTGG
3' GGTCACCCC<u>AAAAA</u>GGGGTGACCGATGA

The 5 Adenines on the two strands (underlined) will not pair. Instead they will form a bubble in the middle of this double stranded molecule. There are now free 3' ends to which DNA polymerase can add nucleotides as well as templates to guide the addition. The product will be the same as that from the template with the hairpin loop.

Section 6.5 – Recombination at the DNA Level

6-28. The coinfection of bacteria with c^+ and c bacteriophage allowed recombination to occur between the two genotypes of bacteriophage. How can a single progeny bacteriophage produce both c^+ and c progeny in the second round of infection? How can one bacteriophage with a single double stranded chromosome contain two kinds of genetic information? Remember that a key intermediate step in recombination is the formation of a heteroduplex region (**Feature Figure 6.24 step 5**). Normally, the heteroduplex regions are corrected by DNA repair systems. However, **in rare cases the heteroduplex is not corrected and a single bacteriophage particle can be generated with a chromosome that contains the mismatch. One strand of DNA would be c^+ and the other strand would be c.** When such a bacteriophage infects a new bacterial host cell, DNA replication creates some molecules in which both strands are c^+ and other DNAs in which both strands are c.

   As an aside, you might be wondering how you can actually isolate a single bacteriophage and look at its progeny. This will become clear in the next chapter, when methods for handling bacteriophage are explored in more detail.

6-29. A mutant *recA E. coli* **strain will not be able to undergo recombination.** Any event requiring recombination would not occur normally (**Feature Figure 6.24 step 3**).

6-30. The numbers of *B* and *b* alleles are not in the 2:2 ratio predicted from meiosis of an *A B C / a b c* diploid. The 3:1 ratios indicate that gene conversion occurred. Gene conversion results from the DNA repair of the nucleotide mismatches of heteroduplex DNA that forms during branch migration in the process of recombination (**Feature Figure 6.24 step 5**). Because the B gene shows the 3:1 ratio of alleles this gene must have been in the heteroduplex region.

Recombination events can be resolved in two ways when the flanking markers are considered - genes A and C in this case. In one resolution the flanking markers do not show recombination (**Feature Figure 6.24 steps 6-8**). These are called aborted crossovers. Tetrad I is the aborted crossover type. The A and C markers show the non-recombinant configuration of AC and ac. In this tetrad the correction of the nucleotide mismatches in the two regions of heteroduplex DNA was to *B* (**Figure 6.23**) giving spores of the genotypes *A B C, A B C, a B c* and *a b c*.

In the other resolution of the recombination event the flanking markers do show recombination. In tetrad II the flanking markers show tetratype genotypes (*A C, A c, a C* and *a c*) indicating a single cross over between these genes. Again the heteroduplex DNA must have included the region of the B gene. In this case the heteroduplexes were both repaired to the b allele giving spores of the genotypes *A B C, A b c, a b C* and *a b c*.

6-31. The process of recombination involves the breakage, exchange, and rejoining of strands to make the Holliday intermediate (**Feature Figure 6-24**, especially **steps 2-4**). This is followed by a final resolution involving cutting and rejoining the strands. **If the same two strands of the DNA molecule that were cut during the initiating event are cut during the resolution, there will be no crossing over.** (Crossing over is defined as the genetic recombination of markers outside of the gene conversion region.) **There is an equal likelihood that the other strands will get cut and genetic recombination is the result.** Regardless of which strands are cut during resolution, those resulting in crossing-over or those not leading to crossing-over, **mismatches within the heteroduplex region that are generated early in the process can be corrected to the same allele, resulting in gene conversion** (also **Figure 6.23**). Therefore, gene conversion occurs equally as frequently as recombination of genetic markers.

6-32.

a.   **The crossover initiated either between *e* and *f* or between *f* and *g*.**

b.   **If you said the initiating cut was between *e* and *f* then the resolving cut occurred between *f* and *g*. If you said the initiating cut was between *f* and *g* then the resolving cut occurred between *e* and *f*.** In either case, gene *f* must have been included in the heteroduplex region to get the resulting octad (because gene *f* does not segregate 4:4).

c.   **The chromosome that ended up with the f^+ - f mismatch in the heteroduplex region was corrected by mismatch repair to f f. The e gene was not within the heteroduplex region and therefore it was not affected by the mismatch repair, so the alleles segregated 4 e^+ : 4 e^-.**

6-33. During strand invasion an enzyme similar to *E. coli* recA opens the nearby double helix and which allows the 3' end of the invading strand to search for a homologous sequence (**Feature Figure 6-24 step 3**). **If many short repeats are present in the double helix at the point where the invading strand is pairing, it is likely that the invading strand will not line up perfectly.** This incorrect alignment can either generate extra repeats or eliminate some copies of the repeats.

Chapter 7   Anatomy and Function of a Gene: Dissection Through Mutation

Synopsis:

This chapter is about MUTATIONS! They are the heart and soul of genetics - the basis of genetic variation, the raw material for evolution, and the basic tool of genetic analysis. Now that you know the structure of DNA, you can understand the molecular nature of different kinds of mutations, how errors arise that can result in mutation (**Figure 7.10**), and how errors can be corrected. You can also understand some of the implications of various mutations on protein structure and function.

The functions of proteins produced within an organism determine phenotype. The connection between genes and what the gene products do is apparent in the consequences of a defect (mutation) in a gene encoding a protein that is needed in the pathway. The order of genes acting in an enzymatic pathway is determined based on the ordering of intermediate compounds and the enzymes that catalyze the conversion of one compound to another.

Two of the most important tools for genetic analysis are complementation tests and deletion mapping. Complementation analysis is used to determine if mutations that result in the same phenotype are in the same gene. In this way, the number of genes required for a particular process can be determined. In complementation analysis, gametes containing chromosomes with two different mutations are combined. If a gamete containing a mutation in gene A, but not gene B is combined with a gamete with a mutation in gene B but not A, wild type gene products for both A and B will be made in the resulting zygote and the zygote will be wild type, not mutant. Thus, these mutations *complement* each other. But if the mutations are in the same gene, there is no wild type copy of the gene available so there is no functional gene product for that gene and the cell is still mutant. Thus the mutations are *non-complementing*.

Deletion analysis is used to determine the location and order of genes. It is a quicker way to determine map order than doing linkage analysis on each pair of mutations (or genes) along a chromosome.

Recombination can occur within a gene, even between adjacent nucleotide pairs.

Significant Elements:

After reading the chapter and thinking about the concepts, you should be able to:

♦ Identify different types of mutations that can occur in DNA: frameshift (deletion and insertion), transition, transversion, and complex mutations like inversions and reciprocal translocations (see Chapter 13 for a thorough presentation of chromosomal rearrangements). Understand the natural processes that change DNA sequence like depurination, deamination and the effects of UV and X rays.

♦ Understand how mutagens alter DNA (**Figure 7.10**).

♦ Understand how these various mutations change the structure and function of the proteins coded by the mutated genes.

♦ Understand how DNA is repaired by the various repair systems (**Figures 7.11, 7.12, 7.13** and **7.14**).

♦ Know how to set up a complementation test; a genotype with the mutations in trans is a complementation test; a genotype with a mutation in cis is a test for dominance/recessiveness.

$$\frac{a^+ \quad b^+}{a \quad b} \qquad \frac{a^+ \quad b}{a \quad b^+}$$

cis genotype   trans genotype

♦ Determine the number of genes represented by the results a of complementation test.

♦ Understand the difference between complementation and recombination analysis. Complementation is a test for function and does not require any interaction between the DNA molecules. Recombination requires physical exchange between 2 homologous chromosomes. This occurs during meiosis in eukaryotes and during DNA replication in prokaryotes and bacteriophage. Make sure you understand which process is being analyzed.

♦ Use deletion mapping to position mutants on a map (see problem 7-25).

♦ Use fine structure analysis to order mutations and calculate recombination frequencies.

♦ Determine the order of intermediate compounds in an enzymatic pathway using data on the ability of mutants to grow on intermediate compounds (see problem 7-30).

♦ Determine the order that genes act in a pathway composed of several enzymatic steps (see problem 7-30).

♦ Understand how to use mutations to dissect a complex biosynthetic process like the assembly of the bacteriophage T4 (see **Figure A in the Fast Forward box**).

Problem Solving Tips:

♦ Many of the problems in this chapter involve cross schemes - remember to apply the 3 Essential Questions to help with assigning genotypes - see Chapter 5 'Problem Solving - How to Begin' on page 66 of this Study Guide / Solutions Manual.

- A complementation test is a test for function. Lack of complementation means the mutations are in the same gene. If a wild type phenotype is seen in all cells then the mutations are in different genes (are complementing).

- Group the negatives (i.e. mutations that are in the same gene) when analyzing complementation data.

- Remember that dominant mutations cannot be mapped to a single gene based on complementation data.

- Recombination between the mutant chromosomes and reversion of the mutation on one of the mutant chromosomes are both rare events which regenerate a wild-type gene. Therefore, a small number of wild-type cells when most of the cells in the complementation test are mutant are due to recombination or reversion.

- Deletions do not revert and can be recognized by this characteristic. Deletions do not complement mutations that are located in the deleted region.

- Deletions can also be defined by their behavior in fine structure recombination mapping. A deletion is a mutation which does not recombine with 2 other mutations which do recombine with each other (consider deletion #3 in problem 7-25).

- Fine structure recombination analysis can be used to order closely linked mutations. When 2 closely linked mutations are in trans, the rare single recombination events between these mutations will give one product which is wild type for the phenotype. By examining the genotypes of flanking markers it is possible to order the original mutations (see problem 7-24). Recombination frequency = 2(# wild type recombinant progeny) / total progeny.

- An enzymatic pathway is a series of steps each catalyzed by a gene product (enzyme). To order the compounds, work from the final product toward the beginning of the pathway. Remember that the final product is the compound on which all the mutants will grow. If there is a mutation early in the pathway, providing later intermediate compounds will allow growth, because the subsequent enzymes are available. The block point has effectively been bypassed in these cases.

Solutions to Problems:

Vocabulary

7-1. This answer has been grouped 2 different ways. First, as the terms (numbered column) that apply to each mutational change (lettered column): a. **1, 5**; b. **1, 2, 5**; c. **2**; d. **8**; e. **9**; f. **3, 6, 9**; g. **4, 7**; h. **1, 5**; i. **6, 8, 9**; j. **3, 4**; k. **6, 7**. The following groupings show the mutational changes (lettered column) that apply to each term (numbered column): 1. **a, b, h**; 2. **b, c**; 3. **f, j**; 4. **g, j**; 5. **b, h**; 6. **f, i, k**; 7. **g, k**; 8. **d, i**; 9. **e, g, i**.

Section 7.1 - Mutations

7-2. We are told the trait in the pedigree is very rare. The trait is first seen in individual III-1 and shows a vertical pattern of inheritance between generations III and IV, suggesting a mutation that is dominant to the wild-type. The mutant gene cannot be X-linked because male IV-1 inherits it from his father. Thus the trait is autosomal dominant, yet it is not seen in generations I and II.

It is possible that the mutant allele is incompletely penetrant. In this case one individual in generation I and one individual in generation II (either II-2 or II-3) have the mutant allele but do not display the trait.

A second possibility is that a mutation occurred in the germ line of either II-2 or II-3 producing a gamete with the dominant autosomal mutation. This gamete was then used to form III-1 (the propositus).

The data in the pedigree does not exclude either of these possibilities. Incomplete dominance is probably less likely. The trait is transmitted to half the progeny of the propositus, suggesting complete dominance. However there is no evidence that anyone in the first two generations has the trait.

There are two potential ways to discriminate between these hypotheses. First, more information about this trait from a more extended family pedigree or from pedigrees of other families might indicate whether the trait is or is not completely dominant. Second, if you could obtain DNA sequence information for the gene in question from II-2, II-3 and III-1 you could see if one of the alleles of the gene in the propositus has a mutation not found in his parents.

7-3. Each independently derived mutation will be caused by a different single base change. When you find a base that differs in only one of the sequences, it is the mutation. Determine the wild-type sequence by finding the base that is present at that position in the other two sequences. The wild-type sequence is therefore:

5' ACCGTA_G_TCGAC_T_GGT_A_AACTTTGCGCG

7-4. Dominant mutations can be detected immediately in the heterozygous progeny who receives the mutant gamete (see problem 7-5). **Recessive mutations can only be detected when they are homozygous. To detect the appearance of new recessive alleles you must test cross with a recessive homozygote.** This can be done in mice, where the researcher can control the mating, but it cannot be done in humans!

7-5. Of the achondroplasia births observed, 23/27 are due to new mutations because there is no family history of dwarfism. Achondroplasia is an autosomal dominant trait, so it will be expressed in the child that receives the mutant gamete. There were 120,000 births registered, so there were 240,000 parents in which the mutation could have occurred during meiosis, leading to a mutant gamete. **The mutation rate = 23 mutant gametes/240,000 gametes = 9.5 x 10^{-5}. This rate is higher than 2 to 12 x 10^{-6} mutations per gene per generation which is the average mutation rate for humans**.

7-6. To ensure that the mutants you isolate are independent you should follow procedure #2. If you follow procedure #1, a mutation causing resistance to the phage could have arisen several generations before the time when you spread the culture and several of the colonies you isolate could be clones of the same cell and thus have the identical mutation. Different mutations in the same gene often give different information about the role of the gene product. Therefore, geneticists generally strive to find many independent and thus different mutations in the same gene in order to understand as much about the gene as possible.

7-7. Kim's hypothesis is that the bacteriophage is able to induce resistance in ~1 in every 10^4 bacteria. If she is right, then several (~10 if there are 10^5 bacteria on each plate) of the colonies on each of the replica plates should contain resistant cells that will continue to grow after exposure to phage. **These resistant cells and the surviving colonies that grow from them will be randomly distributed over the three plates**. **Maria**'s hypothesis is that the resistant cells are already present in the population of cells they plated on the original plate. If she is right then some of the colonies on the original plate (about 10 of them) should have contained resistant bacteria which will give rise to colonies after treatment with the phage. **These resistant colonies will appear at the same locations on all three of the replica plates (Figure 7.5b)**.

7-8. Since the mutation is not found in the somatic tissue (blood) from any of the parents in generation II **the mutation must have occurred in the germ line of one of the parents. This parent must be II-2** since the two affected children had the same father but different mothers. However mutations are rare, so **how can III-1 and III-2 have exactly the same mutation? The mutation must have occurred in the germ line of II-2 at a point before the formation of separate sperm**. In the male germ line cells called spermatogonia divide mitotically to make many meiotic cells (primary spermatocytes; see Figure 4.18). **It is likely that a mutation occurred in an early spermatogonial cell leading to a large proportion of spermatocytes and therefore a large proportion of sperm**

that carry the same mutation. **Thus this pedigree suggests that the testes of II-2 exhibit germ line mosaicism** - some of the spermatocytes carry the mutation while others do not. Note that this is similar to the pattern seen in Hypothesis 2 of the Luria-Delbrück fluctuation test (Figure 7.4): **if a mutation occurs early in the growth of a bacterial colony (or in the mitotic division of the germ line) then many of the cells in the colony (or in the germ line) can have the mutation**. Germ line mosaicism in which a large proportion of the germ line carries a new mutation is likely to be rare because the mutation must occur early in the multiplication of the germ line. At this early stage there are fewer cells than later, so there are fewer chances for mutations to occur.

7-9. Diagram the crosses (InX^{Bar} = X chromosome with inversions and the mutant allele of Bar eyes, X* = mutagenized X chromosome:

X / InX^{Bar} ♀ x X* / Y → F$_1$ InX^{Bar} / X* ♀ x X / Y → 1/2 InX^{Bar} / Y (Bar ♂) : 1/2 X* / Y

(wild type ♂). These F$_1$ females must also have had daughters: 1/2 InX^{Bar} / X (Bar ♀) : 1/2 X* / X (normal ♀). However, there were 3 F$_1$ females who did NOT give these ratios of sons: ♀A, ♀B and ♀C.

F$_1$ InX^{Bar} / X^{A}* (♀A) x X / Y → 1/2 ♀ : 1/4 Bar ♂ : 1/4 white ♂

F$_1$ InX^{Bar} / X^{B}* (♀B) x X / Y → 2/3 ♀ : 1/3 Bar ♂

F$_1$ InX^{Bar} / X^{C}* (♀C) x X / Y → 4/7 ♀ : 2/7 Bar ♂ : 1/7 normal ♂

Remember that each F$_1$ daughter inherited a different mutagenized X chromosome from their mutagenized male parents. Thus, ♀A, ♀B and ♀C all are each heterozygous for a different mutagenized X chromosome. Any unusual result obtained with these females is due to some alteration of the X that underwent the mutagenesis. **The mutagenized X chromosome carried by ♀A must have a recessive white eye mutation.** There is no effect on viability, but the sons who inherited mutant (non-*Bar*) X chromosome (1/2 of the total sons) have white eyes. In ♀B, there is a reduction in the number of males by half. This indicates that **♀B is heterozygous for a mutagenized X chromosome with a new recessive lethal mutation**, and the sons that inherit that X chromosome die. Female C produced fewer sons than expected among those that inherited the mutagenized X chromosome; however, some of the sons which inherited that chromosome survived. Thus, **the mutagenized X chromosome from ♀C may contain a recessive mutation that causes lethality but is not completely penetrant, or she could be a mosaic** where some of her cells are of one genotype, while other cells are a different genotype. In the case of mosaicism, ♀C inherited a mutagenized X chromosome from her father in which one strand of DNA contained a lethal mutation and the other strand had the wild-type sequence. This scenario implies that DNA replication or repair mechanisms have not yet acted to make sure both strands were completely complementary in sequence. As a result, some of her tissues will have the wild type chromosome from her father and others will have the mutant homolog. All of the cells in ♀C will have a second X chromosome, the InX^{Bar} chromosome that she inherited from her mother. Inheritance of these chromosomes could cause the germ line of ♀C to be a mosaic that contains a mix of gametes: InX^{Bar}, wild-type and the lethal mutation.

7-10. Diagram the cross; * denotes the mutagenized fly:

wild type ♀ x wild type* ♂ → F₁ ♀ x *y cv ct sn m* → 1/3 wild type ♀ : 1/3 *ct sn* ♀ : 1/3 wild type ♂

a. X-rays cause breaks in DNA. The *ct sn* ♀ in the second generation suggests that **the mutagenized wild type X chromosome is missing the *ct, sn* region**. Thus this mutagenized chromosome has a deletion that removes both *ct*$^+$ and *sn*$^+$ and therefore uncovers the recessive alleles on the other homolog.

b.

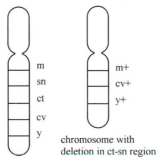

chromosome with
deletion in ct-sn region

c. *ct sn* ♀ (from 2nd generation of cross above) x wild type ♂. The genotype of this female is *y cv ct sn m / y*$^+$ *cv*$^+$ (del) *m*$^+$. Therefore, **this female will make a variety of non-recombinant and recombinant gametes**. The table shows the reciprocal pairs of parental gametes and the SCO classes: region 1 between *y* and *cv*; region 2 between *cv* and one end of the deleted region and region 3 between the other end of the deleted region and *m*):

female gametes	male gamete y^+ cv^+ ct^+ sn^+ m^+	male gamete Y
y cv ct sn m (parental)	*y cv ct sn m* / y^+ cv^+ ct^+ sn^+ m^+ (wild type)	y cv ct sn m / Y (y cv ct sn m ♂)
y^+ cv^+ (del) m^+ (parental)	y^+ cv^+ (del) m^+ / y^+ cv^+ ct^+ sn^+ m^+ (wild type)	y^+ cv^+ (del) m^+ / Y (dead)
y cv^+ (del) m^+ (SCO region 1)	*y* cv^+ (del) m^+ / y^+ cv^+ ct^+ sn^+ m^+ (wild type)	y cv^+ (del) m^+ / Y (dead)
y^+ *cv ct sn m* (SCO region 1)	y^+ *cv ct sn m* / y^+ cv^+ ct^+ sn^+ m^+ (wild type)	y^+ cv ct sn m / Y (cv ct sn m ♂)
y^+ cv^+ *ct sn m* (SCO region 2)	y^+ cv^+ *ct sn m* / y^+ cv^+ ct^+ sn^+ m^+ (wild type)	y^+ cv^+ ct sn m / Y (ct sn m ♂)
y cv (del) m^+ (SCO region 2)	*y cv* (del) m^+ / y^+ cv^+ ct^+ sn^+ m^+ (wild type)	y cv (del) m^+ / Y (dead)
y^+ cv^+ (del) *m* (SCO region 3)	y^+ cv^+ (del) *m* / y^+ cv^+ ct^+ sn^+ m^+ (wild type)	y^+ cv^+ (del) m / Y (dead)
y cv ct sn m^+ (SCO region 3)	*y cv ct sn* m^+ / y^+ cv^+ ct^+ sn^+ m^+ (wild type)	y cv ct sn m^+ / Y (y cv ct sn ♂)

The female will also make rarer DCO classes of recombinants. Note that half of all the sons inherit the deletion, which is lethal when hemizygous because many essential genes are missing.

7-11.

a. *Essential* genes are those that produce a recessive lethal phenotype when mutant - if they don't function because of a mutation, the animal does not live. H.J. Muller's control results showed that 0.3% of the females produced only wild type (non-Bar) sons. Thus these females had a recessive lethal mutation on the X chromosome with the Bar mutation. In other words mutations in essential genes on the X chromosome occur at a rate of 3×10^{-3} mutations per gamete. The average spontaneous mutation rate for *Drosophila* genes is 3.5×10^{-6} mutations per gene per gamete. Therefore, 3×10^{-3} X-linked lethal mutations per gamete / 3.5×10^{-6} mutations per gene per gamete = **857 essential (lethal) X-linked genes**. This estimate depends on the accuracy of the estimate for the average mutation rate per gene in *Drosophila*.

b. The estimated 857 essential genes on the X chromosome / 2279 total genes on the X chromosome
    x 100 = **37.6% of the genes on the X chromosome are essential**. Although this is a rough figure,
    it matches fairly well to other estimates made with other techniques suggesting that only 1/4 - 1/3
    of the genes in the *Drosophila* genome are essential. In other words, 66-75% of the genes in the
    genome could be disrupted by mutation and the flies could still survive.

c. Clearly the mutation rate has increased with exposure to X-rays, demonstrating that X-rays are
    mutagenic. Genetics became much easier to do because scientists could increase the chances they
    would find rare mutations. (H.J. Muller received a Nobel Prize for this finding.) You can calculate
    the average mutation rate per gene upon exposure to this dosage of X rays if you use the answer to
    part a indicating the number of essential genes. There are 857 essential genes on the X
    chromosome x X-ray induced mutation rate/gamete = 0.12 X-linked lethal mutations/gamete. Thus
    the X-ray induced mutation rate = 0.12/857 = 1.4×10^{-4}. In other words, this high dosage of X-
    rays has **increased the mutation rate about 40-fold** over the spontaneous rate (1.4×10^{-4} / 3.5 x
    10^{-6}).

7-12. All of the mutagens listed in Figure 7.10 cause **transitions**.

7-13. A perfect reversion is a very rare event, even when induced by two-way mutagens. Such a
reversion demands one particular change in the exact same base pair that was originally mutated.
These kinds of reversion events can normally be seen only in experimental organisms like bacteria or
bacteriophage where it is possible to examine so many individuals that rare events can be detected.

a. Figure 7.10 shows how the base analog 5-bromouracil (5-BU) can cause a T:A to C:G substitution.
    5-BU can sometimes behave like thymine and sometimes like cytosine. In the figure, 5-BU
    behaves like T (because it base pairs with A), but then shifts to a state where it behaves like C, so
    that during replication, DNA polymerase will incorporate a G in the newly forming strand. **5-BU
    is a two-way mutagen**. A reversion can occur if 5-BU is incorporated into DNA in the C-like state
    the DNA will have a 5-BU:G base pair. If the 5-BU shifts into its alternative form during
    replication, it will act like a T, so the result will be a 5-BU:A base pair; following the next round
    of DNA replication, the result will be T:A.

b. **Hydroxylamine is a one-way mutagen** because it changes C to hydroxylated C (C* in the
    figure), and C* can only pair with A. This will yield a T:A pair after the next round of replication.
    Hydoxylamine can not mutagenize this.

c. **Ethylmethane sulfonate (EMS) is a two-way mutagen** because it can modify either G or T.
    Figure 7.10 shows that ethylated G (G*) pairs with T, resulting in a G:C to A:T substitution. If

EMS now ethylates T forming T* (this part is not shown in the figure), then the modified T will pair with G. This will result in an A:T to G:C substitution which is an exact reversion.

d. Nitrous acid changes C to U and also A to hypoxanthine. Figure 7.10 shows how this mutagen can cause both a C:G to T:A substitution as well as a reverse T:A to C:G substitution making it a **two-way mutagen**.

e. Proflavin can cause either an insertion or deletion of a base, thus it is a **two-way mutagen**. If a chromosome with an insertion of a base caused by proflavin is now treated with proflavin a deletion of this additional base may result, restoring the original sequence.

7-14.

a. EMS induced new mutations in the *dumpy* gene in the sperm of the treated males. However, as shown in Figure 7.10, EMS-induced mutations are fixed in both strands of the DNA only after a few rounds of DNA replication. **The DNA in the *dumpy* gene of a sperm just treated with EMS would have one DNA strand with the normal G and the other DNA strand with hydroxylated C (C*).** This sperm now fertilized a dumpy egg. **After several rounds of DNA replication and mitosis, some cells will have the normal C:G base pair, while other cells will have a *dumpy* mutant A:T base pair.** An animal in which different cells have different genotypes is called a mosaic and will be described in more detail in other chapters. **The phenotype of the fly that grows from this fertilized egg would depend upon which particular cells in the body had the wild type or mutant sequence of the *dumpy* DNA.** This explains why one wing might be short and the other long. **Whether or not the new dumpy mutation in the F_1 animals can be transmitted to future progeny depends upon whether some of the cells in the germline carried the mutant A:T base pair.**

b. If the progeny of the second cross have short wings then the germline cells of their F_1 male parent carried the mutant A:T base pair. Many rounds of DNA replication are involved in producing the male germline from the zygote. Therefore **these progeny are not mosaics because they have received an A:T base pair from their F_1 parent in which both DNA strands are mutant so these progeny will have two short wings.**

7-15.

a. DNA in most organisms is not exposed to high levels of aflatoxin B_1. It is therefore unlikely that most cells would have evolved genes for enzymes that could directly remove the aflatoxin B_1 from guanine, or a glycosylase that could specifically remove the adduced guanine-aflatoxin B_1 base (as in base excision repair). **The repair system most likely to repair this is the nucleotide excision repair**. This system would treat the adduct in a similar fashion to a thymine-thymine dimmer, which is also a bulky group that distorts the double helix. **It is also possible that the SOS-type error-prone repair system will repair aflatoxin B_1 damage**. Some other repair systems make less sense. Methyl-directed mismatch repair is unlikely because the mutagen works independently of DNA replication. Double-strand break repair is not involved because this mutagen does not produce double-strand breaks.

b. This new information changes the picture somewhat. Apurinic (AP) sites are intermediaries in base excision repair, so **AP endonuclease and other enzymes in the base excision repair system could remove the damage**. A glycosylase would not be required because the adduct in effect removes itself from the sugar. It is also known that **SOS repair systems can work at AP sites, adding any of the 4 bases at random** across from the AP site during DNA replication.

7-16. The mutagen initially mutates somatic cells, not a gamete producing cell. These somatic cell mutations give rise to the tumor cells. **When the tumor cells are injected into a new mouse, they will divide in an uncontrolled manner and cause a tumor to develop. The somatic mutation that caused the original cell to become cancerous is not present in the germ cells of the mouse.** Thus, the cancer phenotype cannot inherited in a Mendelian fashion.

7-17. Yes. The rat liver supernatant contains enzymes that convert substance X to a mutagen, and his^+ revertants occur. Our livers contain similar enzymes that process substances, converting them into other forms that cause mutation and can lead to cancer.

Section 7.2 - Mutations and Gene Structure

7-18. Do a complementation test by mating the two mice. If the mutations in each mouse are in the same gene all the progeny will be mutant (albino). If the mutations causing albininsm in each mouse are in different genes, all the progeny will be wild-type (the genes will complement).

7-19.

a. **Each complementation test involves taking pollen from a plant homozygous for one mutation and using it to fertilize ovules from a plant homozygous for another mutation. This creates plants heterozygous for the two mutations**. The diagonal represents self-fertilization; a recessive mutation cannot complement itself. A **- indicates a lack of complementation and the resulting plants have serrated leaves. A + means that the two mutations complemented each other, so the resulting plant had wild type leaves with smooth edges. Some of the boxes in the table are colored because they represent crosses that were already done**, but the parents were switched. For example a cross with mutant 1 pollen and mutant 2 ovules will produce the same kinds of progeny as a cross with mutant 2 pollen and mutant 1 ovules.

b. If a box has a - then the two mutants must be in the same complementation group or gene. If the box has a + the two mutations are in different complementation groups or genes. From the results given mutation 1 is in the same complementation group as mutation 3 (-). Mutation 1 is in a different complementation group than mutation 2 (+). Therefore mutation 2 must be in a different complementation group than mutation 3 (+). **The completed boxes are circled**.

	1	2	3	4	5	6
1	-	+	-	(-)	+	(+)
2		-	(+)	(+)	(+)	-
3			-	-	(+)	(+)
4				-	(+)	(+)
5					-	+
6						-

c. There are **3 complementation groups or genes. One includes mutations 1, 3, and 4. A second group has mutations 2 and 6. The third consists of mutation 5.**

7-20.

a. Assuming that the man and woman represented in the pedigree in Figure 3.19c each have a recessively inherited form of albinism then **this is a complementation test**. In this example, **the two albinism-causing mutations are in different genes** (complement each other) and none of the children is albino. If the two mutations were in the same gene, then all the children should be albinos, as shown in Figure 3.16b.

b. Complementation tests upon demand are not possible in humans. Therefore, the most straightforward way to tell if a trait can be caused by mutations in more than one gene (exhibits locus heterogeneity) is to **map the mutations in a number of independent families**. Mapping involves recombination analysis of the gene causing the mutant trait with other markers. If the mutations in different families map to different chromosomes or are far apart on the same chromosome then they must be in different genes. If the mutations map close to each other on the same chromosome they may or may not be in the same gene. For example, the two mutations could be in adjacent, closely related genes such as the red and green photoreceptor genes shown in Figure 7-28c. In a case like this researchers can determine the DNA sequence of these closely related genes in various mutant individuals to see if the mutation affects the same gene. This approach will be discussed in later chapters.

c.   Again, as explained in part b above, you can try to **map the dominant mutations or analyze the DNA of candidate genes** you think might be involved in the conditions.

7-21.

a.   Deletion mutations can be identified in a couple of ways. **Deletions are mutations that never revert to wild type. Deletions are also mutations that don't recombine with 2 other mutations that do recombine with each other**. For example, mutations a and b <u>do</u> recombine with each other, but mutation c does not recombine with mutation a nor mutation c; this implies that mutation c is a deletion (see problem 7-26b below).

b.   The length of the T4 chromosome in micrometers predicts the number of nucleotide pairs because the physical size of the molecule is known from the Watson and Crick model. Recombination analysis with many mutants distributed over the T4 chromosome suggests the total map units in the T4 genome. Thus, **Benzer could estimate the number of map units/nucleotide pair. He then compared this to the smallest distance he could measure between 2 mutations by recombination**.

c.   *rII⁻* mutants in the same nucleotide pair cannot recombine with each other to produce *rII⁺* phage (again, see problem 7-26b below).

7-22.

a.   **The starting tube (call it tube A) contains 5 ml of bacteriophage with 1.5×10^{10} phages. Take a 1 μl (0.001 ml) sample of tube A, corresponding to 3×10^6 phages, and add it to 999 μl (in practice, 1 ml) of diluent in tube B. This step is a 2×10^{-4} dilution (1 μl / 5000 μl). Repeat this step with 1 μl of tube B (3×10^3 phages) and mix it with 999 μl of diluent in tube C (2×10^{-7} dilution). Next take 10 μl of tube C (30 phages) and mix it with bacteria (about 100x more cells than phage = a low multiplicity of infection (MOI)). Allow the phages to infect the cells, then add cells to a top agar and pour on an agar plate. Repeat the infection/top agar step with 100 μl of tube C (300 phages) and plate. There should be 30 plaques on the first plate** and about 300 plaques on the second plate. This describes only one of many possible protocols: other dilution steps, such as 10^{-2} dilutions or 10^{-1} dilutions, could also be employed.

b. **To figure out the total number of phage, you need to look at a particular dilution in the electron microscope and count all the phage particles. The ratio of plaques to total phage is the plating efficiency. In part a, it is fair to assume that only one phage initiated each plaque because of the very low MOI. Because there were many more bacterial cells than phage, the chances are very high that any individual bacterial cell could have been infected only by a single phage.**

7-23.

a. **There are two complementation groups and therefore two genes**.

b. **The complementation groups are (1, 4) and (2, 3, 5)**.

7-24.

a. Diagram the cross:

$Ly^+ ry^{41} Sb / Ly ry^{564} Sb^+$ ♀ x ry^{41} / ry^{41} ♂ → 8 $Ly ry^+ Sb$ and lots of ry progeny

Recombination within the *rosy* gene, between the two *ry* mutations in the heterozygous female, generates the eight offspring with wild type eyes. Such a recombination event will give one recombinant gamete with the wild type sequence for both rosy mutations, thus giving a ry^+ phenotype in the progeny of this cross. The reciprocal recombinant gamete will be a double mutant, $ry^{41} ry^{564}$, which will yield *ry* progeny indistinguishable from the parental type *ry* progeny. We assume the $ry^{41} ry^{564}$ progeny are found in equal numbers to the ry^+ recombinants, so the **recombination frequency = (8 + 8) / 100,000 = 0.00016 = 0.016%. The distance between ry^{41} and ry^{564} = .016 mu.**

b. The wing and bristle phenotypes of those eight recombinant offspring are a consequence of the order of the ry^{41} and ry^{564} with respect to the flanking markers, the *Ly* and *Sb* genes. Try both orientations of the two *ry* mutations with respect to the flanking markers to see which order produces *Ly, ry⁺ Sb* as the result of a crossover between the rosy mutations.

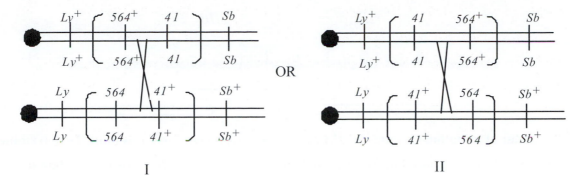

I   II

Orientation II produces the *Ly ry⁺ Sb* recombinants obtained so the order must be *Ly ry⁴¹ ry⁵⁶⁴ Sb*.

7-25.

a. Based on the complementation data (the first table), mutation 6 is almost certainly a deletion, because it doesn't complement with any of the other mutations. The second table gives the recombination results. A '+' in this table is the result of recombination that occurred in the *E. coli* B host to generate wild-type phage that can now grow in *E. coli* K (λ). The '-' designation indicates that a few phage were able to form plaques. These are revertants. **Remember that point mutations can revert to wild type, while deletion mutations cannot.** Mutants 3, 6, and 7, did not form any plaques here (as designated by '0'), so these three must be deletions. **Thus, mutants 3, 6, and 7 are deletion mutants (non-reverting).**

b. The first table, based on the coinfection of *rII* mutants into *E. coli* K(λ), gives the results of complementation analysis and lets us place mutations in the two *rII* complementation groups. Notice that deletion 6 does not complement any of the mutants so it must delete at least part of each of the two *rII* genes The two complementation groups are (1, 2, 5) and (4, 8, 9). The second complementation group is the *rIIB* gene as defined by the problem. Next use the recombination data to order the mutations with respect to the deletions. Deletion 6 does not recombine with mutation 1 (*rIIA*) nor with mutations 8 and 4 (*rIIB*), so these mutations must be near each other because deletion 6 spans both genes. Deletion 3 allows you to order 8 (outside deletion 3), then mutation 4 (overlaps both deletions 6 and 3), then mutation 9 (only overlaps deletion 3). Mutations 2 and 5 cannot be ordered relative to each other, except to say that they do not overlap deletion 6, and so they are surrounded by {}. **The map is:**

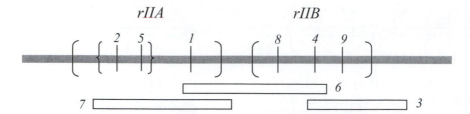

c.   **The order of mutations 2 and 5 in *rIIA* cannot be determined form this data. To determine the order, you would need to use other deletions that occur in *rIIA* in recombination testing, or cross mutants 2 and 5 and test for linkage to appropriate genetic markers in genes that lie to either side of the *rIIA/B* loci.**

7-26. Mutations 5 and 6 do not revert, so they are deletions. Deletions are not included in the complementation groups shown in part a below. All the rest of the mutations do revert, so they are point mutations.

a.   When analyzing complementation data, group the mutants that do NOT complement as these are mutations of the same gene. Mutation 1 does not complement mutation 8, but these 2 mutations do complement all the other point mutations. Therefore, mutations 1 and 8 make up one complementation group. Mutation 2 complements every other mutation; therefore it is the sole mutation in another complementation group. Mutation 3 does not complement 4 or 7, so these form a third complementation group. **In total there are three complementation groups: (1, 8); (2); and (3, 4, 7).**

b.   A '-' result in the recombination data in part b means that the two mutations in the diploid cannot recombine. Therefore the mutations must affect the same nucleotide, since recombination occurs between adjacent nucleotides (**Figure 6.24 step 1**). Diploids that were generated by mating the same mutation (e.g., *1* x *1*) have the mutation in the same position on both homologs. All point mutations recombine with all other point mutations, so no two point mutations affect the same nucleotide. Note that mutations *5* and *6*, which are known deletions (do not revert) do not recombine with various point mutants. In the complementation data, deletion *5* acts like any of the point mutations in the complementation group (*4, 3, 7*). However, it acts differently in the recombination data. It does not recombine with (overlaps) point mutations *4* and *3*, but it does recombine with *7*. Thus it is possible to genetically define mutation 5 as <u>a deletion = a mutation</u> <u>that does not recombine with 2 other mutations that DO recombine with each other</u>. If two point mutations recombine, they must affect different nucleotides, and because the deletion fails two recombine with the two mutations, it must remove more than one nucleotide. Deletion 6 does not recombine with mutations *1* nor *4*. Because *1* and *4* are in different genes (complementation groups), this deletion must span the distance between these 2 genes as well as 'uncover' the point mutations themselves. Deletion *6* <u>does</u> recombine with mutation *8*, which places *8* on the far side of its gene from the gene containing mutation *4* (**Figure 7.21a**). Combining this information with the complementation data from part a allows you to **draw the map below**:

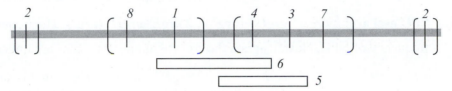

The location of gene 2 (to the right or left side) cannot be determined from this data.

c. The diploid cells from part a are allowed to undergo meiosis. If the two mutations in the heterozygous diploid are in the same gene (mutations *1* and *8* or mutations *4*, *3* and *7*), or if the two mutations are in different genes on the same chromosome (the *1* and *8* mutations must on the same chromosome as mutations *4*, *3* and *7* because deletion mutation *6* overlaps both of these genes), then **recombination can occur between two mutations producing prototrophic (lys$^+$) spores**. The example below shows the result for mutations *1* and *8* which are in the same complementation group. After the recombination event shown, the 4 spores will be 1/4 *1*$^+$ *8* (lys$^-$) : 1/4 *1*$^+$ *8*$^+$ (lys$^+$) : 1/4 *1* *8* (lys$^-$) : 1/4 *1* *8*$^+$ (lys$^-$) = **3/4 lys$^-$: 1/4 lys$^+$**. In the case of genetic linkage, the numbers of tetrads showing the 3/4 lys$^-$: 1/4 lys$^+$ ratio will depend on the distance between the mutations.

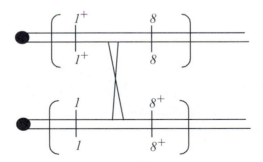

Mutations 1 and 2 are in different complementation groups. The original cross was mutant 1 x mutant 2. If these genes are on separate chromosomes the mutations will assort independently and **half of the meiosis will yield Parental Ditype asci (2 *1* *2*$^+$: 2 *1*$^+$ *2*) with no prototrophic spores, and half of the meiosis will yield Non-Parental Ditype asci (2 *1* *2* : 2 *1*$^+$ *2*$^+$) with 1/2 prototrophic spores.**

Section 7.3 - Mutations and Gene Function

7-27.

a. Diagram the cross, assuming the genes are unlinked (**Figure 5.15a-c**):

$argE^-$ $argH^+$ x $argE^+$ $argH^-$ → $argE^-/argE^+$; $argH^-/argH^+$

When this diploid is sporulated, the PD asci are: 2 $argE^-$ $argH^+$ (arg⁻) : 2 $argE^+$ $argH^-$ (arg),

and the NPD asci are: 2 $argE^+$ $argH^+$ (arg⁺) : 2 $argE^-$ $argH^-$ (arg⁻). The frequency of PD

spores = frequency of NPD. Next, diagram the cross assuming the genes are linked (**Figure 5.17a**

and **f**): $argE^-$ $argH^+$ x $argE^+$ $argH^-$ → $argE^-$ $argH^+$ / $argE^+$ $argH^-$. In this case, the PD ascus

is the same as above: **2 $argE^-$ $argH^+$ (arg⁻) : 2 $argE^+$ $argH^-$ (arg⁻). The NPD asci are the same**

as above also: 2 $argE^+$ $argH^+$ (arg⁺) : 2 $argE^-$ $argH^-$ (arg⁻), however there would be many

more PD asci than NPD asci. In all cases the distribution of spores will show MI segregation for

both genes (**Figure 5.21**). In other words, the center of the ascus corresponds to the ':' in the ratios

above.

b. **The $argE^-$ $argH^+$ spores of the PD will grow when you supplement the media with ornithine,**

citrulline, arginosuccinate or arginine. For the 2 $argE^+$ $argH^-$ spores, only arginine itself in

the media allows growth. In the case of the 2 $argE^-$ $argH^-$ spores of the NPD asci, only

arginine allows growth. The two 2 $argE^+$ $argH^+$ spores are prototrophs that grow on

minimal medium without supplementation.

7-28.

a. Diagram the cross:

orange x black → F_1 brown

The problem says that orange is caused by one autosomal mutation and black is caused by another.

This implies that they are in different genes. Therefore, if the parents are true breeding then the F1

is doubly heterozygous. Thus, the underlying ratio in the F2 will be: *BB oo* (orange) x *bb OO*

(black) F_1 *Bb Oo* (brown) → F_2 9 *B- O-* : 3 *B- oo* : 3 *bb O-* : 1 *oo bb*. If orange and black are

two intermediates in the pathway to brown (for instance, orange → black → brown), with the O^+

gene product carrying out the conversion from orange to black and the B^+ gene product catalyzing

the conversion from black to brown, **the F_2 would have 9 brown, 3 black, 4 orange.** In other

words, epistasis would be seen. (Note: We don't know the order of orange and black in this

pathway. **If the order were black → orange → brown, a 9 brown : 3 orange : 4 black ratio**

would be seen.)

b. If there are only two pathways, one producing orange and the other black, then there would be four different phenotypes in the F$_2$ generation: **9 brown (*O- B-*) : 3 black (*oo B-*) : 3 orange (*O- bb*) : 1 nonpigmented (*oo bb*)**.

7-29. Designate the genes and alleles. The *W* gene product converts a colorless (white) pigment to green. The *G* gene product converts green to blue flowers; the mutant allele is *g*. Either of two gene products *B* or *L* can convert blue to purple flowers; *b* and *l* are the mutant alleles. Diagram the cross. Note that both parents are *WW*. All progeny will be *WW*, and it will not affect the array of phenotypes in the progeny. For this reason, it is not considered in this cross:

   gg BB LL (green) x *GG bb ll* (blue) → F$_1$ *GgBbLl* → F$_2$:

   3/4 *G-* x 3/4 *B-* x 3/4 *L-* = 27/64 *G- B- L-* (purple);

   3/4 *G-* x 1/4 *bb* x 3/4 *L-* = 9/64 *G -bb L-* (purple);

   3/4 *G-* x 3/4 *B-* x 1/4 *ll* = 9/64 *G- B- ll* (purple);

   3/4 *G-* x 1/4 *bb* x 1/4 *ll* = 3/64 *G- bb ll* (blue);

   1/4 *gg* x 3/4 *B-* x 3/4 *L-* = 9/64 *gg B- L-* (green);

   1/4 *gg* x 1/4 *bb* x 3/4 *L-* = 3/64 *gg bb L-* (green);

   1/4 *gg* x 3/4 *B-* x 1/4 *ll* = 3/64 *gg B- ll* (green);

   1/4 *gg* x 1/4 *bb* × 1/4 *ll* = 1/64 *gg bb ll* (green)

The ratio is 45 purple : 16 green : 3 blue. You can see why the problem specified that the green parent was mutant in only a single gene, as *gg bb LL* or *gg BB ll* plants would still be green yet would yield a very different ratio of phenotypes in the F$_2$.

7-30. First, order the compounds from final product to first one in the pathway. The final compound is the one on which all of the mutants in the pathway will grow. The compound before that (E in this example) is the one that allows all the mutants except one class to grow. Continue working toward the beginning of the pathway in this manner. Next, order the mutants. Again, you can do this by working backwards from the final product through the intermediates, look for the mutant which grows only when supplied with G. In this problem it is mutant 2. The mutation must be in the gene encoding the enzyme catalyzing the last step synthesizing compound G. Then look for the mutant that grows only when supplied with G or one other intermediate. Mutant 7 can grow only when supplied with intermediate E or with G. This verifies our earlier assignment of E as the intermediate that precedes G, and it also tells us that the gene in which mutation 7 is located encodes the enzyme that allows the synthesis of E. In this way, continue working back through the pathway to get the answer.

$$X \xrightarrow{6} F \xrightarrow{1} D \xrightarrow{5} A \xrightarrow{3} C \xrightarrow{4} B \xrightarrow{7} E \xrightarrow{2} G$$

7-31.

a. Diagram the crosses:

1. blue x white → purple → 9 purple : 4 white : 3 blue

2. white x white → purple → 9 purple : 7 white

3. red x blue → purple → 9 purple : 3 red : 3 blue : 1 white

4. purple x purple → purple → 15 purple : 1 white

3EQ #1 - there are 2 genes controlling the phenotypes in each cross, because all four crosses show epistatic modifications of the 9:3:3:1 ratios; **3**EQ #2 - in all 4 crosses the purple phenotype corresponds to the "A- B-" class; **3**EQ #3 - none of the genes are X-linked. Assign genotypes in all the crosses:

Cross 1. *AA bb* (blue) x *aa BB* (white) → *Aa Bb* (purple) → 9 *A- B-* (purple) : 4 *aa --* (white) : 3 *A- bb* (blue)

Cross 2. *AA bb* (white) x *aa BB* (white) → *Aa Bb* (purple) → 9 *A- B-* (purple) : 7 *aa --* + *- - bb* (white)

Cross 3. *AA bb* (red) x *aa BB* (blue) → *Aa Bb* (purple) → 9 *A- B-* (purple) : 3 *A- bb* (red) : 3 *aa B-* (blue) : 1 *aa bb* (white)

Cross 4. *AA bb* (purple) x *aa BB* (purple) → *Aa Bb* (purple) → 15 *A- --* + *-- B-* (purple) : 1 *aa bb* (white)

b. 1. colorless $\xrightarrow{A}$ blue $\xrightarrow{B}$ purple

2. colorless1 $\xrightarrow{A}$ colorless2 $\xrightarrow{B}$ purple

3. colorless $\xrightarrow{A}$ red

   colorless $\xrightarrow{B}$ blue   (red pigment + blue pigment = purple)

4. colorless $\searrow^{A}$
                     purple
   colorless $\nearrow_{B}$

c. **Cross #2** is compatible with a single-step pathway in which genes *A* and *B* encode two different subunits of a multimeric enzyme that catalyzes the step. In such a case, enzyme activity would result only if at least one allele of each gene were the dominant allele specifying subunit production.

d. Assuming that "tightly linked" means the distance between the A and B genes is 0 mu, we can rewrite all four of the crosses in the form: $A\,b\,/\,A\,b$ x $a\,B\,/\,a\,B$ → $A\,b\,/\,a\,B$ (selfed) → $1/4\,A\,b\,/\,A\,b : 1/4\,A\,b\,/\,a\,B : 1/4\,a\,B\,/\,A\,b : 1/4\,a\,B\,/\,a\,B$. The phenotypes for each cross:

Cross 1. 2 purple ($A\,b\,/\,a\,B + a\,B\,/\,A\,b$) : 1 blue ($A\,b\,/\,A\,b$) : 1 white ($a\,B\,/\,a\,B$)

Cross 2. 1 purple ($A\,b\,/\,a\,B + a\,B\,/\,A\,b + A\,b\,/\,A\,b$) : 1 white ($a\,B\,/\,a\,B$)

Cross 3. 2 purple ($A\,b\,/\,a\,B + a\,B\,/\,A\,b$) : 1 red ($A\,b\,/\,A\,b$) : 1 blue ($a\,B\,/\,a\,B$)

Cross 4. all purple ($A\,b\,/\,a\,B + a\,B\,/\,A\,b + A\,b\,/\,A\,b + a\,B\,/\,a\,B$)

Note that if there is any recombination between the two genes, then recombinant $a\,b$ gametes will be produced, allowing the emergence of white F_2 plants ($a\,b\,/\,a\,b$) in crosses 3 and 4. The greater the distance between the two genes, the more the ratios will resemble those in part a.

7-32. To solve this problem consider first only those mutants that are only defective in biosynthesis of one amino acid. The mutants defective in only the proline pathway are those that grow when given proline in the media but not when given glutamine. Mutants 2, 6, and 1 are of this type. There is no intermediate that allows the growth of mutant 2 so the defect must be in the final enzyme that produces proline. Working backward from this point in the pathway, mutant 6 grows when supplied with intermediate A, so A is the final intermediate and mutant 6 is blocked in the step that leads to A. Mutant 1 grows when supplied with intermediates E or A, indicating that E is prior to A in the proline pathway. Now conduct the same analysis for glutamine. Mutants 7 and 4 are only defective in glutamine biosynthesis. Mutant 4 grows only on glutamine, while mutant 7 grows when supplied with B or glutamine, indicating that mutant 7 is blocked in the production of B, and mutant 4 cannot convert intermediate B to glutamine. Now look at the mutants that area defective in both glutamine and proline biosynthesis. Mutants 5 and 3 are of this type. Mutant 3 grows only if given intermediate C, so it must be blocked just prior to this step. Mutant 5 grows if given C or D, so it is blocked prior to the D intermediate. This represents the first part of the pathway that is used both in proline and glutamine biosynthesis. Putting all of this information together, we have the following branching pathway:

$$X \xrightarrow{5} D \xrightarrow{3} C \xrightarrow[1]{7} \begin{array}{c} B \xrightarrow{4} gln \\ E \xrightarrow{6} A \xrightarrow{2} pro \end{array}$$

7-33.

a.   This problem is solved in the same way as <u>problem 7-30</u>.

$$X \xrightarrow{18} D \xrightarrow{14} B \xrightarrow{9} A \xrightarrow{10} C \xrightarrow{21} \text{thymidine}$$

b.   **Double mutant 9 and 10 blocks at the B to A step first, and so it accumulates intermediate B. Similar reasoning predicts that double mutant 10 and 14 would accumulate intermediate D.**

7-34. The data represent complementation experiments done at the biochemical level in the test tube. The results suggest that **there are 2 different X-linked genes that cause hemophila when mutant**. Individuals 1 and 2 are mutant in one gene (call it gene A) and individuals 3 and 4 are mutant in the other gene (gene B). Individuals 1 and 2 thus lack the function of one factor needed for clotting, while 3 and 4 lack a different factor. When the two kinds of blood complement (as in the mixture of blood from individuals 1 and 3), clotting occurs because the blood of each patient supplies the factor lacking in the other patient.

The results exclude a pathway in which the product of one of these genes (say the *A* gene) is required for the synthesis of the protein encoded by other gene (*B*). In such a case, the blood of a patient mutant for gene *A* would have neither protein A nor protein B, so the mixture of mutant bloods would have no source of compound B. Another excluded scenario is one in which protein A is a substrate for a reaction catalyzed by enzyme B, and this reaction could not take place in the test tube (for example, if protein A were rapidly degraded if not immediately converted into something else by the action of enzyme B). Many other linear, convergent, or divergent pathways are still consistent with the results.

As an interesting historical sidelight, the cited article in the *British Medical Journal* was published in the December 27th (Christmas) issue of 1952. The first patient whose blood could complement that of most other hemophiliacs in the test tube (thus indicating the existence of two different kinds of X-linked hemophilia) was a 5 year old boy whose family name was Christmas. Because of these facts, the rarer form of the disease, usually called hemophilia B, is still often called Christmas disease.

7-35. Remember that the von Willebrand factor is necessary at fairly high levels to stabilize factor VIII.

a.   This should be a **successful treatment** because the normal plasma contains both vWF and factor VIII. The **effect should be immediate** because both factors are present **and prolonged** because vWF stabilizes factor VIII.

b.   This treatment should **not be successful** because vWD plasma has neither vWF nor factor VIII.

c. This should be a **successful treatment** because the hemophilia A plasma contains vWF even though it does not have factor VIII. The **effect should be delayed** because the patient has no factor VIII and the added vWF will only stabilize factor VIII newly synthesized by blood cells in the patient, and this takes time. The **effect should also be prolonged** because vWF stabilizes factor VIII.

d. This should be a **successful treatment** because the normal blood contains both vWF and factor VIII. The **effect should be immediate** because both factors are present **and prolonged** because factor VIII in the transfused plasma is already stabilized.

e. This **treatment will not be successful**. The vWD plasma has neither vWF nor factor VIII, and the patient cannot synthesize any factor VIII.

f. This **treatment should not be successful**. Neither the patient's blood nor the transfused plasma has any factor VIII.

g. This treatment **should be successful, but only after a delay** to allow the patient's blood cells to synthesize enough factor VIII that can be stabilized by the injected vWF. The **effects should be prolonged** because of the stabilization.

h. This treatment should be **unsuccessful** because no factor VIII can be made.

i. This treatment **should be successful immediately** because you are injecting factor VIII, but the **effects will be only very short-term** because the injected factor VIII will be degraded in the absence of vWF.

j. This treatment **should be successful immediately, and the effects will be prolonged** because the patient's blood has vWF which can stabilize the injected factor VIII.

7-36. This problem is similar to an ordered enzymatic pathway except the gene products are a series of nonenzymatic proteins that make up a structure that is assembled in a particular order. The loss of one protein due to mutation will prevent all the subsequent proteins from being added. The loss of the first protein at the surface would prevent all others from being at the cell surface. The mutant that fits this description is E. Mutants A and C have a similar pattern in which only E and C or A respectively are at the surface, so genes A and C encode the two proteins that form the dimer structure shown second from the embryo surface. The logic is continued to place the remaining three gene products in their order.

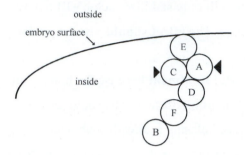

7-37.

a.   **Two loci are needed**, one for the α globin polypeptides and one for the β globin polypeptides.

b.   Assuming that both alleles of both genes are expressed at the same levels, then you would see 1/2 α1 : 1/2 α2 for the 2 forms of the hemoglobin α subunit and 1/2 β1 : 1/2 β2 for the forms of the β subunits. The α subunits will form the following sorts of dimers: 1/4 α1α1 : 1/2 α1α2 : 1/4 α2α2. The β subunits will assemble into dimers in the same way, giving a genotypic monohybrid ratio of: 1/4 β1β1 : 1/2 β1β2 : 1/4 β2β2. In order to figure out the types of hetero-tetramers and their frequencies, apply the product rule to the 2 monohybrid ratios: **1/16 α1α1 β1β1 : 1/8 α1α2 β1β1 : 1/16 α2α2 β1β1 : 1/8 α1α1 β1β2 : 1/4 α1α2 β1β2 : 1/8 α1α2 β2β2 : 1/16 α1α1 β2β2 : 1/8 α1α2 β2β2 : 1/16 α2α2 β2β2**.

7-38. See **Chapter 7, Fast Forward Figure A**.

a.   You will see: **i) the base plate** - this is the structure just to the left of the number 19 in the Figure; **ii) a completely formed head filled with DNA; and iii) completely formed tail fibers**. None of these should be attached to each other.

b.   You will see: **i) an immature head, not filled with DNA, ii) replicated phage genomes not inserted into heads; iii) completely formed tail in which the sheath is on top of the base plate** (the structure to the right of numbers 3 and 15); and **iv) completely formed tail fibers**. None of these should be attached to each other.

c.   Because these two mutations are in different genes, the two strains should complement each other and you should see **completely formed new progeny T4 bacteriophage** in the electron microscope.

d.   You will see: **i) nearly complete tail assemblies** (the figure just to the left of numbers 3 and 15); **ii) mature heads filled with phage DNA; iii) and iv) two separate halves of the tail fiber** (notice that the final tail fibers have a kink between the 2 halves). All four parts should be separate from each other.

Section 7.4 – Mutations change genotype and affect phenotype

7-39. In unequal crossing-over the homologous chromosomes align out of register, and instead use homology between related but different genes. **The result of crossing-over is a homolog with a duplication, β δ/β δ, and a homolog with a deletion, β/δ.** The genes with a slash indicate hybrid genes. Even though these genes are very similar, they do have differences. One of the more important differences between the δ and β genes is the time of expression. In these hybrid genes the regulatory region of the β gene may have been replaced with the regulatory information of the δ gene, or vice-versa, so the hybrid gene may be expressed inappropriately (that is, during the wrong time in development). In addition, the polypeptides formed from the hybrid genes may affect the affinity of hemoglobin for oxygen in unpredictable ways.

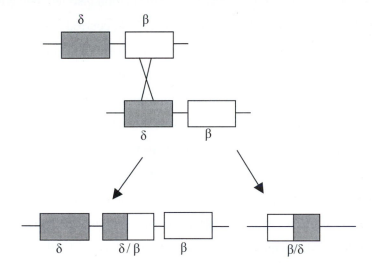

7-40.

a.   Most mammals, including New World primates, are dichromats, while Old World primates are trichromats. Primates diverged from mammals 65 million years ago (65 Myr), while Old World primates diverged from each other about 35 Myr ago.

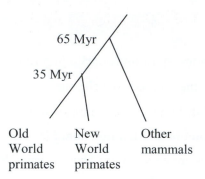

Compare this with **Figure 7.28d**. You can see that the final gene duplication event – the one eventually giving rise to different red and green photoreceptor genes – most likely occurred in the lineage leading to the trichromatic Old World primates (like humans) but not in the lineage leading to the dichromatic New World primates (like marmosets). This dates the final gene duplication event to some time subsequent to 35 Myr. The information also lets you conclude that the two previous gene duplication events shown in Figure 7.28 occurred at some time prior to 65 Myr because all mammals have rhodopsin plus at least two color photoreceptors.

b. Assuming there is only one allele for the autosomal receptor gene, female monkeys will have 2 alleles of the X-linked receptor while males will only have one. **Males will be dichromats**, as they are hemizygous for the X-linked gene. However, **there will be 3 classes of males that see colors in different ways**. **Females can be dichromats** if they are homozygous for an allele of the X-linked gene, again there will be 3 classes. **Females can also be trichromats** if they are heterozygous for this gene, and there are 3 different combinations: 1-2, 1-3 and 2-3.

c. The fact that 95% of our receptors work best in low light conditions suggests that **early mammals were active in low light situations**, for example in forests at night!

Chapter 8   Gene Expression: The Flow of Genetic Information from DNA to RNA to Protein

Synopsis:

This chapter describes how the information in DNA is converted into usable machinery (proteins) in the cell via the processes of transcription and translation. This flow of information is part of the central dogma of genetics. You need to become very comfortable with using the terms transcription and translation <u>accurately</u>. In <u>transcription</u>, DNA information is converted into RNA information. In <u>translation</u>, RNA information is converted into protein information. Work on developing some mental pictures for yourself so you can see the process occurring when you speak the words. The three letter DNA code and the correspondence between DNA sequence and protein sequence is described in this chapter. Tie together your knowledge of transcription/translation and the genetic code. This chapter contains many new vocabulary terms. The best way to know you have a good grasp on the terms is use the terms while pretending you are describing transcription, translation and the genetic code to another person.

Begin to introduce more inquiry into your learning process. For example, think about the components involved in transcription, RNA processing, and translation. How could they be affected by alterations (mutations) in any one of the components? Start thinking about how we know what we know and what evidence supports a particular view of how a process occurs.

Significant Elements:

After reading the chapter and thinking about the concepts, you should be able to:

◆   Identify open reading frames in a DNA sequence.

◆   Assign 5' and 3' end designations to DNA or RNA sequences.

◆   Describe the parameters for transcription - describe the enzymes (RNA polymerase) and proteins that play a role and what information in the DNA (promoter and terminator sequence) is important for their ability to function (**Feature Figure 8.11**).

◆   Identify the mRNA-like strand in the double-stranded DNA sequence either by knowing the sequence of the mRNA transcript or by looking for open reading frames.

◆   Describe the parameters for translation - describe the RNAs (rRNAs, tRNAs), proteins (aminoacyl synthetases) and RNA-protein complexes (ribosomes with the A, P and E sites, **Figure 8.23**) that are important and the information in the mRNA (ribosome binding site, start codon, stop codon) that is important for their ability to function (**Feature Figure 8.25**).

◆   Understand that the nearly universal genetic code consists of 64 codons. Of these, 61 specify amino acids, while the other 3 (5' UAA, 5' UAG and 5' UGA) are nonsense or stop codons.

♦ Use a codon table (**Figure 8.3**) to do a virtual translation of DNA into protein sequence and reverse translate protein into DNA sequence.

♦ Explain the steps in processing a eukaroytic primary transcript into an mRNA: addition of 5' methyl CAP (**Figure 8.13**), addition of 3' poly-A tail (**Figure 8.14**) and splicing out of introns (**Figure 8.15**).

♦ Answer questions that require you to know the roles of the nucleic acids and proteins in transcription, translation, and RNA processing.

♦ Understand the ways that different types of mutations affect gene expression (see **Figure 8.28** for an overview): silent mutations, missense mutations, nonsense mutations (and nonsense suppressors, see **Figure 8.32**), frameshift mutations (**Figures 8.5** and **8.6**) and mutations outside of coding sequences that alter signals required for transcription, processing of the mRNA or translation. Many types of mutations are classified by their effects on protein function (**Table 8.2**).

♦ Understand and can explain the differences in gene expression between eukaryotes and prokaryotes (**Table 8.1**).

Problem Solving Tips:

♦ The convention is to write DNA sequences with the top strand running 5' to 3' left to right.

♦ Transcription occurs in a 5' to 3' direction along the template DNA strand. The complementary strand of DNA is the mRNA-like strand. The mRNA-like DNA strand has the same polarity and sequence (with Ts instead of Us) as the mRNA.

♦ Ribonucleotides are added to a 3' end of the growing mRNA strand.

♦ Information in the 5' portion of a coding region of the mRNA will be information in the NH_2 terminal portion of the protein.

♦ There is mRNA sequence at the 5' and 3' ends of the transcript which is NOT translated into protein. These sequences are called the 5' and 3' untranslated regions (UTRs).

♦ By definition an open reading frame (ORF) is an RNA sequence with no stop (nonsense) codons in the adjacent groups of 3 nucleotides.

   If you are given an mRNA sequence, there are 3 possible reading frames to examine to see if they are open or closed (have an in-frame stop codon). Remember that the ribosome reads the mRNA from 5' to 3', so you check for ORFs by starting at the 5' end of the mRNA sequence. The first reading frame begins with the first nucleotide, the second frame begins with the second nucleotide and the third frame begins with the third nucleotide. When you begin with the fourth nucleotide you are back in the first reading frame.

If you are given a DNA sequence to examine, there are 6 possible reading frames! Remember that the template strand of DNA is transcribed 3' to 5' to synthesize a complementary mRNA that is 5' to 3'. Use the complementary mRNA-like strand of DNA as a stand-in for the mRNA. Read the mRNA-like strand of DNA from its 5' end, checking the 3 possible reading frames. Then assume the other strand of DNA is the template, and repeat the process.

♦ Codons are found in mRNAs and anticodons are found in tRNAs. To avoid confusion, be consistent in how you write them. In the Study Guide the codons are written 5' to 3' and the anticodons are written 3' to 5'.

Solutions to Problems:

Vocabulary

8-1. a. **5**; b. **10**; c. **8**; d. **12**; e. **6**; f. **2**; g. **9**; h. **14**; i. **3**; j. **13**; k. **1**; l. **7**; m. **15**; n. **11**; o. **4**; p. **16**.

Section 8.1 - The Genetic Code

8-2. a. **4**; b. **6**; c. **1**; d. **2**; e. **3**; f. **5**.

8-3.

a. **GU GU GU GU GU or UG UG UG UG UG.**

b. **GU UG GU UG GU UG GU UG GU.**

c. **If you start with the first base: GUG UGU GUG. If you start with the second base: UGU GUG UGU.**

d. **GUG UGU GUG UGU GUG UGU GUG UGU.** This is the result of reading the first 3 nucleotides then going back to the second nucleotide and reading a codon, etc. **There are other possibilities**, such as reading codons starting on 1, 2, 4, 5, etc. Overlapping codes will always give more coding information for the same number of bases compared to the non-overlapping code.

e. **GUGU GUGU or UGUG UGUG.**

8-4.

a. **Comparing the mutant to the wild-type sequence you can see where insertions, corresponding to + mutations, and deletions, corresponding to - mutations, occurred,** see part b.

b. The amino acids in the wild-type and mutant protein are shown:

<pre>
                       Lys Ser Pro Ser Leu Asn Ala
     wild-type: 5' AAA AGT CCA TCA CTT AAT GCC 3'
                          (-)   (+)
     mutant: 5' AAA GTC CAT CAC TTA ATG GCC 3'
                       Lys Val His His Leu Met Ala
</pre>

The five amino acids in between the - and the + mutations are different from wild-type.

c. **The substitutions of amino acids between the - and + mutations in the mutant must not alter the structure of the protein significantly enough to alter protein function.**

8-5. Glutamic acid can be encoded by either GAA or GAG. In sickle cell anemia this amino acid is changed to valine by a single base change. Valine is encoded by GUN with N representing any of the four bases. Therefore, the second base of the triplet was altered from A to U in the Hb^S allele. In Hb^C the glutamic acid codon (GAA or GAG) is changed to a lysine (AAA or AAG). The change here is in the first base of the codon. **The mutation causing the Hb^C allele therefore precedes the Hb^S mutation in the sequence of β-globin gene** when reading in the 5' - 3' direction that the RNA polymerase travels along the gene.

8-6.

a. **If the Asn6 (5' AAC) is changed to a Tyr residue, the nucleotide change is to a UAC. In protein B this means that the Gln (5' CAA) at position 3 becomes a Leu (5' CUA).**

b. **Leu (5' CUA) at position 8 is changed to Pro (5' CCA). In protein B the Thr (5' ACU) at position 5 is still a Thr residue even though the codon is different (5' ACC).**

c. **When Gln (5' CAA) at position 8 in protein B is changed to a Leu (5' CUA), the Lys codon (5' AAG) at position 11 in protein A is changed to a stop codon (5' UAG).** This would cause the production of a truncated form of protein A only 10 amino acids long.

d. This is a thought question that involves some speculation; the following are two reasonable possibilities. (1) As seen in parts a-c above, **a mutation in the region of overlap has a high probability of causing alterations in both of the proteins simultaneously. Any change in this DNA sequence has the potential to affect two proteins instead of just one, and an organism would be less likely to tolerate mutations affecting the production of two proteins** than mutations affecting the production of a single protein. **There would thus be strong evolutionary selection against overlapping reading frames. (2) If a region of DNA evolved so as to encode a protein with a stable three-dimensional conformation, it is very unlikely that a stable protein could be produced by the sequence shifted by one nucleotide.** This is because the alternate reading frame is likely to have a stop codon every 3 codons out of 64. This means that in reality, among the very few examples of overlapping reading frames that exist, either one or both of the proteins is very small, composed of only a few amino acids.

8-7. Note that the nucleotide sequences below are written as mRNA sequences. From this you can easily convert to the DNA sequence of the gene. Remember that proflavin causes frameshift mutations (single base insertions and deletions) <u>in the DNA</u>. **All mutagens work at the level of the DNA, even though the changes are often written at the level of the mRNA!** Notice that there are several ambiguous bases in the wild-type sequence (any one of four bases possible is indicated by N; other amino acids are encoded by two different codons). Line up the invariant bases in the mutant with the wild-type and it is clear that a single base insertion occurred in the mutant. Knowing the amino acid sequence of the mutant and therefore the nucleotide sequence, all but one of the third base ambiguities in the wild type can be resolved. The mRNA sequence of the wild-type gene would be:

	Gly	Ala	Pro	Arg	Lys
wild-type mRNA:	5' GGN	GCN	CCN	AGA/G CGN	AAA/G 3'
mutant mRNA:	5' GGN	CAU/C	CAA/G	GGN	AAA/G 3'
	Gly	His	Gln	Gly	Lys

After comparing the wild type and mutant sequences you can see that the first nucleotide of the second codon was deleted to make the mutant sequence. The deduced DNA sequence of wild-type is:

5' GGN GCA CCA AGG AAA 3'

8-8. Nierenberg and Leder used an *in vitro* translation system to determine that 5' CUC is the leucine codon and 5' UCU codes for serine. **The basis of the assay is that the combination of a synthetic triplet RNA codon, matching charged tRNA, and ribosome bound together would be too large to pass through a filter.** They set up 20 reactions, each containing 5' CUC, one radioactive amino acid attached to its tRNA, and the other 19 non-radioactive amino acids attached to their tRNAs. **In the mixture containing the radioactive amino acid leucine that corresponds to the codon 5' CUC, the radioactivity would be trapped on the filter.** The same experiment was done for the 5' UCU triplet; in this case, serine was the radioactive amino acid that was trapped on the filter.

8-9. A nonsense mutation is a single nucleotide change that turns a sense codon (one coding for an amino acid) into one of the three stop codons. The change can occur at any of the three positions in the codon. The easiest way to identify such sense codons is to begin with each nonsense codon and systematically change each position to all other possible nucleotides. This gives nine different possible codons. Use the coding table (page 257) to translate the resulting codons. Not all of these changes lead to sense codons – some of them will result in nonsense (STP) codons.

Stop Codon / Change	UAA		UAG		UGA	
1st position	AAA	**Lys**	AAG	**Lys**	AGA	**Arg**
	CAA	**Gln**	CAG	**Gln**	CGA	**Arg**
	GAA	**Glu**	GAG	Glu	GGA	**Gly**
2nd position	UUA	**Leu**	UUG	**Leu**	UUA	**Leu**
	UCA	**Ser**	UCG	**Ser**	UCA	**Ser**
	UGA	STP	UGG	**Trp**	UAA	STP
3rd position	UAU	**Tyr**	UAA	STP	UGU	**Cys**
	UAC	**Tyr**	UAC	**Tyr**	UGC	**Cys**
	UAG	STP	UAU	**Tyr**	UGG	**Trp**

8-10. The original sequence in each case is: Met Asn Asn Ala Pro Glu Glu Ala Asp

The mutant sequences are:

(a) Met Asn Asn Arg Ala Gly Gly Ala Asp;

(b) Met Asn Lys Arg Gly Glu Ala Asp for the three single base deletions and

Met Asn Asn Gly Ala Arg Gln Glu Ala Asp for the three single base insertions;

(c) Met Asn Lys Arg Arg Arg Lys Arg for the frameshift due to a single base deletion and

Met Asn Asn Gly Ala Gly Ser Gly for the frameshift due to a single base insertion.

8-11.

a.   The protein would terminate after the His codon due to a nonsense mutation. **The Trp codon (UGG) could have been changed to a either a UGA or a UAG codon**. These stop codons result from changing the second or third base of the Trp codon to an A.

b.   Reverse translate the amino acid sequence:

```
             N Ala  Pro  His     Trp  Arg     Lys  Gly  Val  Thr C
mRNA     5' GCN  CCN  CAU/C  UGG  CGN     AAA  GGN  GUN  ACN
                                          AGA/G
```

If you restrict the possibilities to mutations that substitute one base pair for another, there are four possible ways to generate a nonsense mutation from this sequence. (i) A UAG stop codon will result from a change of the second base of the Trp codon to A. (ii) If the third base of the Trp codon UGG changes to A, a UGA stop codon will result. (iii) If the Lys codon was AAA, and there is an A to T substitution at the first position, a UAA stop codon would be produced. (iv) If the Gly codon is GGA, a mutation of G to T at the first position will generate a UGA nonsense codon.

There are many other changes that could cause the premature termination of the protein encoded by this sequence, for instance, the insertion or deletion of a single base pair causing a frameshift mutation. As just one example, if the Arg codon is CGU, a single base insertion in the DNA before or within this codon would lead to a UAA codon in the mRNA, with the U would come from the former Arg codon and AA from the Lys codon.

8-12. The extra amino acids could come from an intron that is not spliced out due to a mutation in a splice site. The genomic DNA sequence in normal cells should contain this sequence. An alternative is that the extra amino acids could come from the insertion of DNA from some other part of the genome, such as the insertion of a small transposable element. In this case, the normal allele of the gene will not have this sequence. Note that the defect in this case could not be caused by a mutation that changes the stop codon at the end of the open reading frame to an amino-acid-specifying codon. If this were the case, the extra amino acids would be found at the C-terminus of the longer protein, not in the middle.

8-13. Both strands of a double-helical DNA molecule are possible template strands (see for example problem 8-19, page 159). There are three possible reading frames on both the top and bottom strands. If a stop codon is found in a frame, then it is not an open reading frame (that is, the entire sequence cannot not code for a protein). Assume first that the bottom strand is the template strand, so that the top strand is the mRNA-like strand. Treat it as the equivalent to an mRNA, which means it would be translated in the 5' to the 3' direction and that you would substitute T for U to allow translation from the genetic code. Thus, the first potential reading frame begins with the first nucleotide, the second reading frame with the second nucleotide, etc. In the following sequence, an 'x' above the nucleotide means the reading frame is closed (that is, it contains a stop codon), while an 'o' means it is open. The stop codons in the top and bottom strands are shown in bold. Scan the sequence looking for stop codons (or their direct DNA equivalents: TAA, TAG and TGA). Repeat this process assuming the top strand of the DNA is the template strand and the bottom strand is the mRNA-like strand, with 3 possible reading frames beginning at the 5' end of the sequence. **When you analyze this DNA**

sequence you find that the top strand has two open reading frames in the top strand (reading from 5' to 3' marked with an o above the first nucleotide of the open frame) and one closed reading frame (a stop codon in the frame marked with an x above the first nucleotide; these stop codons are shown in bold type). The bottom strand has one open reading frame while the other two frames are closed (reading 5' to 3').

```
     xoo
5' CTTACAGTTTATTGATACGGAGAAGG 3'
3' GAATGTCAAATAACTATGCCTCTTCC 5'
                         oxx
```

8-14.

a.   **The physical map is based on the number of base pairs**. In <u>Figure 8.4a</u> the physical map is represented by the red bar. On this map of the *trpA* gene in *E. coli* the distance between amino acids 1 and 10 will be the same as the distance between amino acids 51 and 60. **The genetic map, shown in purple in this figure, is based on numbers of crossovers. If different regions in the gene (or genome) have different rates of crossing over, the genetic map will vary proportionally**. Notice that amino acids 15 and 22 (21 nucleotides apart) are about the same genetic distance apart as amino acids 22 and 49 (81 nucleotides apart).

b.   Regions with a high rate of crossing over will have a proportionally larger genetic map relative to the physical map. Regions with a low crossover rate will appear smaller relative to the physical map. A comparison of the maps shown in <u>Figure 8.4a</u>, show that **the N terminal 1/3 and the C terminal 1/2 of the gene have relatively high recombination rates, while the central portion of the gene has a lower rate of recombination**.

8-15.

a.   The mutant nucleotide is marked in bold. The corresponding change to the amino acid sequence is shown below the RNA sequence.

wild type	5' AUG ACA CAU CGA GGG GUG GUA AAC CCU AAG
	Met Thr His Arg Gly Val Val Asn Pro Lys
mutant 1	5' AUG ACA CAU **CCA** GGG GUG GUA AAC CCU AAG
transversion	Pro
mutant 2	5' AUG ACA CAU CGA GGG **U**GG UAA ACC CUA AG
deletion	Trp STOP
mutant 3	5' AUG AC**G** CAU CGA GGG GUG GUA AAC CCU AAG
transition	Thr (no change)
mutant 4	5' AUG ACA CAU CGA GGG GU**U** GGU AAA CCC UAA G
insertion	Val Gly Lys Pro STOP
mutant 5	5' AUG ACA CAU **U**GA GGG GUG GUA AAC CCU AAG
transition	STP
mutant 6	5' AUG ACA **UUU ACC** ACC CCU CGA UGC CCU AAG
inversion	Phe Thr Thr Pro Arg Cys Pro Lys

b.   **The frameshift mutations 2 and 4 (single base insertions and deletions) can be reverted with proflavin**. Mutations 1, 3 and 5 are single nucleotide substitutions, either transitions (changes from one purine to the other or from one pyrimidine to the other) or transversions (changes a C-G base pair to a G-C base pair). **EMS causes transitions (Figure 7.10b), so it can revert mutations 3 and 5 (transition mutations) back to the original DNA sequences**. Mutation 1 is a transversion, and EMS can not change the G-C back to C-G. **Often "reversion" is used in a more general sense, meaning simply a restoration of wild type protein function.** In this case, it is possible that another, non-wild type amino acid at the mutant position will restore function. An EMS-induced transition in the mutant codon could give a functional protein. With this second meaning of "reversion", **all three point mutations (1, 3, and 5) could be reverted by EMS.**

8-16. Proflavin creates frameshift mutations. Thus the mutant protein is expected to have the normal amino acid sequence up until the point of the frameshift mutation (+1 insertion or -1 deletion). At the site of the frameshift mutation the amino acid sequence of the mutant protein will change because the one of the other two reading frames in the mRNA is begin translated. A second equal (but opposite) frameshift in the gene would be expected to restore the correct frame and suppress the original mutation. This is how the intragenic suppressors located between the N-terminus and the original mutation restore a functional polypeptide.

Notice that the original mutant protein is shorter (110 amino acids) than the normal protein (157 amino acids). Thus **the initial frameshift mutation changes the reading frame and there is a stop codon in this alternate reading frame**. This terminates translation early and generates a shorter polypeptide. The second frameshift mutation of the opposite sign cannot restore function if it occurs after the premature stop codon. Since no suppressors can be identified after the original mutation (in other words, between the original mutation and the premature stop codon), **we can predict that the stop codon is likely to be very close to the site of the original frameshift mutation**.

For example: 5'...GUG GCA AUA GAC... Unmutated sequence

5'...GUG GCA **UAG** AC....   original frameshift mutation (-1 A)

5'...GUG GCA UAG AC....

In this example only an insertion in the underlined region, within three base pairs of the original deletion, would be able to restore the original, normal open reading frame.

Section 8.2 - Transcription

8-17. In transcription, complementary base pairing is required to add the appropriate ribonucleotide to a growing RNA chain.

8-18. DNA replication is a permanent event – the new daughter cells will each get only one copy of the DNA molecule. That one copy must serve as the template for all transcription and the template for further cell divisions. This is not true for transcription, and consequently there are several reasons why the higher level of error during transcription is tolerated:

(i) **Transcription produces a transient product**. Even if a particular mRNA molecule contains critical errors that produce nonfunctional proteins, the cell has many other mRNA molecules transcribed from that same gene and the majority of them do not contain errors in critical locations.

(ii) Further, **the degeneracy of the genetic code limits the effects of errors in transcription** because many single errors will produce silent changes, and others will lead to conservative substitutions.

(iii) In most cases, portions of the primary RNA transcript produced by transcription will be spliced out of the molecule. For instance, only 0.6% of the dystrophin initial transcript will be present in the mature mRNA; the range for this value in different genes varies from 100% to 0.6%. **Mistakes that are made in transcription of the untranslated regions (5' and 3' UTRs and introns) will have no effect on the amino acid sequence of the popypeptide.**

(iv) Lastly, **the body produces much more protein than is actually required for most genes**. Many disorders caused by lack of protein function (e.g. cystic fibrosis, muscular dystrophy, tyrosinase negative albinism, hemophilia and other clotting disorders) are recessive. This means an individual with one non-functional, mutant allele and normal, functional allele of these genes will only produce about one half the normal levels of functional protein. Yet these individuals have a completely unaffected phenotype. Therefore a small percentage of abnormal mRNA molecules will not give rise to enough abnormal protein to affect the phenotype of the individual.

8-19. DNA sequences are generally written with the 5' to 3' strand on top and the 3' to 5' strand on bottom. If the protein coding sequence for **gene F** is read from left (N terminus) to right (C terminus), then top strand is the RNA-like strand which is read from 5' to 3', and the **template strand must be the bottom strand of DNA.** The template for gene G is the opposite since the coding sequence is read in the opposite direction (right to left). **The template strand for gene G is the top strand.** Note that this means the enzyme RNA polymerase moves from left-to-right along the DNA in transcribing gene F, and from right-to-left in transcribing gene G.

8-20. The heavier lines in the figure below represent mRNAs from 2 different genes, I and II; two genes are portrayed to show that the results would be slightly different depending upon whether or not the gene has introns. In general the DNA strands form a double stranded DNA structure, except where their pairing is interrupted by the presence of the mRNA/DNA heteroduplex. Gene II pairs with the template strand of the DNA, forcing the other strand to loop out. Gene I has an intron which has been processed out of the mature mRNA. When the mRNA pairs with the template strand of DNA there is a loop-out of the DNA in the region corresponding to the intron.

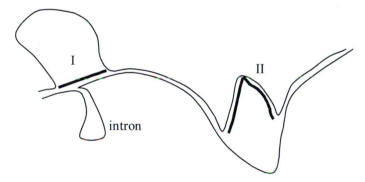

Section 8.3 - Translation

8-21. In translation, complementary base pairing between the codon in the mRNA and the anticodon in the tRNA is responsible for aligning the tRNA that carries the appropriate amino acid to be added to the polypeptide chain.

8-22. See **Figure 8.19** for tRNA structure, **8.21** for codon/anticodon pairing, **8.23** for ribosome structure and **8.25** for translation. Assume that the complete gene sequence is shown. For part e only one codon is labeled. Part q is only present if this figure represents an mRNA from a eukaryotic cell. **The following items are not found in this figure: a, d, g, i, j, n and t.**

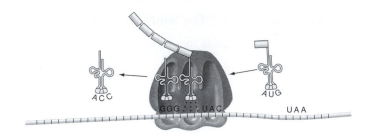

For part e. only one codon is labeled, but any of the 9 codons in this gene could have been labeled. Part q. is only present if this is figure represents an mRNA from a eukaryotic cell. **The following items are not found in this figure: a, d, g, i, j, n and t.**

8-23. Refer to the figure in the answer to problem 8-22 above.

a.   This figure represents **translation**.

b.   **The next anticodon is 5' GUA which is complementary to the codon 5' UAC which codes for tyrosine. The tyrosine is added to the carboxy-terminal (C-terminal) end of the growing polypeptide chain. The protein will be 9 amino acids long when completed.**

c.   There are 2 other building blocks with known identities. **The amino acid just placed on the carboxy-terminus of the growing polypeptide chain is tryptophan** (anticodon of departing tRNA is 5' CCA which is complementary to the codon 5' UGG which codes for tryptophan). This would be the fourth amino acid from the protein's N terminus. There is also **an assumed building block with a known identity – the initiation codon (5' AUG) that starts off protein synthesis, placing the amino acid methionine at the amino-terminal (N) end of the protein**.

d.   **The first amino acid at the N terminus would be f-Met in a prokaryotic cell and Met in a eukaryotic cell. The mRNA would have a cap at its 5' end and a poly(A) tail at its 3' end in a eukaryotic cell but not in a prokaryotic cell. If the mRNA were sufficiently long, it might encode several proteins in a prokaryote but not in a eukaryote.**

Section 8.4 – Gene Expression in Eukaryotes and Prokaryotes

8-24. Eukaryotic genes contain intervening sequences (introns) that do not code for proteins. Because they are spliced out of the primary transcript and thus are not included in the mature mRNA, introns do not have to contain open reading frames. In fact, introns almost always contain stop codons that would halt all possible reading frames. Thus, the reading frames of almost all eukaryotic genes are interrupted by introns that contain stop codons in that frame.

8-25.

a. **The minimum length of the coding region is 477 amino acids x 3 bases/codon = 1431 base pairs** (not counting the stop codon). The gene could be longer if it contained introns.

b. Look at both strands of this sequence for the open reading frame (remember that the given sequence is part of a protein-coding exon, so there must be an open reading frame). It occurs on the bottom strand, starting with the second base from the right (x = closed reading frames, o = open reading frames). The direction of the protein is N-terminal to C-terminal going from right to left in the coding sequence on the RNA-like strand (that is, the bottom strand of the DNA).

xxx

template strand   5' GTAAGTTAACTTTCGACTAGTCCAGGGT 3'

mRNA-like strand 3' CATTCAATTGAAAGGTGATCAGGTCCCA 5'

xox

mRNA   5' ACCCUGGACUAGUCGAAAGUUAACUUAC 3'

c. The amino acid sequence of this part of the mitotic spindle protein is: **N...Pro Trp Thr Ser Gly Lys Leu Thr Tyr...C.** Notice that the open reading frame begins with the second nucleotide. You cannot determine the amino acid corresponding to the A nucleotide at the 5' end of the mRNA because you don't know the first two nucleotides of the codon.

8-26. First look at the sequence to determine where the Met Tyr Arg Gly Ala amino acids are encoded. The top strand clearly does not encode these amino acids. On the bottom strand, there are Met Tyr codons on the far right (reading 5' to 3') and Arg Gly Ala much farther down the same strand. Why aren't these codons adjacent? There could be an intron in the DNA sequence which is spliced out of the primary transcript. Thus, the mature mRNA would encode this short protein.

a. **The bottom strand is the RNA-like strand, so the top strand is the template. The RNA polymerase moves 3' to 5' along the template.**

b. The sequence of the processed nucleotides in the mature mRNA is shown below. The junction between exons is marked by a vertical line:

5' CCC AUG UAC AG|G GGG GCA UAG GGG 3'

The sequence of this mRNA contains nucleotides prior to the AUG initiation codon and subsequent to the UAG stop codon to emphasize that the mRNA does not begin and end with these codons: remember that mRNAs contain both 5' and 3'- untranslated regions (UTRs).

In fact, this problem has a simplified DNA sequence to allow the sequence to fit on the page and to facilitate your analysis of the sequence in a reasonable amount of time. If you look very carefully at the sequence, you can see that there are canonical splice donor and splice acceptor sequences at the borders between the intron and the two exons that flank it. However, the intron does not contain a canonical branch site (see **Figure 8.16b** to review the nature of the three sequences needed for splicing, and then try to verify these statements yourself). In reality, the shortest introns are about 50 bp long, instead of the 24 bp in the problem, and must contain branch sites.

c. **A Thr residue at this position could occur if the G base on the bottom strand that just precedes the junction between the intron and the first exon (underlined in the answer to part b above) was mutated to a C. This base change would also alter the splice donor site, so splicing does not occur. The next codon after the ACG for Thr is a UAA stop codon, so the polypeptide encoded by the unspliced RNA is only three amino acids long.**

8-27. Mitochondria do not use the same genetic code! In yeast mitochondria, the codon 5' CUA 3' codes for Thr, not Leu as it does in yeast or human nuclear genes. If you want to ensure that the correct protein will be made by the yeast cell, **you should mutate all the 5' CUA 3' codons in the mitochondrial gene to 5' ACN 3' before putting the gene into a chromosome in the yeast nucleus.** This ensures that the cellular translation machinery will put a Thr at all positions it is required in the protein.

8-28.

a. The differences in transcription and translation prokaryotes (bacteria) and eukaryotes (humans) mean that the prokaryotic transcription and translation systems can not produce a functional insulin protein from the human gene. For instance, the **promoters for the human insulin gene may not work in *E. coli*,** thus blocking transcription of the gene. One of the main problems is that the **human insulin gene has introns** which are transcribed into the primary transcript and then spliced out by the spliceosome. **Bacterial cells do not have introns and thus have not evolved the machinery necessary to remove them from the RNA.** Other eukaryotic post-transcriptional modifications such as **addition of the 5'-cap are found in eukaryotes but are absent in prokaryotes.** Also, the **human insulin mRNA may not have sequences like the Shine-Delgarno box** needed for efficient initiation of translation at the AUG. Finally, it is also possible **correct folding of the polypeptide and other post translational modifications may not occur in the same ways in prokaryotic cells**.

b.   To make these bacterial insulin factories, you would have to transform *E. coli* cells with a composite gene in which some parts would be from the human insulin gene and other parts from a highly expressed *E. coli* gene. **The only parts of the human insulin gene that would be needed are the protein-coding sequences in the exons. These must be properly spliced together**. The easiest way to get these sequences is to make a DNA copy of an insulin mRNA molecule (known as a cDNA), a method that will be explained in <u>Chapter 9, Figure 9.8</u>. **All of the DNA sequences controlling gene expression (promoter, Shine-Dalgarno sequence, transcription terminator) should come from an *E. coli* gene that is transcribed at very high levels**.

Section 8.5 – Computerized Analysis of Gene Expression in C. elegans

8-29. The order of these elements in the gene is: **c; e; i; f; a; k; h; d; b; j; g**.

8-30.

a.   **All of these elements are in the RNA (that is, the terms are abbreviated) except the promoter (c) and the transcription terminator (g)**. These 2 are the only structures in this list that the RNA polymerase enzyme recognizes on the DNA. RNA processing and translation occur post-transcriptionally.

b.   **a, e, f and i are found partly or completely in the first exon; a, h and k are found partly or completely in the intron** (note that the splice donor site shown on the splicing figure includes nucleotides in both the upstream exon and the intron); and **b, d, j and g are found partly or completely in the second exon**.

Section 8.6 – Mutations and Gene Expression

8-31. Nonsense or frameshift mutations that affect codons for amino acids near the C-terminus are usually less severe than nonsense or frameshift mutations in codons for amino acids near the N-terminus because less of the protein that will be affected.

a.   **very severe effect** as there will be no functional protein

b.   **probably mild effect** if none of the last few amino acids are important for enzyme function

c.   **very severe effect**, see part a

d.   **probably mild effect**, see part b

e.   **no effect**, by definition a silent mutation maintains the same amino acid

f.   **mild to no effect** as the replacement is with an amino acid with similar chemical properties

g. **severe effect** as the mutation is likely to destroy the enzyme activity

h. **could be severe** if it affects the protein structure enough to hinder the action of the protein, **or could be mild** if the enzyme's function can tolerate the substitution

8-32.

a. The deleted homolog has a known null activity allele for all the genes within the deletion (because the genes are not present at all). Therefore, **if a *mutant / deletion* genotype has the same mutant phenotype as the *mutant / mutant* genotype, this suggests that the mutant allele has the same level of activity as an allele with known zero activity (the deletion). One limitation of this assumption is that some phenotypes have a threshold level of enzyme activity. In other words, the mutant phenotype is seen as long as the level of enzyme activity is below some critical threshold**. Once the level of enzyme activity rises above this level, the phenotype becomes wild type. For example, imagine a situation where any individual has the mutant phenotype if the enzyme activity is <30%. Suppose allele *m* of an autosomal gene produces an enzyme with 20% of the activity as the enzyme encoded by the wild-type allele. Thus, an *m / m* individual would have 20% the enzyme activity of wild-type homozygotes, and an *m / deletion* heterozygote would have only 10% of normal enzyme activity. Because of the threshold, both of these individuals would have the same mutant phenotype, yet the *m* allele is clearly not null.

b. You can determine that a mutant allele of a gene is a true null activity allele **if an assay for enzyme activity of the encoded protein shows none present in individuals homozygous for that allele (or in *mutant / deletion* heterozygotes)**. In some cases, antibodies can be used to determine the amount of gene product present in individuals of these genotypes. Thus, **if the antibody does not detect any protein, you can safely assume there is no enzyme activity = null allele**. Remember that the converse of this statement is not true: the presence of protein as detected with an antibody does NOT mean there is enzyme activity, as even the change of a single amino acid in a large protein can completely abolish its function.

8-33. The size of the protein is 2532 amino acids. Thus the mRNA must be 2532 x 3 = 7.6 kb. To be detectable, any changes must be more than 1% of normal size or amount. The answers below are presented **in the order of mRNA size; mRNA amount; protein size; protein amount**. A '+' means there will be a >1% change, while a '-' means there will be no change. We assume in the answers below (excepting part j) that mutant mRNAs or proteins have normal stability in the cell, though this is not always true in practice.

a. - - - -. This is a non-conservative amino acid substitution changing the identity of only a single nucleotide in the mRNA and a single amino acid in the protein, so it will probably affect the ability of the protein to function normally, but it will not affect the size or amount of mRNA or protein.

b. - - - -. This is a conservative amino acid change, so it probably won't affect the function of the protein either.

c. - - - -. This is a silent change, so it won't affect any detectable parameters.

d. - - + -. This is a nonsense mutation, so it won't affect mRNA size or amount, nor amount of protein. It will affect the size of the protein.

e. - - + - or +. Met1Arg could mean that no protein is made, as the ribosome can't initiate translation at an Arg codon. However, it is possible that a protein could be made if there is another downstream, in frame 5'AUG codon that can be used to initiate translation. This second possibility would result in a smaller protein.

f. - + - +. A mutation in the promoter would most likely make it a weaker promoter, so fewer RNA polymerase molecules would bind. With less mRNA, less protein would be produced.

g. - - + -. This change would not be detectable in the mRNA, but it will cause a frameshift mutation in the protein which will obviously affect the size of the protein.

h. - - - -. A deletion of 3 bases (a codon) would only remove 1 amino acid / 2532 amino acids; this is a <1% change in size of the mRNA or protein.

i. + - + -. This mutation causes alternate splicing to occur, removing exon 19. Depending on the size of exon 19, this could easily have a >1% affect on the size of the mRNA and protein.

j. + + - +. Lack of a poly(A) tail could make the mRNA noticeably smaller. It could also decrease the stability of the mRNA. If the mRNA is degraded faster then less protein will be translated.

k. - - + or - -. This substitution in the 5' Untranslated Region will not affect the mRNA. However, it could have an affect on the amount of protein made if it affects ribosome binding, for example. Conversely, it may have absolutely no affect on how well the protein is translated.

l. - - - -. If this insertion is into an intron, then it should be spliced out of the primary transcript, producing a wild type mature mRNA.

8-34.

a. Null mutants are those with no enzyme activity. **The following changes could lead to loss of enzyme activity: a (if this amino acid substitution blocks protein activity); b (if this amino acid substitution blocks protein activity); d (would destroy most of the protein activity); e (if there is no translation initiation); f (if promoter does not function); g (frameshift would alter most of the amino acid sequence of the protein); h (it is possible that the deletion of 1**

amino acid blocks protein function or alters the tertiary structure of the protein); i (will
produce an altered protein that might be unable to function); j (could make the mRNA
completely unstable); and k (losing exon 19 could inactivate the protein). In practice,
mutations like j and k seem to cause less severe reductions in gene expression.

b. Any mutation that makes an altered protein with impaired function or lower levels of otherwise
functional protein is likely to be recessive to wild type. Most null or hypomorphic alleles are in
fact recessive, loss-of-function mutations. Thus, **any of the mutations in the list (except c and l)
could be recessive. Those listed in the answer to part a are also the most likely to be
recessive.**

c. **Any of the mutations, other than c and l, could potentially be dominant to wild type.** For
most of these, if the mutation was null or strongly hypomorphic there could be a **dominant effect
from haploinsufficiency**, where one wild type allele of the gene does not express enough gene
product for a wild type phenotype. This is true for any mutation that alters the protein product in
size or amount. Any of the mutations that change the size of the protein, as well as a, b and h
(which alter the protein to a lesser degree) could potentially have **dominant negative or
neomorphic dominant effects** that would depend on the protein (for example, whether it is
multimeric) and the exact consequences of the mutation. Finally, it is possible that a promoter
mutation, as in f, could turn on a gene in the wrong tissue or at the wrong time, leading to an
ectopic dominant phenotype. However, you should remember that most of the scenarios leading
to dominant phenotypes are generally much more rare than recessive effects due to loss of
function.

8-35. Cross each of the reversed (reverted) colonies to a wild type haploid to generate a heterozygous
wild type diploid. Then sporulate the diploid and examine the phenotype of the spores. **If the met$^+$
phenotype is due to a true reversion, then the cross was:** *met$^-$* x *met$^+$* → *met$^+$* / *met$^-$* → 2
met$^+$: 2 *met$^-$*. **If there is an unlinked suppressor mutation in another gene, the suppressor (*su$^-$*)
and original *met$^-$* mutation should assort from each other during meiosis:** *met$^-$ su$^-$*
(phenotypically met$^+$) x *met$^+$ su$^+$* (wild type) → *met$^-$ / met$^+$* ; *su$^-$ / su$^+$* → 1/4 *met$^-$ su$^-$* (met$^+$) :
1/4 *met$^-$ su$^+$* (met$^-$) : 1/4 *met$^+$ su$^-$* (met$^+$) : 1/4 *met$^+$ su$^+$* (met$^+$) = 3/4 met$^+$: 1/4 met$^-$. If the two
genes are linked, you will still get the four classes but the proportions of the recombinant gametes

(*met⁻ su⁺* being the one you can detect as it is the only one with a met⁻ phenotype) will increase proportionate to the distance between the 2 genes.

8-36. See Figures 8.29, 8.30 and 8.31.

a.   **Null mutations have no functional gene product. Hypomorphic mutations have a lower level of protein activity while hypermorphic mutations have a higher level of activity. Dominant negative mutations interfere with the functioning of the normal polypeptide made by the other allele or they interfere with the functioning of other proteins that interact with the gene product. Neomorphic mutations exhibit a new phenotype.**

b.   **Null and hypomorphic mutations would usually be recessive** to wild-type, unless the phenotype is particularly sensitive to decreases in the amount of a gene product (as in the cases of incomplete dominance or of haploinsufficiency). **Hypermorphic and neomorphic mutations would probably be dominant**, but it is possible that in rare cases, the phenotype might not be sensitive to the extra or new function unless the mutant allele were present in two doses. **Dominant negative mutations are dominant by definition.**

8-37.

a.   The anticodon is complementary to the codon. **The anticodon in this nonsense suppressor tRNA is 3' AUC 5'.** Because the 5'-most nucleotide in the anticodon is C, the wobble rules predict that it can pair only with a 5' UAG 3' stop codon and not with any other triplet.

b.   The wild-type tRNA^{Gln} recognizes either a 5' CAA 3' or a 5' CAG 3' codon. Only a single nucleotide was changed to turn the wild type tRNA into the nonsense suppressor. The suppressor tRNA recognizes a 5' UAG 3' codon, so the wild type tRNA must have recognized the 5' CAG 3' codon if only one base change occurred. The anticodon sequence in the wild-type tRNA is therefore 3' GUC 5'. The template strand employed to produce the tRNA is complementary to the tRNA sequence itself, so **the sequence of the template strand of DNA is 5' CAG 3'.**

c.   In any wild type cell of any species there is a minimum of one tRNA^{Gln} gene. This one has to be able to recognize both of the normal Gln codons, 5' CAA 3' and 5' CAG 3'. The wobble rules say that a tRNA with a 3' GUU 5' anticodon can recognize both Gln codons, so this would be the minimum tRNA gene in any organism (**Figure 8.22**). However, *B. adonis* is a species that can harbor the Gln nonsense suppressing tRNA discussed in part a. **Therefore there must be two tRNA^{Gln} genes in a wild-type *B. adonis* cell. One would code for the tRNA (anticodon of 3' GUC 5') that was changed into a nonsense suppressor, and the other gene would code for the tRNA (with an anticodon of 3' GUU 5') that recognizes both of the normal Gln codons.**

8-38.

a. There is a progression of mutations from Pro (5'CCN, wild type) to Ser (5'UCN) to Trp (5'UGG, strain B) codons. **The original wild type proline codon must have been 5' CCG, so the sequence of the DNA in this region must have been: 5' CCG**

    3' GGC

b. The wild type amino acid is Pro (proline). The amino acid at the same position in strain B is Trp (tryptophan). Although these are not the same amino acids, the phenotype of strain B is wild type. This means that **Trp at position 5 is compatible with the function of the enzyme encoded by the gene.** This implies that the change from Pro to Trp is a conservative substitution. In fact, if you look at **Figure 7.24**, you can see that Pro and Trp are both amino acids with nonpolar R groups. The original mutation changed Pro to Ser (serine). This mutant is non-functional, so the **Ser at position 5 is not compatible with enzyme function - it is a non-conservative substitution**. **Figure 7.24** shows that Ser has an uncharged polar R group, so its chemical properties are likely to be very different than those of Pro.

c. **Strain C does not have any detectable protein, so it is likely to be a nonsense mutation that stops translation after only a few amino acids have been added.** One way in which thus could have happened is that the codon for Ser in the mutant (5' UCG) could have been changed by mutation to a 5' UAG. However, the nonsense mutation could also have occurred at other locations in the gene's coding region.

d. Strain C-1 is either a same-site revertant (5' UAG to sense codon) or a second site revertant (nonsense suppressing tRNA mutation). **Because the reversion mutation does not map at the enzyme locus, it must be a nonsense suppressing tRNA.**

8-39. A missense suppressing tRNA has a mutation in its anticodon so that it recognizes a different codon and inserts an inappropriate amino acid. Problem 8-37 dealt with the change of the tRNA^{Gly} anticodon from 3' GUC 5' to 3' AUC 5', making a nonsense suppressor tRNA. Imagine instead that the tRNA^{Gln} anticodon had mutated to 3' GCC 5'. This mutant tRNA will respond to 5' CGG 3' codons, thus putting Gln into a protein in place of Arg. This is a missense suppressor.

a. Consider here the effect of the presence of a missense or nonsense tRNA on the normal proteins in a cell that are encoded by wild-type genes without missense or nonsense mutations. **A missense suppressing tRNA has the potential to change the identity of a particular amino acid found in many places in many normal proteins.** Using the example above of a missense suppressor that replaces Arg with Gln, most proteins have many Arg amino acids, so a missense suppressor could potentially substitute Gln for Arg at many sites in any single protein. **In** contrast, **a nonsense suppressing mutation can only affect a single location in the expression of any wild-type gene (that is, the stop codon that terminates translation), making a normal protein longer**. The longer protein will still have all of the amino acids comprising its normal counterpart. Together, these considerations mean that the presence of a missense suppressing tRNA has more likelihood of damaging proteins synthesized in the cell than the presence of a nonsense suppressing tRNA.

b.   In addition to the situation described above in which a mutation in a tRNA gene would change the anticodon to recognize a different codon, there are other possible ways to generate missense suppression, including: **(i) a mutation in a tRNA gene in a region other than that encoding the anticodon itself, so that the wrong aminoacyl-tRNA synthetase would sometimes recognize the tRNA and charge it with the wrong amino acid; (ii) a mutation in an aminoacyl-tRNA synthetase gene, making an enzyme that would sometimes put the wrong amino acid on a tRNA; (iii) a mutation in a gene encoding either a ribosomal protein, a ribosomal RNA or a translation factor that would make the ribosome more error-prone, inserting the wrong amino acid in the polypeptide; (iv) a mutation in a gene encoding a subunit of RNA polymerase that would sometimes cause the enzyme to transcribe the sequence incorrectly**.

8-40. If a tRNA was suppressing +1 frameshift mutations, then **it must have an anticodon that is complementary to 4 bases, instead of 3**.

8-41. Use the wobble rules (**Figure 8.22**) to help solve this problem.

a.   The nonsense codons that differ only at the 3' end are 5' UAG and 5' UAA. **A tRNA with the anticodon 3' AUU 5' could recognize both of these nonsense codons** because the U at the 5' end of the anticodon could pair with G or A at the 3' end of the nonsense codons.

b.   **This nonsense suppressing tRNA would suppress 5' UAA 3' and 5' UAG 3'**.

c.   This question asks which wild type tRNAs could have their anticodons changed to 3' AUU 5' with a single nucleotide change. It is easier to answer this question from the perspective of the mRNA sequences. In other words, what sense codons could have mutated to become 5' UAA/G? **These are: 5' CAA/G = Gln, 5' GAA/G = Glu, 5' AAA/G = Lys, 5' UCA/G = Ser, 5' UUA/G = Leu, 5' UAU/C = Tyr, and 5' UGG = Trp**. However, there is an additional consideration that excludes Trp as an answer. There is only one possible anticodon for a Trp-carrying tRNA. This anticodon is 3' ACC 5'. A normal Trp tRNA could not have an anticodon of 3' ACU 5' because wobble would allow such a hypothetical tRNA to recognize a 5' UGA 3' stop codon. Thus two mutational changes would be required to change the only possible normal tRNA^{Trp} with a 3' ACC 5' anticodon to a nonsense suppressing tRNA with a 3' AUU 5' anticodon.

8-42. In the second bacterial species where the isolation of nonsense suppressors was not possible, there must be only a single tRNA^{Tyr} gene and a single tRNA^{Gln} gene. Thus, if either gene mutated to a nonsense suppressor, it would be lethal to the cell, as there would not be any tRNA that could put Tyr or Gln where they belong. In this scenario, the single tRNA^{Tyr} would have to have an anticodon of 3' AUG 5' to recognize the two Tyr codons of 5' UAU 3' and 5' UAC 3' based on the wobble rules. The single tRNA^{Gln} would have to have an anticodon of 3' GUU 5' to recognize both 5' CAG 3' and 5'CAA 3' Gln codons.

Chapter 9   Digital Analysis of DNA

Synopsis:

This chapter introduces you to many of the recombinant DNA techniques that have provided a powerful new approach for studying the mechanisms of inheritance and functions of specific genes. Restriction enzymes, cloning DNA, making libraries, identifying clones of interest, DNA sequencing and PCR amplification are now just a part of the toolkit that all biologists (not just geneticists) use. These techniques will be referred to over and over throughout this textbook (and probably in your other biology courses as well) so it is worthwhile to get a solid understanding of these techniques from this chapter.

As you read about the various techniques and apply them to solve problems, try to keep in mind which techniques are done in solutions in test tubes (restriction enzyme digests, ligating fragments together, PCR, DNA sequencing, making cDNA) and which techniques involve analyzing or manipulating DNA in cells (transformations, screening libraries, preparing large amounts of cloned DNA, total genomic DNA or cellular RNA). This should help your understanding of the techniques and their uses. Hybridization of nucleic acids is central to many techniques but is often challenging to understand. The basis of hybridization is *complementarity* of bases in forming double stranded nucleic acids. A probe DNA or RNA molecule is used to locate a specific sequence (on a nitrocellulose or membrane based blot after electrophoresis in a gel, as a clone inside a cell, or in a chromosome squash) based on hybridization. A probe contains a recognizable radioactive or fluorescent tag that makes it possible to identify the place where the probe found a complementary sequence.

Significant Elements:

After reading the chapter and thinking about the concepts, you should be able to:

- Describe the essential steps in cloning.
- Describe the basic components and uses of different types of cloning vectors.
- Make a map of restriction enzyme sites.
- Read and interpret DNA sequencing gels (**Feature Figure 9.13**) and automated DNA sequencing results (**Figure 9.14**).
- Design PCR primers.
- Determine which technique(s) you must use to achieve a desired goal. There is often more than one way to reach a goal. However, there is usually one most efficient, preferred way to solve a problem.
- The technique used determines what is being examined and limits the interpretation of the data. For instance, probing a genomic library will give you a clone that is homologous to the probe, but

this clone probably won't be transcribed and translated in *E. coli*. Probing a cDNA library will give you a clone which can be translated and transcribed in *E. coli*.

Problem Solving Tips:

Essential Steps in Cloning:

Cloning is basically a straightforward process that has lots of options and variations that can be used depending on what is desired. Basic components are insert DNA and vector. There are relatively few sources for the insert DNAs. However there are many, many types of vectors that have been developed for various purposes.

Types of insert DNA

♦   cDNAs contain only the regions of genes that are present in processed (spliced) transcripts synthesized in the cell from which they were isolated (**Figure 9.8**).

♦   genomic DNAs are digested fragments of the genomic DNA of an organism, and so contain all of the DNA (genes and non-coding regions) from the cells.

Basic vector criteria

♦   vectors must have an origin of replication so they can be replicated in the host organism, usually *E. coli*.

♦   vectors must have a selectable marker(s) so you can determine that they are present in the host organism; the selectable marker is often an antibiotic resistance.

♦   vectors also often have multiple cloning sites with known restriction sites and ways to detect the presence of an insert DNA after cloning. One example of an insert detection system is the β-galactosidase / X-gal detection system. Insertion of a fragment into the middle of the *lacZ* gene inactivates the gene. Cells carrying an insert within the *lacZ* gene are unable to cleave a lactose-like substrate (X-gal) and are phenotypically Lac⁻. They are recognized as white colonies while colonies that received intact copies of the vector (no insert interrupting the *lacZ* gene) can cleave the substrate, turning the cells blue.

Types of vectors/purpose of cloning (**Table 9.2**)

♦   plasmid vectors accept small pieces of insert DNA (10 kb or less). Plasmid vectors may be used to amplify large amounts of specific DNA sequences. Specialized plasmid vectors called expression vectors allow transcription and translation of cloned genes; must be used with cDNA inserts (Genetics and Society, Recombinant DNA Technology and Pest-resistant Crops **Figure A**). Use your knowledge of the requirements for transcription and translation when considering if genes cloned into expression vectors will be expressed in the host cell.

♦   BAC vectors (bacterial artificial chromosomes) accept very large inserts of 300 kb.

Cloning

- after restriction enzyme digestion, mix insert and vector DNAs and ligate together sticky ends that have complementary overhanging single-stranded bases can be. It may be helpful to draw out the 5' and 3' ends generated (including the individual bases of the recognition site) when a double stranded DNA is cut by a restriction enzyme (**Figure 9.2**).

- transform the ligation mix into the host cells, usually *E. coli.*

- select for presence of vector (may also be able to isolate those vectors that you know have an insert).

- grow up a large amount of the clone(s).

Identifying the desired clone

- often you must identify a particular desired clone from a large variety of different inserts; this usually involves probing, or hybridization with a labeled DNA.

Other Techniques

- gel electrophoresis separates DNA fragments according to their size (**Feature Figure 9.4**).

- blotting is the process of transferring the material in the gel to a nitrocellulose filter or a nylon membrane and covalently binding the material from the gel to the filter or membrane. A Southern blot has DNA on the membrane (a genomic Southern has genomic DNA), a Northern blot has mRNA on the membrane and a Western blot has protein on the membrane (**Feature Figure 9.11**).

- Restriction mapping is part science and part art, like putting together a jigsaw puzzle. Use a pencil and an eraser. Be patient. The first step is usually ascertaining if you began with a linear or a circular piece of DNA. Usually this is gotten out of context - a plasmid clone is circular, for instance. Begin the map by examining a single digestion lane on the gel and determining the total size of the DNA (the sum of all the fragments) and the number of restriction sites for that enzyme (2 fragments when you digested a circular piece of DNA means there were 2 restriction sites; 2 fragments when you digested a linear piece of DNA means there was only 1 restriction site). Next, look at the double digestion lane. Determine which bands from the single digestion are left undigested in the double enzyme digestion. The fragments from the single enzyme digestion that disappear in the double digestion must have a restriction site for the second enzyme within them. Figure out which smaller fragments they have been broken into, then begin mixing and matching various combinations of bands until you find one that gives you an order that will give the correct pattern of bands when you digest the DNA with the second restriction enzyme alone (see problems 9-5 and 9-6). Make sure the final sites you put on a map are consistent with results from all digests.

- DNA sequencing provides the ultimate description of a cloned fragment of DNA. Make sure you can explain the Sanger sequencing method (dideoxy sequencing) to a friend (**Feature Figure 9.13**).

♦ PCR rapidly purifies and amplifies a single DNA fragment from a complex mixture (**Feature Figure 9.12**). In order to do PCR you must know something about the DNA sequence of 2 short stretches of the DNA to be amplified. The DNA fragment to be amplified is defined by a pair of oligonucleotide primers that are each complementary to one of the strands of the DNA template. These primers are extended at their 3' ends. The size of the final product of the PCR reaction is determined by the distance between the 5' ends of the primer pair.

Solutions to Problems:

<u>Vocabulary</u>

9-1. a. **10**; b. **1**; c. **9**; d. **7**; e. **6**; f. **2**; g. **8**; h. **3**; i. **5**; j. **4**.

<u>Section 9.1 – Sequence-Specific DNA Fragmentation</u>

9-2.

a.   *Sau*3A recognition sites are 4 bases long and are expected to occur randomly every 4^4 or 256 bases. The human genome contains about 3×10^9 bases, one would expect $3 \times 10^9/256 = 1.2 \times 10^7$ **~12,000,000 fragments**.

b.   *Bam*HI recognition sites are 6 bases long and would be expected every 4^6 or 4096 bases. $3 \times 10^9/4,100 = 7.3 \times 10^5$ **~700,000 fragments are expected**.

c.   The *Sfi*I recognition site is 8 specific bases. The N indicates that any of the four bases is possible at that site and therefore does not enter into the calculations. Recognition sites would be expected every 4^8 or 65,536 bases; $3 \times 10^9/65,500 = 4.6 \times 10^4$ **~46,000 fragments are expected**.

9-3. See **Feature Figure 9.4** and the section in the chapter 'Gel electrophoresis distinguishes DNA fragments according to size.' The rate at which a piece of DNA moves through a gel is dependent on the strength of the electric field, the gel composition, the charge density and the physical size of the molecule. When electrophoresing DNA the only variable is the size of the molecule - all the rest of the variables are the same for each molecule. **Longer DNA molecules take up more volume and therefore bump into the gel matrix, slowing down the molecule's movement**. Shorter molecules can easily slip through many pore sizes in the gel matrix.

9-4. When you digest a circular DNA one fragment indicates that the DNA has 1 restriction site for the enzyme. Thus, *Bam*HI and *Eco*RI each cut the plasmid once. The double digest gives information about the relative positions of these two sites. The 2 restriction sites are at two different positions on the plasmid. The *Eco*RI site is 3 kb away from the *Bam*HI site and it is 6 kb around the rest of the circle back to the *Eco*RI.

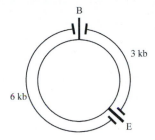

9-5.

a. Remember the Problem Solving Tips at the beginning of this chapter! If there is one restriction site then digesting a circular molecule results in one fragment, while digesting a linear molecule generates two fragments. Digestion of a circular molecule will always result in one fewer restriction fragments than the digest of a linear molecule. **Sample A is therefore the circular form** of the bacteriophage DNA.

b. The length of the linear molecule is determined by adding the lengths of the fragments from one digest. 5.0+3.0+2.0 kb = **10.0 kb**. (This size is not realistic - λ DNA is, in fact, about 50 kb in length.)

c. The circular form is the same length - **10.0 kb**.

d. Comparison of the circular and linear maps gives you information on which fragments contain the ends of the linear molecule. The 5.0 kb *Eco*RI fragment is present in the circular but not the linear digest so the 4.0 and 1.0 kb fragments must be joined in the circular map while they are at either end of the linear molecule. Begin drawing a picture of the molecule for yourself at this point. The same logic applies to the 2.7 kb *Bam*HI fragment – it is present in the circular but not the linear digest so the 2.2kb and 0.5 kb pieces must be at the ends of the linear molecule. If the 0.5 kb *Bam*HI fragment was at the end where the *Eco*RI 1.0 kb fragment is, the 1.0 kb *Eco*RI fragment would have been cut by *Bam*HI in the double digest. However, the 1.0 kb fragment is still in the double digest, so the 0.5 kb fragment must be within the 4.0 kb *Eco*RI fragment. The remaining *Eco*RI site is placed based on the double digests. The 2.0 kb *Eco*RI fragment is not cut by *Bam*HI but the 3.0 kb fragment is, so place the site within the 3.0 kb. Now double check that all the *Bam*HI+*Eco*RI fragment sizes are as seen in the different double digests.

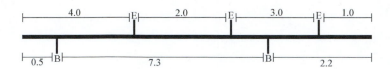

9-6. Plasmids are circular pieces of DNA, thus the *Eco*RI and *Sal*I digests indicate that there is one site for each of these enzymes. *Hin*dIII, in contrast, cuts the molecule at three sites. Draw a circle showing the three *Hin*dIII sites. In the *Sal*I+*Hin*dIII digest the 4.0 kb *Hin*dIII fragment is cut into 2.5 and 1.5 kb fragments. The *Sal*I site is therefore 1.5 kb from one end or the other in the 4.0 kb *Hin*dIII fragment. Similarly the *Eco*RI+*Hin*dIII double digest splits the 1.0 kb *Hin*dIII fragment into 0.6 and 0.4 kb fragments, but the orientation of the *Eco*RI site within the 1.0 kb *Hin*dIII is ambiguous. Try placing the *Eco*RI site in the two different positions in the 1 kb *Hin*dIII fragment. In each case see how this fits with the *Eco*RI+*Sal*I digestion results. The orientation that works places the 0.4 kb *Hin*dIII-*Eco*RI fragment adjacent to the 2.5 kb *Sal*I-*Hin*dIII fragment.

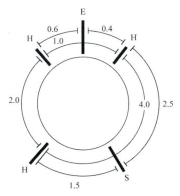

9-7.

a. The full length is represented by the largest fragment in the partial digestion, as this **1.83 kb** fragment is the one that has not been digested.

b. After digestion only one of the products is visualized on the autoradiograph. Because the fragment is only labeled at the *Eco*RI end, the *Bam*HI end 'disappears.' Thus you see a set of nested bands with the longest one representing digestion at the enzyme site furthest from the *Eco*RI end, the next longest digested at the next site closer to the *Eco*RI end, etc. If you start with the biggest fragment and subtract each subsequent fragment size from the preceding one you will generate a map that begins at the unlabeled or *Bam*HI end of the fragment. For *Hha*I (from left to right): 0.42, .09, 0.93, 0.26, 0.13; for *Sal*I: 0.26, 0.64, 0.35, 0.10, 0.24, 0.24.

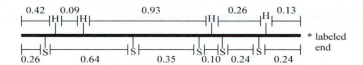

c.   The complete double digest will yield the following fragments: 0.26, 0.16, 0.09, 0.39, 0.35, 0.10, 0.09, 0.15, 0.11 and 0.13. All of these fragments will be seen on an ethidium bromide stained gel. Only the fragment that includes the labeled *Eco*RI end will be seen on the autoradiograph. The 0.13 kb fragment contains the labeled end. It is rightmost on the map.

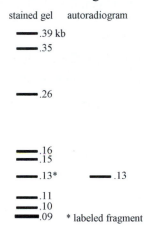

Section 9.2 – Cloning Fragments of DNA

9-8. An artificial chromosome is **a vector which can accept very large pieces of foreign DNA** – up to 1Mb in the case of yeast artificial chromosomes. Commonly used artificial chromosome vectors include BACs (bacterial artificial chromosomes) from *E. coli* and YACs (yeast artificial chromosomes) from yeast. **Such vectors make it possible to clone large pieces of genomic DNA**, so they are useful for making genomic libraries of organisms with large genomes, like humans. This sort of vector must be able to act as a chromosome in the host cell (*E. coli* for BACs and yeast for YACs) so they must include an origin of replication, a selectable marker and an origin of replication. YACs must also include a centromere and telomeres at the end of the vector.

9-9. Selectable markers in vectors provide a **means of determining which cells in the transformation mix take up the vector**. These markers are often drug resistance genes so a drug can be added to the media and only those cells that have received and maintained the vector will grow.

9-10. The study of genes often involves studying mutations in the genes and the phenotypes (or diseases) associated with these mutations. If you are interested in studying mutations and diseases then you want to focus on the protein-coding part of the genes. **Eukaryotic genes are often very large**. However **the majority of this DNA consists of intronic sequences** which do not end up in the mRNA. For example the human dystrophin gene in humans is 2,500 kb (2.5 Mb, see Figure 8.15). The gene has more than 80 introns which are spliced out to give an mRNA that is 14kb long. Therefore 2,486 kb of the dystrophin gene is introns! Thus, **most of the DNA in eukaryotic genomic libraries does not code for proteins**. It can be difficult to figure out which sequences of the genomic DNA are actually part of the mRNA so it can be difficult to figure out which gene sequences are important to the protein and which are unimportant. **cDNA libraries, which are made from the mRNAs, allow you to ignore all of these intronic sequences**. All eukaryotic mRNAs have polyA tails at their 3' end and this is used to make cDNAs. The process begins by isolating mRNAs from an organism or a tissue in an organism and then using polyT primer with reverse transcriptase (**Figure 9.8**).

In prokaryotes most of the DNA in the genome codes for mRNA – there is very little non-transcribed DNA. Prokaryotes also lack introns, so without processing the transcript is the same thing as the mRNA. In general the 5' and 3' UTRs are small, so most of the mRNA consists of coding sequences. It would also be difficult to make cDNA libraries **in prokaryotes because there is no polyA tail nor any other common sequence between all mRNAs**.

9-11. First, work through the digestion and ligation of the DNA fragments and the vector. The vector is cut with *Bam*HI, leaving the following ends:

```
5' —G          GATCC—
3' —CCTAG          G—
```

The insert DNA is cut with *Mbo*I, leaving the following sticky ends:

```
5' —           GATC—
3' —CTAG           —
```

The ligation of an *Mbo*I fragment to a *Bam*HI sticky end will only occasionally create a sequence that can be digested by *Bam*HI. It depends on the exact base sequence at the ends of the *Mbo*I fragment. The 'X' in the sequence below indicates this ambiguity. In all cases the following sequence will be found: The sequences from the inserted *Mbo*I fragment are in bold.

5' —GGATCX————————XGATCC—
3' —CCTAGX ————————XCTAGG—

a. **100%** of the junctions can be digested with *Mbo*I

b. A junction that can be digested with *Bam*HI must have a C at the 3' end of the *Mbo*I recognition sequence. This would occur 1/4 or **25% of the time**.

c. **None** of the junctions will be cleavable by *Xor*II.

d. The first five bases fit the recognition site for *Eco*RII. The final position must be a pyrimidine (C or T). There is a **1/2 chance** that the junction will contain an *Eco*RII site.

e. For the restriction site to be a *Bam*HI site in the human genome it must have had a G at the 5' end. This G was in the vector sequence in the clones created. The chance that the 5' end was NOT a G=**3/4**.

9-12.

a. The **genomic library** is based on the most inclusive and complex starting material, so it would consist of the greatest number of different clones.

b. **All of these libraries would overlap each other to some extent**. The genomic library contains all the DNA sequences, while the other libraries are made up of subsets of the genomic sequences. All cells express a common subset of genes (housekeeping genes). These genes would result in some overlap of clones, although the cDNA libraries will each contain some unique sequences. Although introns often have repeated DNA, the transcribed and translated portions of sequences are usually unique, so the library of unique genomic sequences will overlap with the cDNA libraries as well.

c. **The genomic library uses the total chromosomal DNA inside the cell. The repetitive sequences in the genomic DNA would have to be removed to create the unique DNA library. The cDNA libraries all start with the mRNA present in the cells and thus represent therefore the expressed genes in these cells.**

9-13.

a. You need **4-5 genome equivalents** to reach a 95% confidence level that you will find a particular unique DNA sequence.

b. The number of clones needed depends on the total size of the genome of your research organism and the average insert size in the vector. BAC inserts can be 500kb while plasmid vectors normally have inserts smaller than 15 kb. **Divide the number of base pairs in the genome by the average insert size then multiply by five** to get the number of clones in five genome equivalents.

9-14.

a. An intact copy of the whole gene would be on a fragment larger than 140 kb and would therefore have to be cloned into a **BAC vector**.

b. The entire coding sequence of 38.7 kb could be cloned into a **BAC vector** (30-45 kb inserts) as a cDNA copy of the gene.

c. Exons are usually small enough to clone into a **plasmid vector** (<15 kb inserts).

9-15. When the vector (pWR590) is digested with *Eco*RI you get one 2.4 kb fragment. When the vector is digested with *Mbo*I there are 3 fragments - 0.3, 0.5 and 1.6 kb. The somatostatin insert was cloned into the vector at the *Eco*RI site. There is also an *Eco*RI site very near one end of the insert DNA. Therefore**, after digestion of the recombinant plasmid with *Eco*RI, a small *Eco*RI insert fragment of 49 bp and the vector fragment of 2.4 kb will be generated**. Next, consider the *Mbo*I restriction pattern. The insert fragment contains an *Mbo*I site 5 bp from one end. The insert fragment could ligate into the vector in either of 2 possible orientations. **In one orientation the *Mbo*I site in the insert is nearest the 700 bp *Mbo*I vector fragment, so digestion with *Mbo*I produces 705, 300, 500 and 944 (formed from the 900 bp vector fragment + the rest of the insert) bp fragments. In the other orientation, the *Mbo*I digest produces 905, 500, 300 and 744 bp fragments.**

9-16. Draw the recombinant plasmid to help you determine the fragment sizes before sketching the gel.

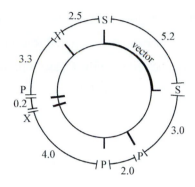

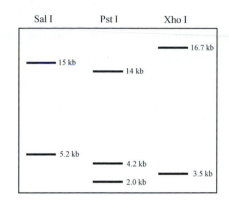

9-17.

a. The goal of a ligation is to generate clones which have attached one piece of frog DNA to one vector molecule. A ligation mixture consists of linear double stranded vector DNA with complementary *Eco*RI sticky ends (**Figure 9.2b** and **Figure 9.6**) at both ends and linear double stranded frog DNA with complementary *Eco*RI sticky ends at both ends. Ligase simply attaches a 3'OH (hydroxyl) group to a 5'P (phosphate). There are three different products that will occur in a ligation mix. (i) The desired ligation is vector/frog (intermolecular ligation). (ii) Ligase will also join vector/vector (intramolecular ligation which yields reconstituted vector molecules with no inserts) and (iii) frog/frog (intramolecular ligation, giving chains of insert DNA with no vector). In order to encourage the desired result you add more vector than insert – the vector DNA is easier to come by. This decreases the likelihood of chains of the insert DNA and increases the probability that any vector molecule that is ligated to an insert is only ligated to one insert molecule. However adding more vector increases the likelihood of reconstituted vector with NO inserts. To decrease the amount of reconstituted vector you treat the linear, digested vector with alkaline phosphatase. **Alkaline phosphatase removes the 5'-phosphate groups** on the linear DNA molecule – see * below. Remember that this represents the digested vector, so the DNA strands are contiguous except for the boxed area. This continuity is represented by the dashes at the ends of the lines   . The boxed area represents the sticky ends created by EcoRI.

After the treatment with alkaline phosphatase **ligase can not join a hydroxyl group to the de-phosphorylated 5' ends**. Therefore the 2 ends of the vector can not be ligated to each other and this treated molecule will remain linear. If insert DNA is added then the ligase will join the 3'OH on the vector with the 5'P on the insert. In effect this will ligate the left end of the top strand of the vector shown above to the insert. The left end of the bottom strand can not be ligated to the insert leaving a nick in the bottom strand at this point. On the right end the bottom strand ligates to the insert and the top strand at the right end can not ligate leaving another nick. The ligation mix is then transformed into *Escherichia coli*. These nicks in the phosphate backbone of the cloned DNA are repaired after the ligated DNA enters the cells.

Plasmid vectors are constructed so that they contain the *lacZ* gene with a restriction site right in the middle of the gene. If the vector reanneals to itself without inclusion of an insert, the *lacZ* gene will remain uninterrupted; if an insert has been cloned into the vector the *lacZ* gene will be interrupted. The ligation mix is transformed into *E. coli* cells such that about one cell out of 1,000 cells takes up a plasmid. The transformed cells are plated on media containing ampicillin. Only the cells with a plasmid will grow, thus removing the intramolecular ligation products that consist of inserts. The media also contains X-Gal. This is a substrate for the β-galactosidase protein that is coded for by the *lacZ* gene. The β-galactosidase enzyme cleaves X-Gal and produces a molecule that turns the cell blue. Those cells that took up an intact, re-circularized vector with no insert will produce β-galactosidase and form blue colonies. The bacterial cells that took up a vector + insert (clone) will not be able to produce functional β-galactosidase and will form white colonies.

The ligation with the non-phosphorylated vector reanneals to itself at a high frequency, leading to 99/100 blue colonies. The phosphorylated vector formed 99/100 white colonies, showing that almost all of the vectors had an insert.

b. **Yes**, the suggestion was a good one. **The dephosphorylation of the vector increased the number of clones (vector + insert) 100 fold.**

c. The choice of whether to dephosphorylate the vector versus the insert DNA is based on an understanding of the mechanics of the bacterial transformation that is carried out after the ligation. If the vector is dephosphorylated it cannot self-ligate. The insert can self-ligate. The self-ligated inserts do not have any vector DNA, so they do not have a bacterial origin of replication (ORI) nor do they have a gene encoding antibiotic resistance. Therefore, these recircularized

DNA's will not allow the transformed bacteria to grow on the selective media. **If the insert were dephosphorylated, it will not self-ligate, but the vector WILL self-ligate. The vector has the antibiotic resistance gene and ORI, so the "empty" vector will be propagated in _E. coli_, generating a high level of "background."**

Section 9.3 – Hybridization

9-18. One order is: **c, j, f, a, k, i, g, b, e, h d**. However, there is some flexibility in the order. Steps c and j can be switched, but they must be done before j; steps c, j and f must be completed before a; c, f, j and a must occur before k and all of these must happen before i. The next step must be the g, b, and e group, although these 3 can be changed among each other, then step h and lastly step d.

9-19.

a. **(1) 3.1, 6.9 kb; (2) 4.3, 4.0, 1.7 kb; (3) 1.5, 0.6, 1.0, 6.9 kb; (4) 4.3, 2.1, 1.9, 1.7 kb; (5) 3.1, 1.2, 4.0, 1.7 kb**.

b. The **6.9 kb fragment in the _Eco_RI+_Hin_dIII digest; the 2.1 and 1.9 kb fragments in the _Bam_HI+_Pst_I, and the 4.0 kb fragment in the _Eco_RI+_Bam_HI digest** will hybridize with the 4.0 kb probe.

9-20. If the gene has been isolated from other organisms, you could use **a probe of the gene from another organism**. Because the protein sequence of the ozonase enzyme is known, the **amino acid sequence could be 'reverse translated' into potential DNA** sequences that could encode that protein. A mixture of **oligonucleotides that could encode that peptide sequence could be used as a probe to hybridize to a library of genomic or cDNA clones.** Furthermore, **the cDNA clones can be cloned into an expression vector** (a vector that allows E. coli to transcribe and translate the gene(s) cloned into the vector). **The library clones can then be screened for expression of the ozonase enzyme by screening with the antibody**.

9-21. Probes need to be at least 15 nucleotides to effectively anneal to DNA. In this experiment short probes are desirable, because the longer the probe the greater the degeneracy. Thus, this type of experiment is usually done with **probes between about 15 and 18 nucleotides long**. The design of degenerate probes is based on reverse translation, and there are a few considerations to keep in mind: (i) if you know the amino acid sequence of the protein in one species then you can make some guesses about the amino acid sequence of the corresponding gene in the second species. You hope that the amino acid sequence of a particular, small region of the protein will be identical in the two species. Since there are 20 different amino acids even one amino acid difference would make it hard to design a probe. **If you knew the sequence of the protein from several bacterial species you could choose a very highly conserved region on which to base a probe**. If the amino acids are identical in several different species then they might be identical in *Beneckea nigripulchritudo.* (ii) If you don't know anything about the amino acid sequence of the protein in other species of bacteria then you would **find a region of 5 or 6 contiguous amino acids with low degeneracy** - that is amino acids that are encoded by the lowest possible number of codons. The best choices are Met and Trp which are each encoded by only a single codon. Unfortunately, it is highly unlikely that a region of 5 or 6 amino acids would be composed solely of Met and Trp. The next best choices are Phe, Tyr, Cys, His, Gln, Asn, Lys, Asp, or Glu, which are each coded for by 2 codons. The worst choices would be Leu, Arg, and Ser (6 codons). If you had a 5 amino acid region composed only of these three amino acids, then the number of different molecules in the degenerate probe would be $6^5 = 7776$.

9-22.

a.   These enzymes all recognize sites that are 6 bp long (**Table 9.1**). Therefore, in DNA of random sequence they will create **DNA fragments that are 4096 bp long on average. Fragments of this size are best resolved by electrophoresis in agarose gels**. Such fragments are too large to be resolved on polyacrylamide gels - they will all bunch up at the top of the gel (**Feature Figure 9.4**).

b.   Remember that total human genomic DNA was digested. The size of the haploid human genome is 3 billion base pairs (3×10^9 bp). Thus, digestion with *Eco*RI would produce roughly 700,000 fragments of DNA. Although the average size of the digestion fragments is ~4000 bp (4×10^3 bp) there is a wide range of sizes. Upon electrophoresis, these fragments would form **a broadly distributed smear centered around 4 kb** (see the agarose stained gel in Feature Figure 9.4).

c.   In problems 9-11, 9-12, 9-15 and 9-6 the restriction patterns being examined were from a small linear or circular molecule. This molecule was in the form of a cloned of DNA. Thus each digestion gave rise to very few discrete fragments, and all of these fragments were detected directly with ethidium bromide. Since all the restriction digests were of the same discrete piece of DNA, the sum of all the fragments in each restriction digestion must equal the size of the original molecule. In this example **the DNA that is digested is very long – a human chromosome**. The fragments were electrophoresed, subjected to a Southern blot and probed with a 5 kb *Eco RI* fragment. **Some fragments of DNA produced by digestion with restriction enzymes other than *Eco*RI will have either one or both ends at a position beyond the *Eco*RI sites defining the probe**. The location of these ends can vary, so the sum of the sizes of the fragments homologous to the probe can vary in different digests – see the map below.

d. The dark gray line with the dashed ends represents the chromosome. The line above the chromosome shows the extent of the 5 kb *Eco RI* probe. The numbers represent the distances in kb between adjacent restriction enzyme sites.

e. There are no markers in this experiment that indicate the centromere-to-telomere direction, so **you can not orient the restriction map relative to this direction**. If you had a complete restriction map (or DNA sequence) of the entire chromosome you could orient this small fragment with respect to the centromere and telomere. Chapter 10 explains how genome projects can provide such information.

Section 9.4 – PCR

9-23.

a. The human genome sequence shows the sequence of the normal allele of PKU. **You wish to know whether the PKU syndrome in this patient is caused by a mutation in the phenylalanine hydroxylase gene.** You suspect that there might be such a mutation in this particular exon, so you will sequence the PCR product. If there is a mutation in this 1 kb exon, you want to know exactly what it is, how it affects the enzyme, and perhaps something about the history of this mutation in human populations. For example, if you compare the sequence in many patients and track where the patients are from, you might get an idea of where this mutation arose in time and geographical space. If you do not find a mutation in this 1 kb exon that changes the amino acid sequence of the enzyme, there might still be a mutation in a different exon.

b. One haploid human genome contains 3×10^9 bp. Therefore (3×10^9 bp/haploid genome) x (6.6×10^2 g/mole) x (mole/6.02×10^{23} bp) = 3.3×10^{-12} g/haploid genome. In other words, one haploid genome weighs 3.3×10^{-12} g or 3.3 picograms. Each haploid genome will contain only one phenylalanine hydroxylase gene to be used as the template for the PCR reaction. You start the PCR reaction with 1 ng (1×10^{-9} g) of human DNA. Therefore (1×10^{-9} g DNA) x (1 haploid genome/3.3×10^{-12} g) x (1 template molecule/1 haploid genome) = 0.3×10^3 template molecules = **300 template molecules in 1 ng of DNA**.

c. You begin the PCR with 300 template molecules. If the PCR runs for 25 cycles then this number of molecules doubles exponentially 25 times. Therefore you will end up with 300 molecules x $2^{25} = 10^{10}$ or about 10 billion molecules. This result explains the power of PCR: you started with only 300 template molecules and end up with 10 billion copies of the region you are amplifying. In practice the yields are not quite as high because not all potential template molecules get amplified each cycle. However the amplification is still substantial. The PCR product is 1 kb long, so (10^{10} molecules of PCR product) x (10^3 bp/molecule of PCR product) x (mole/6.02 x 10^{23} bp) x (6.6 x 10^2 g/mole) = 1.1 x 10^{-8} g = 110 ng. You started with 1 ng of the whole genome and ended up with **110 ng of a 1 kb section of the genome after the PCR!**

9-24. Primers have to be 5' to 3' and have the 3' end toward the center so DNA polymerase can extend into the sequence being amplified. Only **set b.** satisfies these criteria.

9-25.

a.   Both of the primers in set b in <u>problem 9-24</u> are 18 nucleotides long. If (i) human DNA is assumed to be a random sequence of equal proportions of A, G, C, and T (this is not entirely accurate, but it is close enough for this discussion), and (ii) no mismatches are allowed between the primer and the genomic template (again, this is not entirely accurate as seen in parts b and c below, but again, it is close enough) then **the chance that one of the two primers will anneal to a random region of DNA that is not the targeted CFTR exon would be $(1/4)^{18}$, or about 1 chance in 7 x 10^{10}.** In other words, **an 18 base sequence will be present once in every 70 billion nucleotides**. Since the human genome is 3 billion nucleotides long it is extremely unlikely that even one of the primers will anneal anywhere else than the desired target. The probability is much lower that <u>both</u> of the primers will anneal to other stretches of DNA that happen to be close enough together to allow the formation of a PCR product. This latter number is hard to calculate exactly because of the variation in the possible distance between the primers.

b.   (i) The lower limit on the size of the primers is governed by two main factors. First, the PCR amplification must be specific, so the primers should be long enough to guarantee this specificity. As in part a, **the chance probability of a 16 base sequence in random DNA is $(1/4)^{16}$, or 1 chance out of 4 x 10^9.** Therefore, two 16 base pair primers allow a comfortable margin for specificity. More importantly the primers must anneal to the genomic DNA to be amplified. As discussed in Chapter 9, hydrogen bonding between 15 or 16 nucleotides of contiguous base pairs is required to allow DNA to remain double stranded. (ii) If the primers are too long, several potential problems arise. First, **the longer the primers the more expensive they are to synthesize**. Second, **the longer the primers the more likely they are to anneal with each other**, or for a single primer to anneal to itself and form a hairpin loop, and the less likely the primers are to anneal with the template. Third, and most importantly, **if the primer is too long it**

can hybridize with DNA with which it is not perfectly matched. Internal mismatches are tolerated and hybridization can occur as long as there are enough surrounding base paired nucleotides, especially at the 3' end of the primer. Thus, **longer primers might anneal to other regions of the genome than the region you actually want to amplify**.

c.   You would be more likely to obtain a PCR product if the mismatch were at **the 5'-end**. The 3'-end of a primer is its business end - that is where DNA polymerase adds additional nucleotides to the chain. **Mismatches at the 3'-end would prevent DNA polymerase from adding any new nucleotides to the chain**. (You might remember that some DNA polymerases have a 3'-to-5' exonuclease that could potentially remove the mismatch, now allowing further polymerization. This is true of *E. coli* DNA polymerase, but many of the DNA polymerases used in PCR come from thermophilic bacteria and these DNA polymerases do not have this exonuclease activity.) A mismatch at the 5'-end of the primer does not matter as long as there is enough base-pairing between the primer and genomic template to allow annealing.

9-26.

a.   The *Eco*RI and the *Sal*I restrictions sites are both found in the pMore vector sequence shown in the problem. The *Eco*RI site is nearer the 5' end and the *Sal*I site is nearer the 3' end of the pMore sequence shown. This region of pMore is at the C-terminal end of the maltose binding protein (MBP). Therefore your cloning will insert the CFTR DNA sequence into the DNA sequence that codes for the C-terminal end of the MBP protein. In other words, **the N-terminus of the fusion protein contains most of the MBP protein sequence**. The MBP sequence ends at the 8th amino acid from the C-terminus of MBP where the *Eco*R1 site cuts the MBP DNA. **The next part of the fusion protein contains the CFTR protein encoded by the PCR product**. Note that the PCR amplifies the last protein coding exon of the CFTR gene. Therefore **the C-terminal end of the fusion protein will contain the C-terminal end of CFTR**. Remember that the N-to-C orientation of the CFTR protein must be the same as that of the fusion protein as a whole. Further details of the fusion protein will be discussed in part c below.

b.   When you use two different restriction enzymes, **the CFTR gene can only be inserted into the vector with the desired orientation** yielding the fusion protein you described in part a. Thus the N-to-C orientation of the CFTR protein will be the same as the MBP protein. If the vector was only cut with *Eco*R1 and the PCR product had *Eco*R1 sites at both ends, then the PCR product could be inserted into the vector in two equally likely orientations, only one of which is the one you desire. A second advantage is that cutting with two enzymes minimizes unwanted products of the ligation in which ends of the same molecule come together (see problem 9-17 a and b).

c.   There are many things to take into consideration here. First, you can use the <u>set b</u> PCR primers you designed in your answer to <u>problem 9-24</u> in order to amplify the entire CFTR exon. Second, the CFTR exon does not have sites for *Eco*R1 and *Sal*I so you need to add nucleotides to the 5'-ends of the two primers that will contain appropriate sites for the two restriction enzymes. These sites cannot be exactly at the 5'-ends of the PCR primers – you must also add 5 more nucleotides beyond the restriction sites to enable the restriction enzymes to bind to their recognition sequences and digest the DNA. The sequence of these 5 nucleotides is not important. Third, the two parts of the fusion protein must end up being in frame. Because the PCR product encodes the C terminus of the fusion protein, there are fewer constraints on the identity of the additional nucleotides added to the second (backwards) primer. The answer below is just one of many possible solutions. The sequence of the critical part of the pMore vector is reproduced here. The dots at the left and right ends of this sequence represent the continuity of the DNA - this was a circular plasmid before the digestion.

```
5'...AGGATTTCAGAATTCGGATCCTCTAGAGTCGACCTGTAGGGCAA...3'
3'...TCCTAAAGTCTTAAGCCTAGGAGATCTCAGCTGGACATCCCGTT...5'
```

The vector is digested with *Eco*RI and *Sal*I to generate these sticky ends:

```
ArgIleSerGluPh
5'...AGGATTTCAG          TCGACCTGTAGGGCAA...3'
3'...TCCTAAAGTCTTAA          GGACATCCCGTT...5'
```

The PCR product using the set b primers (<u>problem 9-24</u>) is shown below. Remember that this PCR product contains the last protein coding exon of the CFTR gene. The left hand primer only has one open reading frame with the amino acid sequence shown below. The right hand primer contains the DNA sequence coding for the last four amino acids at the C-terminal end of the CFTR protein, as shown in the problem. The stop codon (STP) is underlined. Therefore the amino acids are

```
    LeuArgSerGluPheSerGlu.....TrpAlaIleMet
5' GGCTAAGATCTGAATTTTCCGAG...TTGGGCAATAATGTAGCGC 3'
3' CCGATTCTAGACTTAAAAGGCTC...AACCCGTTATTACATCGCG 5'
```

Now you need to add an *Eco*R1 site to the 5' end of the left primer and a *Sal*I site to the 5' end of the right primer – the restriction sites are underlined below. These sites cannot be directly at the ends of the DNA sequence, so you need 5 random nucleotides added to each of the primers. Furthermore, you must maintain the continuity of the ORF (open reading frame) between the MBP and the CFTR proteins after the vector and insert are digested and ligated. Therefore two more nucleotides (note the two G:C pairs, italicized) were added to the left primer between the restriction site and the beginning of the CFTR ORF. Also, the region between the vector and the insert cannot have any in-frame stop codons. The PCR product using these primers is:

```
                    LeuArgSerGluPheSerGlu  TrpAlaIleMet
5' CCCCCGAATTCGGGCTAAGATCTGAATTTTCCGAG...TTGGGCAATAATGTAGCGCGTCGACCCCCC 3'
3' GGGGGCTTAAGCCCGATTCTAGACTTAAAAGGCTC...AACCCGTTATTACATCGCGCAGCTGGGGGG 5'
```

Upon digestion of the PCR product with *Eco*R1 and *Sal*I, you will get:

```
                    LeuArgSerGluPheSerGlu....TrpAlaIleMet
5' AATTCGGGCTAAGATCTGAATTTTCCGAG...TTGGGCAATAATGTAGCGCG 3'
3'     GCCCGATTCTAGACTTAAAAGGCTC...AACCCGTTATTACATCGCGCAGCT 5'
```

Now you can ligate the vector and the PCR product yielding:

```
ArgIleSerGluPheGlyLeuArgSerGluPheSerGlu......TrpAlaIleMetSTP
5'...AGGATTTCAGAATTCGGGCTAAGATCTGAATTTTCCGAG...TTGGGCAATAATGTAGCGCGTCGACCTGTAGGGCAA...3'
3'...TCCTAAAGTCTTAAGCCCGATTCTAGACTTAAAAGGCTC...AACCCGTTATTACATCGCGCAGCTGGACATCCCGTT...5'
```

The Gly (italicised) is the result of the adjustment to the PCR primer to ensure that the N-terminal part of the CFTR region was in frame with MBP. So in summary, the two PCR primers needed are:

5' CCCCCGAATTCGGGCTAAGATCTGAATTTTC 3' and

3' ACCCGTTATTACATCGCGCAGCTGGGGGG 5'

Again, there are many possible answers that have minor variations, but you must still go through all of these steps to make sure your PCR primers will work properly

d. The fusion protein contains almost all of MBP, so it should also bind to the amylose resin. The cloning described in part b removes only the last 7 amino acids from MBP. **Make extracts of bacterial cells expressing the fusion protein and add these extracts to amylose resin**. The fusion protein should stick on the resin while all the other bacterial proteins in the extract should not. You can **wash the other bacterial proteins away** leaving the fusion protein bound to the resin. **To get the fusion protein off the resin you can add the sugar maltose**. Maltose and amylase will compete for binding sites on the fusion protein. If maltose is in excess then it will "disconnect" the fusion protein from the resin, leaving a solution with purified fusion protein.

Section 9.5 – DNA Sequence Analysis

9-27. In well studied organisms such as *C. elegans*, *D. melanogaster*, yeast and mice the entire DNA sequence of the genomes is now available. All you need to do **in order to study any region in these genomes** is to **design PCR primers based on the genomic sequence** that will amplify the region of interest. If necessary you can then determine the DNA sequence of the amplified region using automated methods. You might do this, for example, if you wanted to know if an individual's gene carried a mutation. These techniques require much less effort on the part of the investigator. Thus **having the genome sequence of an organism increases the importance of PCR.**

Restriction mapping is becoming a rarity even when studying unusual organisms - if you have cloned a gene from your organism you can sequence the DNA. Once you know the DNA sequence you can automatically find the location of the sites for all known restriction enzymes.

However you still need to use restriction enzymes to construct libraries and specific recombinant DNA molecules. **Restriction digestions remain the basis for many important applications of DNA cloning** and also for understanding in the next chapter how scientists were actually able to determine the DNA sequences of entire genomes.

9-28. Notice how many of these processes require the use of DNA polymerase, underlining why it is so important to learn how this enzyme works.

a.   **Enzyme-based; DNA ligase.**

b.   **Enzyme-based; restriction enzymes.**

c.   **Not necessarily enzyme-based.** You **could cut up DNA into smaller random fragments by mechanical shearing** (for example, passing the DNA through a very narrow needle) or **by chemical treatments.** There are also **DNases that cut DNA at random locations** that could be used for this purpose.

d.   **Non-enzymatic; hybridization relies on complementary base pairing.**

e.   **Enzyme-based; DNA polymerase.**

f.   **Enzyme-based; reverse transcriptase for the first strand of cDNA and DNA polymerase for the complementary strand.**

g.   **Enzyme-based; DNA polymerases from thermophilic bacteria.** *E. coli* DNA polymerase would not be very effective for PCR because at each cycle, heat is applied to denature the DNA, and this heat would inactivate the *E. coli* enzyme. This is not true of DNA polymerases from bacteria that live in high temperature conditions.

9-29.

a. The newly synthesized strand is read from the gel beginning with the smallest band which corresponds to the 5' end of this strand. This newly synthesized strand is complementary to the template strand. Reading the sequence from the gel:

newly synthesized strand: **5' TAGCTAGGCTAGCCCTTTATCG 3'**
template strand: **3' ATCGATCCGATCGGGAAATAGC 5'**

b. The sequencing template is the mRNA-like strand, **so the sequence of the mRNA is:**

5' CGAUAAAGGGCUAGCCUAGCUA 3'.

c. Any mRNA has 3 possible reading frames, which begin at the 5' end with the first nucleotide, the second nucleotide and the third nucleotide. **There are stop codons in each frame** (there are no open reading frames or ORFs) **so it is unlikely that this is an exon sequence of a coding region**.

9-30.

a. Synthesis occurs in the 5' to 3' direction, so the smallest fragment would contain the 5' T added to the primer and the next sized product would incorporate the C.

b. First write out the sequence of both strands and scan each strand for stop codons. **The newly synthesized strand has stop codons in all three frames (underlined) and therefore would not be the coding (exon) sequence. On the DNA sequencing template strand the reading frame that starts with the first nucleotide does not contain a stop codon and therefore is the ORF in this RNA-like strand.**

Synthesized strand: 5' TC<u>TAG</u>CC<u>TGA</u>AC<u>TAA</u>TGC 3'
DNA sequencing template: 3' A<u>GA</u>TCGGACTT<u>GAT</u>TACG 5'

c. The peptide sequence begins with the amino terminal end which corresponds to the 5' end of the mRNA-like DNA sequence (the DNA sequencing template) is **N Ala-Leu-Val-Gln-Ala-Arg**.

9-31.

a. In <u>Figure 9.14a</u>, you can see that the fragments of DNA get successively larger by adding nucleotides onto the 3'-end. DNA polymerase synthesizes growing strands in the 5'-to-3' direction. The trace shows a portion of a synthesized single stranded DNA. The green peak at the left end of the trace means that there is a fragment of DNA of a specific length (see part c) that was terminated when a dideoxy-A (ddA) was incorporated into the DNA strand being synthesized. **This terminal ddA, which is linked to a green fluorescent label, therefore becomes the 3' end of this molecule.**

b. **5'...ACCTATTTTACAGGAATT...3'**

c. **"Residue Position" indicates a peak at a specific location in the scan**. Most probably, nucleotide position 1 corresponds to the first nucleotide at the 5'-end of the newly synthesized fragments. You should note that all of the fragments will start at their 5'-end with the same short oligonucleotide primer, since DNA polymerase requires a primer. Thus, nucleotide position 1 is also the 5'-end of the primer used to generate the nested array of fragments. Therefore **the size of the single-stranded DNA fragment is represented by the residue position**.

d. There are two different peaks showing up at the same position. One is a T, the other is a G. **The double peak at position 370 is most likely caused by the fact that the original DNA actually had two different DNA sequences.** This pattern would be seen if the person whose DNA was amplified was actually a heterozygote with **one chromosome carrying a T-A base pair at this location while the homologue had a G-C base pair**. This is in fact the way that PCR amplification and DNA sequencing can be used together to look for heterozygosity anywhere in the genome. Of course this result could also be due to an error either in DNA sequencing or in PCR amplification.

Chapter 10 Genomes and Proteomes

Synopsis:

This chapter describes the tools and goals of genome analysis, including the construction of both large scale <u>long-range physical maps</u> and <u>sequence maps</u>.

The analysis of the human genome began with the formation of linkage maps are ultimately used to locate genes and explore their function. The physical maps are the molecular counterparts of linkage maps. The physical maps are used to chart the features of the chromosomes using various techniques to localize DNA sequences to specific locations. The DNA clones have been ordered into overlapping clusters that span the chromosomes. Linkage maps and physical maps can be roughly calibrated to each other; in humans 1 cM ~ 1Mb. A second approach does not rely on linkage mapping and genetic markers. Instead, cloned DNAs are used in fluorescent *in situ* hybridization (FISH) (**Figure 10.4**).

The highest resolution genomic map is a long-range DNA sequence maps compiled from the sequences of subclones which provides the complete sequence of the entire genome. The subclones may be derived either from previously mapped large insert clones (heirarchical shotgun approach, **Figure 10.5**) or directly from the genome (whole-genome shotgun approach, **Figure 10.6**).

The Human Genome Project has also led to paradigm changes in our approaches to biology and medicine. For the first time it is becoming possible to take a systems approach where all the elements of a system are studied.

This comparative genome work has lead to a variety of major insights of human and model organism genomes:

- there are surprisingly few genes in the human genome, only ~25,000;
- genes fall into 2 major classes - noncoding RNA genes and protein coding genes;
- the collection of proteins in a particular cell is a proteome and the collection of mRNAs is the transcriptome. These are both dynamic and change throughout the time spans of development and physiological responses;
- repeat sequences constitute >50% of the human genome, with some having evolved to become genes or control sequences;
- the genome contains different types of gene organization, including gene families, gene-rich regions and gene deserts;
- evolution can occur by the lateral transfer of genes from one organism to another;
- different human races have very few, if any, uniquely distinguishing genes;
- the sequences of microbes, plants and animals all employ the same genetic code and show a remarkable similarity;

♦ the Human Genome Project has catalyzed the development of high-throughput equipment for the studies of genomics and proteomics.

Significant Elements:

The problems in this chapter use many of the techniques that were introduced in chapter 9, including restriction mapping, cloning, probing, PCR and dideoxy (Sanger) sequencing (**Figure 10.15**). After reading the chapter and thinking about the concepts, you should be able to:

♦ Describe fluorescent *in situ* hybridization (FISH, **Figure 10.4**) which locates a cloned locus to a particular band on a particular chromosome.

♦ Explain microarrays (**Figures 10.16** and **10.17**).

Solutions to Problems:

Vocabulary

10-1. a. **6**; b. **3**; c. **2**; d. **5**; e. **1**; f. **4**.

Section 10.1 – Large Scale Genome Mapping and Analysis

10-2. BAC clones have 200 kb inserts x 15,000 BAC clones = 3,000,000 kb = 3 Gb human genome. However this does not allow for any overlap at the ends of the BAC clones. Such overlaps are necessary to properly align the contigs. **Thus 15,000 BAC clones are not enough to construct a complete physical map of the human genome.**

10-3. The four cosmids are aligned below. Notice that the third cosmid had to be flipped over for the alignment.

10-4. The two cosmids probably contain DNA sequences that are present as repeat DNAs on many different chromosomes including the one from which the inserts in all the other cosmids came from. Some section of the probe DNA from the BAC contains this repeated sequence and therefore cosmids from other parts of the genome hybridized to it.

10-5.

a.   The monkey DNA of all the BACs in the genomic library is joined to the vector at exactly the same place in the vector, and you know (or can easily determine) the sequence of the BAC vector itself. Thus, **you could generate two sequencing primers that would hybridize in all of the clones to the BAC vector just on either side of the inserted monkey DNA**. If the 5' to 3' orientation of these primers is correct, the DNA sequence obtained would be of the monkey DNA. **The same two primers would enable you to find thousands of different monkey STSs.** In the figure below the oval shows the entire BAC clone (the vector DNA is the heavy line and the monkey DNA is the lighter line). The arrows represent the PCR primers with the 3' end of each primer at the arrowhead.

b.   We don't know the extent of any of the monkey sequences in the BAC clones other than a general knowledge of the amount of insert DNA that can be cloned into the BAC vector (see part d., next page). We can only determine the order of the BAC clones and the STSs. One way to begin assembling the map is to **align the BACs that share sequences in common**. The data shows that BAC A contains the STS sequences 4, 5, 7 and 10. BAC A must therefore overlap the ends of BAC B (sequence 4), BAC C (sequence 5) BAC D (sequence 7) and BAC E (sequence 10) as shown below. The map would be equivalent if flipped horizontally with STS 3 at the left end rather than the right end.

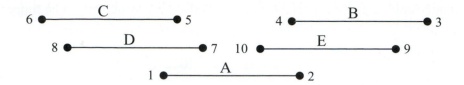

c.   The **minimal tiling path** (the fewest BAC clones needed to cover the entire region) **would be BAC clones C, B, and A.** Clones D and E have redundant sequences that are also found in C, B, or A.

d.   BAC clones can accept DNA inserts between 100 and 500 kb long. The minimal tiling path consists of 3 BAC clones, so **the contig could range from somewhat less than 300 kb long** (because of the overlaps remaining in the minimal tiling path) **to somewhat less than 1500 kb long.**

10-6. There are **two non-overlapping contigs.** The double-stranded sequence of the two contigs are:

5' CAAATAGCAGCAAATTACAGCAATATGAAGAGATCATACAGTCCACTGAA 3'
3' GTTTATCGTCGTTTAATGTCGTTATACTTCTCTAGTATGTCAGGTGACTT 5'
This contig is composed of the first and fifth sequences. The second contig is:

5' GTAGTATCTCCTTTTAAAAATCTCATTTCCTTTAGGGCATTTTCAAATTC 3'
3' CATCATAGAGGAAAATTTTTAGAGTAAAGGAAATCCCGTAAAAGTTTAAG 5'
This contig is composed of the fourth and sixth sequences. The second and third sequences do not overlap with each other or with any of the other sequences.

10-7.

   Recall that you can read approximately 500-800 nucleotides from a single sequencing reaction. If the repetitive sequence of DNA is longer than this size, then all of the data obtained from the reaction will be derived from the repeated sequence. If the repetitive sequence is repeated many times in the genome, you might not be able to tell from where in the genome the copy you have sequenced is derived. Thus, repetitive sequences longer than 500 bp (like transposons) will constitute a problem, as will long tandem arrays of simple DNA like $(AT)_n$. The array type of repetitive sequences pose an additional problem: such sequences are very difficult to clone at all, and the clones that are obtained often contain deletions or other rearrangements.

   Repeats that are dispersed around the genome are not a problem when **using the hierarchical shotgun strategy. This strategy analyzes individual BACs that might contain only a single copy of the repetitive sequence, so shotgun sequencing of an entire BAC could proceed without complications.** The repetitive sequence would simply be a part of the long region contained in the BAC.

The sequencing of such repetitive sequences is more complicated when using the whole genome shotgun approach. Some information can be derived if a particular sequence is partially composed of a repetitive sequence and partially composed of adjacent "single-copy" DNA found only once in the genome. However, more information is needed. The Celera Corporation's **whole genome shotgun strategy involves generating clones with known insert sizes of about 2 kb, 10 kb, and 200 kb. They then sequence the two ends of each clone and retain the information that these two ends are related to each other (that is, one of the sequences is say ~200 kb away from the other sequence)**. Thus, even if one end is repetitive DNA, it can be connected to a single location because of its relationship with the other end.

Both types of sequencing strategies encounter major difficulties in dealing with tandemly repeated sequences. If these stretches are very long then it becomes hard to tell where within the stretch a particular copy of the sequence is located. Fortunately, in most cases the repeats are not exact: for example, in a stretch of $(AT)_n$, a C or G may occasionally appear to provide a guidepost indicating a particular location. But even so, such regions of the genome are refractory to sequencing, particularly given the problems in cloning them in the first place. **Regions around centromeres, for example, may have more than 1 Mb of tandemly repeated simple sequences.** As a result, when scientists claim that a genome project for a particular species is complete, they are not necessarily being completely accurate, because such regions may be excluded from the data. In general, this is not a problem, since regions with simple repeats don't usually contain any genes, but it is a source of frustration for scientists interested in studying centromeres.

10-8. The region may have many **genome-wide repeat sequences**; or the region may be **duplicated at other sites** in the genome. Other **regions of the genome are unclonable** (e.g., centromeric and telomeric sequences) and therefore it is not possible to get the information to design PCR primers. **Long stretches of repeated DNA** cannot be PCR amplified - the primers must be unique sequences (e.g., AT_n can span hundreds of kb of DNA).

10-9. The whole genome shotgun sequencing strategy involves randomly shearing the entire genomic DNA to make a plasmid library with ~2 kb inserts, a plasmid library with ~10 kb inserts and a BAC library with ~200 kb inserts. Sequence the 2 kb and the 10 kb inserts for a 6-fold and 3 fold coverage, then sequence the ends of the BAC clones for 1-fold coverage. **The limitations: some genomic sequences cannot be cloned (e.g., heterochromatin) and some sequences rearrange or delete when cloned (e.g., some tandemly arrayed repeats).**

10-10.

a. The haploid human genome is 3 x 10^9 bp long. There are 23 chromosomes, so the average human chromosome contains about 1.5 x 10^8 bp, or 150 Mb. In Figure 10.4b, the FISH signal occupies (very roughly) about 5% of the length of the chromosome, or about 150 x 0.5 = 7.5 Mb. Therefore **two sequences would have to be approximately 7.5 Mb apart** in order to resolve adjacent independent green and red spots with no overlap. The resolution is actually a bit better in the sense that if the two spots partially overlap you might still be able to see that they don't totally overlap. For example you might be able to tell that the red signal is a bit closer to the telomere than the green signal.

b. Now that the human genome is sequenced **you submit a short stretch of the DNA sequence you want to localize to a computer, which will search the genome to find the location of the sequence. FISH is still useful to help characterize the genomes of organisms whose genome projects are not yet completed. In addition, FISH can still be useful in analyzing the genomes of individual people**. For example, as we will see in Figure 13.10, a probe made to a part of human chromosome 22 will light up only one chromosome in the cells of a DiGeorge syndrome patient, and two chromosomes in a normal individual.

10-11. The FISH protocol for locating a gene on a chromosome **involves a single hybridization** with your labeled gene sequence **to a chromosome spread**. This is especially useful in organisms where the large numbers of matings required for linkage analysis are expensive (e.g., mouse) or not feasible (e.g., humans) and the cytology is good (chromosomes are easily distinguished from one another). **FISH gives results quickly and can be used with any cloned piece of DNA**, while polymorphic alleles are required for linkage analysis. One disadvantage of FISH is the low resolving power.

10-12. Each pair of PCR primers detects a different locus (e.g. the alpha primer pairs detect a 100 bp piece of DNA). Each locus can have 2 alleles – if both primers of the pair anneal then there will be a band (+ band) on the PCR gel OR if one of the primers does NOT anneal because of a single base mismatch with the DNA then there is no band (- band). If the man is homozygous for one allele detected by a given primer pair then all of his sperm would have the same pattern.

a. Because the 200 bp fragment is never amplified, the man may be **homozygous for a mismatched sequence to which the Beta primers anneal**, preventing the fragment from being amplified. (He could also have 1 mismatch on one homolog and a different mismatch on the other homolog.)

b. He is heterozygous for the sequences to which the alpha (a) and delta (d) primers bind because the sequence is amplified in about half of the sperm. If the sequences to which the alpha and delta primers anneal are unlinked the following sperm will be 1/4 a^+d^+ : 1/4 a^+d^- : 1/4 a^-d^+ : 1/4 a^-d^-. The data shows there are 10 a^+d^+ sperm, 11 a^-d^- sperm, 2 a^+d^- sperm and 2 a^-d^+ sperm, showing that the loci are linked. The first 2 classes are the Parentals, so the man's genotype is a^+d^+ / a^-d^-. The last 2 classes are the recombinants, so rf = 4/25 = 16 cM. Thus **the regions to which the alpha and delta primer pairs anneal are 16 cM apart**.

Section 10.2 – Insights from Genomes

10-13. The **human genome contains a large amount of repeated DNA** and intervening sequences **(introns) can be quite large**. The **human genome also has many duplicated genes** (gene families). These factors make the total DNA to protein-coding DNA ratio higher than that in bacteria.

10-14. cDNA clones are made by synthesizing a DNA copy of a processed mRNA. The two different cDNA clones may be due to **alternate splicing** of the same primary RNA (**Figure 10.14**). The same primary transcript splices out different combinations of introns and exons in the middle of the primary transcript. It is also possible that the transcripts are due to different genes. A **gene duplication** originally gave rise to two identical copies of a gene. The center portion of one of the copies was subsequently replaced by a mobile domain in the process known as **domain accretion** (**Figure 10.9**). In either case the center portions of the mRNAs have different protein domains.

10-15. Because cDNAs are made from mRNAs, they **lack regulatory region information and intron DNA**.

10-16. The 4.1 kb mRNA was processed from a primary transcript that extends through all the DNA between A and G bands. The primary transcript could include part of the DNA in bands A and G or the primary transcript could extend further to the left and/or right, off the DNA map shown. **Bands A, D, F and G must contain exon sequences of this 4.1 kb mRNA. There are a minimum of 2 introns - one that includes bands B + C and a second that includes band E.** The 3.4 kb mRNA was produced by a primary transcript that extended from somewhere in fragment B to somewhere in fragment E. **This 3.4 kb gene has a minimum of two exons and one intron in the region covered by fragment D and probably extending into fragments C and E.** The 1.8 kb transcript was produced by a primary transcript that extends from D into G. Because **the mRNA is only 1.8 kb but the total fragment length in this region is 3.5, there must be at least 1.7 kb of intron in this region. This could be 1 intron in fragment E. If not this then there must be a series of small introns because even the small DNA fragments like G have exon sequences.**

10-17.

a. Linkage distances, expressed in centimorgans, are based on recombination frequencies. The difference in genetic distances between the sexes can be explained by **a difference in the amount of recombination in males and females**.

b   **No**, the physical distances between the 2 sexes are the same. **The same chromosomes are passed from generation to generation** (e.g., grandmother to father to daughter).

10-18. One method is to search for transcribed regions in the cloned segment using the DNA as a probe versus RNAs from the tissues **(Northern blot)**. A second method is to use **computational similarity analysis to search the DNA sequence for open reading frames, intron/exon boundaries and codon usage appropriate for the species**. A third method is to **compare complete genome sequences from a related mammalian species such as mouse**. Non-functional regions of the genomes have significantly diverged while functional regions are relatively conserved. Conserved regions of >25 bp are automatic candidates for coding sequences. Look for **homology to STS sequences**.

10-19. (i) The repeat sequences found at **centromeres and telomeres** are thought to be important for the proper function of these chromosome structures. (ii) **Transposon-derived repeats have given rise to at least 47 different human genes**. (iii) **Certain genes have repeated sequences** which lead to mutation when these regions are amplified (e.g., Huntington's disease). (iv) **Transposon-derived repeats have reshaped the genome by aiding the formation of chromosomal rearrangements** like inversions and translocations.

10-20. If pseudogenes arose from cDNAs they would **lack the introns** found in the genomic copy of the functional gene. Pseudogenes would also have **a polyAT tract** at one end of the gene, corresponding to the polyA tail of the original mRNA.

10-21. See <u>Table 10.2</u>. All of these organisms are amenable to genetic analysis and they are already extensively studied and mapped. They are presented in order of increasingly complex genomes.

a. The **genome of yeast is very small** because there is very little repetitive DNA compared with other eukaryotes and therefore the cloning of genes is easier. Any gene involved in basic cell functions would be a good gene to clone in yeast. **Yeast currently has a catalog tryptic peptides related to the genes which code for the as well as a catalog of >7,000 known protein-protein interactions and >500 protein-DNA interactions**.

b. The **nematode is a fairly complex multi-cellular organism** with a genome of ~20,000 genes with a well understood series of developmental steps. *C. elegans* is small relative to humans but worms have several functions that are similar to, yet more primitive than humans, such as a nervous system. Some of the basic functions of nerve cells can be studied this model system.

c. The **mouse is the closest model organism to humans** and genes that are homologous to disease genes in humans can be cloned and analyzed. Mouse and humans diverged from a common mammalian ancestor about 85 million years ago. The mouse genome still shows striking similarities in genes and their orthologous order.

10-22.

a. Once the new gene is sequenced, **the predicted amino acid sequence of the protein product can be examined for protein domains or functional units** (**Figure 10.9**). Tell-tale features of protein sequences are motifs that have been shown to have specific functions in other proteins such as transcription factors. **Using comparisons between sequenced genomes of different species you can find orthologous genes.** You can also **look for paralogous genes** (other members of a gene family) within a species. If a function is known for one member of the gene family it is possible to get an idea of the function of another member.

b. **Northern blot analysis to discover in which tissues the gene is transcribed** provides useful clues about function. **Knocking out a gene and looking for a phenotype** caused by the knock-out is another type of analysis that might lead to information on the gene.

10-23.

a. The zinc-finger motif is a DNA sequence that encodes the amino acids capable of forming the zinc finger in a protein. Such domains allow these proteins to bind to specific DNA sequences, suggestions **suggesting that the protein is a transcription factor**.

b. Similarity with a previously identified gene suggests that **the two genes arose by duplication** of a common ancestor and could be part of a gene family.

10-24.

a. The best consensus would be:

> **5' XXATATAAAAXXXXXX 3'**
> **G   T**

Eight of the positions in the group of four sequences can have any nucleotide. This is denoted with an X. Position 3 can have an A or a G, position 8 can have an A or a T. The remaining nucleotides are invariant – they always have T or A as shown.

b. A consensus sequence **shows the nucleotides that are likely to be the most important to gene or chromosome function**. These nucleotides have been conserved during the evolutionary time since the four organisms last shared a common evolutionary ancestor. Because you can compare any short stretches of random sequences and just by chance see what appear to be conserved nucleotides you must first determine that these sequences might have some functional relationship. **You must choose DNA sequences that are in the same evolutionarily conserved orthologous gene and in the same relative position (like the same exon) in the four species.**

c. If you **compare the amino acids of the proteins encoded by homologous genes, you might be able to find particular conserved amino acids that you can use to define a consensus.** Even if the identical amino acid is not conserved, it is possible that the amino acid might be replaced by another amino acid with similar chemical properties in other species. For example, replacing the acidic amino acid aspartic acid with glutamic acid would be defined as a conservative substitution. This sort of comparison can only be done properly with amino acid sequences because the genetic code is degenerate.

10-25. One reason for the lack of accuracy is that **the definition of 'gene' is somewhat imprecise**. The classical "one gene, one polypeptide" definition is clearly obsolete, since we know that (1) alternative splicing and post-translational modifications like protein cleavage mean that **one gene can actually produce more than one type of protein, although the polypeptides produced by one gene are usually related to each other**, and (2) **many genes are transcribed into RNAs that remain untranslated yet still have important functions** (like rRNAs, to name just one example). At the present time **the most all-encompassing definition of 'gene' is a region of the chromosome that is transcribed into a discrete primary transcript that can subsequently be spliced or otherwise processed.**

A second difficulty with defining the number of genes is technological. With the above definition, scientists can **look for transcribed regions of the genome with techniques such as Northern blots or the sequencing of hundreds of thousands of clones from cDNA libraries. However, if a transcript is found only in very low abundance or if it is very small, these**

techniques do not ensure that it can be detected. In addition, computerized methods for looking for genes are usually pretty accurate for large protein-coding genes that have easily-recognizable features like long open-reading frames. However, these **computer programs are much less successful in identifying genes with very short open-reading frames or the genes that are transcribed into very small non-coding RNAs.**

10-26.

a. **All of these genes, in both humans and chimpanzees would be considered homologous since they would all show some degree of relationship.** However, the degree of homology would differ for different pairs of genes (see part c. below). **The α_1, α_2, β, Gγ, Aγ, δ, ε and ζ genes in humans would comprise one set of paralogous genes**; the α_1, α_2, β, Gγ, Aγ, δ, ε and ζ genes in chimpanzees would also comprise a set of paralogous genes. **The α_1 gene in humans would be orthologous to the α_1 gene in chimpanzees**, the α_2 gene in humans would be orthologous to the α_2 gene in chimpanzees, etc.

b. **Orthologous genes are more likely to have closely related functions, since they have been more recently derived from a common ancestor than paralogous genes.**

c. The **orthologous genes (the human β gene and the chimpanzee β gene) should have a greater degree of nucleotide homology, since they were more recently derived from a common ancestor** than the paralogous human α and β genes.

d. **Most of the gene duplication events depicted in <u>Figure 9.19</u> that gave rise to the different kinds of hemoglobin genes (α_1, α_2, β, Gγ, Aγ, δ, ε and ζ) must have occurred before the human and chimpanzee lineages diverged.** That is, the last common ancestor of these species must have already had all of these genes. In fact, most of these genes must have been present in an ancient species ancestral to all mammals since most of these genes are found in all present-day mammals.

10-27. A cladogram of species (a 'tree of life') is constructed so that the more closely two species are related, the less space (that is, the less time) and the fewer branch points would be found going backwards to the point that they last shared a common ancestor.

a. If genes are **vertically inherited, a cladogram that compared the DNA sequences of the gene should essentially match the accepted species cladogram.** For example, the DNA sequence of a human gene should be closer to the DNA sequence of the corresponding gene in chimpanzees than to that of the corresponding genes in mice. In almost all cases, the sequence of human genes and those in bacteria should be highly dissimilar. However, **if lateral gene transfer occurred**

then a cladogram of the protein sequences would unexpectedly show that the human gene was much closer in DNA sequence to the bacterial gene than to genes in more closely related species like primates or mammals.

b. Imagine that a particular DNA sequence in some closely related species was found to be more similar to a very distant species than to intermediate species. For example, maybe **a gene of bacterial origin was found in humans and chimps, but not in any other primates or mammals. This suggests that the lateral transfer occurred before humans and chimps last shared a common ancestor (estimated to be about 5 million years ago), but before either of these two species last shared a common ancestor with any other current primate species (about 7 million years ago)**.

c. If **a large number of related species had a particular gene but one species in this group** (or a small very highly related subgroup) **did not** have this gene then it is likely that the gene was lost from the genome.

It is sometimes difficult to distinguish between lateral transfer and gene loss. For example, if a gene were found in humans and bacteria but not in yeast and *Drosophila* lateral transfer might have been involve. It is also possible that the gene was lost from the evolutionary lineages for yeast and fruit flies. Such determinations are easier he more genomes that are sequenced: the explanation consistent with the least number of independent events during evolution is probably the correct one. For example, if you could conclude that the gene could have been laterally transferred only once during evolution but that gene loss would have had to have happened many times to explain the species without the gene, lateral transfer would be a more likely hypothesis.

10-28. The two mRNAs present in all tissues are different somehow from the third class found only in the brain. The proteins made by the first two mRNAs could be antigenically different enough that the antibody for the brain protein does not recognize them. **The first two mRNAs could be alternately spliced messages or they could be the transcripts of paralogous, diverged genes. Another possibility is that the two other messages are not translated in any tissues.**

10-29. Not all human genes have been accurately identified: small genes are hard to identify (e.g. small functional RNAs of 25 bp or less); genes expressed only at the RNA level are difficult to identify (tRNAs, rRNAs, snoRNAs and snRNAs); there could be hundred or thousands of genes encoding small peptides that are hard to find. **Genes that are rarely expressed or that have unusual codon usage patterns are difficult to find**. Another factor is that **the human proteome is much more complex** than that of the nematode, and this comes about due to shuffling of functional modules - the human genome evolved many new arrangements that alter the domain architecture. Combinatorial amplification at the DNA level, as seen with the antibody gene segments, and at the RNA level, seen with alternate splicing, also increase the level of complexity of the proteome. **The human genome has more paralogs and chemical modifications of proteins**. Humans have >400 different chemical reactions that affect protein structure, so a human cell can have 20,000 different types of mRNAs and 200,000 different proteins.

Section 10.3 – Global Analysis of Genes and Their mRNAs

10-30.

a.   The gene size is often large due to multiple introns which are sometimes large in size. **You must use cDNAs** as the basis of gene therapy - even for a 100 kb gene the mRNA may only be 1% of that size.

b.   The expression in specific tissues requires **the correct regulatory DNA** so that the therapeutic gene is expressed in the correct tissue. It may also be possible to **target the therapy directly to the correct tissue**.

10-31.

a.   **Black – this gene is either not** transcribed or is transcribed at very low, undetectable levels in both the normal tissue and in the cancerous tissue. **Green – the mRNA for this gene accumulates to higher levels in normal tissue than in the tissue from the tumor** so it is likely that the transcription of this gene is turned down in the cancerous cells. **Red - the mRNA for this gene accumulates to higher levels in cancerous tissue than in normal tissue** so it is probable that the transcription of this gene is turned up in the cancerous cells. **Yellow – the mRNA for these genes accumulates to the same level in both kinds of tissue**.

b.   You would most likely choose **a gene that gives a red signal on the microarray**. If the level of the mRNA goes up in the tumor relative to normal cells the over-expression of the protein product of this gene may contribute to cancer. A drug that inactivates such a protein might be able to stop the cancer from proliferating.

10-32.

a.   The **person** whose PCR products were labeled with rhodamine (red) is **a heterozygote with the M1 sequence on one homologue and the M2 sequence on the other homologue**. The **other** person, whose DNA is labeled with fluorescein, **is heterozygous for M3 and M6**. Remember that the DNA on the chromosomes is double-stranded, not single-stranded as shown for the oligonucleotide sequences.

b. In this particular case, the DNA sequence of the PCR products from both individuals would appear identical. **The sequencing would show that both people have the sequence: 5' ACT(TG)ACCGAGAGA(GA)CCTGCG 3'.** The two nucleotides within a set of parentheses would show up in the same position in the sequencing reactions. When you consider the left set of parentheses **you can tell that both people are heterozygous at this position with the sequence 5' ACT(T)ACC… on one homologue and 5' ACT(G)ACC… on the other homologue. Likewise you would know that both people were heterozygous for G and A at the second position.** This particular case is unusual; the reason these sequences are ambiguous as determined from DNA sequencing of the PCR products is that this short region of the genome has two polymorphisms that are independent of each other. The only way to find the precise genotype of both homologues of both people (for example that the 5' ACT(T)ACC polymorphism is part of the same allele as the 5' AGA(A)CCT polymorphism) is to do the microarray experiment described in part a. above.

c. The results shown in part a demonstrate that **this method of oligonucleotide microarrays can be used to genotype single nucleotide polymorphisms.** Since you can put many oligonucleotides on a single microarray theoretically this technique can be used to simultaneously genotype anyone for a large number of the most important disease-causing polymorphisms,. It is not clear that the method will "scale up" so easily, since you would have to simultaneously PCR amplify many genomic regions (from all the disease genes of interest), and this might contribute to inefficiencies in the PCR or high backgrounds upon hybridization.

Chapter 11   Genome Wide Variation and Trait Analysis

Synopsis:

This chapter deals with a familiar genetic theme: variation. The variation described in this chapter is not the easily visible phenotypes discussed in early chapters, but molecular variation – changes in DNA sequence that can be detected by several different molecular techniques. When DNA is examined, the amount of variation (changes in DNA sequence) between individuals and among individuals in a population is much greater than you may have realized. Differences in DNA sequence seen in a population are called polymorphisms. These can be single base changes (SNPs = single nucleotide polymorphisms) or changes in the number of copies of a small repeated sequence (mini or micro satellites). If the single base variation changes a restriction enzyme recognition sequence (RFLP = restriction fragment length polymorphism), the polymorphism can be identified by restriction enzyme digestion of genomic DNA and hybridization to visualize only that region of DNA. Increases or decreases in repeated DNA can be recognized by changes in the size of restriction fragments or of PCR amplified regions. Changes in base sequence that do not affect a restriction enzyme recognition site or the size of a restriction enzyme fragment can be recognized using allele specific oligonucleotides and hybridization.

Much of the DNA variation seen between individuals is silent. That is, it has no effect on phenotype. A molecular genotype detailing the form (alleles) of DNA markers present in an individual can be used to distinguish between individuals and to map disease genes. The concept of linkage that you learned in chapter 4 is very critical again here. Genes and molecular markers that segregate (are transmitted) together greater than 50% of the time are linked. They are physically close on a chromosome and will only be separated if recombination occurs between them. Polymorphic DNA markers can be used to follow the transmission of a disease locus if the marker is linked to the disease gene.

Positional cloning begins with a phenotype and works to identify the gene(s) responsible for the phenotype. This approach is similar to the classical approach you studied in earlier chapters except in earlier analyses a mutation defined a gene as a factor in determining a phenotype, but the actual function of the gene product was harder to understand. In genome analysis, the goal is to get to the sequence of a gene through mapping and use that information to discover how the gene product works. In the second approach you begin with a gene sequence and try to establish what phenotype the gene causes and therefore the role the gene plays. This bottom-up approach has increased in importance as more and more sequence data has been obtained for several organisms.

The characteristics and phenotypes geneticists are studying now are often far more complex than the traits that were studied in early experiments on peas and humans. Genome analysis has provides a wealth of data that allow us to ask more complex questions. Incomplete penetrance, phenocopies and polygenic inheritance all provide challenges in establishing the relationship between gene(s) and phenotype.

Significant Elements:

After reading the chapter and thinking about the concepts, you should be able to:

♦   Recognize, describe and explain the effects of the four different types of DNA polymorphisms: SNPs, InDels (deletions/duplications/insertions into non-repeated loci), SSRs (simple sequence repeats, microsatellites and minisatellites), CNV/CNP (copy number variation/polymorphism) (**Table 11.1, Figures 11.3, 11.8, 11.9, 11.11** and **11.14**).

♦   Explain the techniques used to detect the different polymorphisms (**Figures 11.4, 11.5, 11.6, 11.8, 11.10, 11.12, 11.14** and **11.15**).

♦   Distinguish different forms of a DNA marker and assign heterozygote and homozygote genotypes.

♦   Determine if a particular marker is informative for mapping a disease gene - is it linked and is it polymorphic in the family you are studying?

♦   Follow inheritance of a disease allele of a gene using a linked molecular marker.

♦   Determine the likelihood that a particular individual carries a disease allele based on the molecular marker data and the degree of linkage between the marker and the disease gene.

♦   Describe the steps in positional cloning from phenotype to gene clone (**Figure 11.17**).

♦   Describe steps for understanding the function of a gene (the phenotype affected) starting with a cloned gene.

♦   Interpret data for locating genes in a DNA sequence (**Figure 11.17**).

♦   Know how to analyze genes involved in a complex quantitative trait.

♦   Define haplotype.

Problem Solving Tips:

♦   Remember that most DNA markers are <u>not</u> in a disease gene (these markers are anonymous markers). Instead the markers are DNA polymorphisms that are near a disease gene, are therefore linked, and are used to track the disease gene.

♦   RFLP alleles are also referred to as forms of a polymorphism (**Figure 11.4**).

♦   In any problem, first establish the different forms (alleles) that are present in the population being considered.

♦   A form (allele) of a DNA marker and the allele of a linked gene are transmitted together in a family unless there is recombination.

♦   A different allele of a marker can be associated with the same disease locus in two different families.

♦   Genomic libraries include all DNA in the genome; cDNA libraries include only expressed (transcribed) genes and lack intron sequences.

♦   Genetic maps are based on recombination frequencies.

♦   The problems in this chapter are more representative of what is actually done in a molecular biology laboratory. Put yourself in the shoes of a researcher; be inquisitive and THINK!

Solutions to Problems:

Vocabulary

11-1. a. **5**; b. **3**; c. **8**; d. **6**; e. **2**; f. **7**; g. **1**; h. **4**.

Section 11.1 – Genetic Variation

11-2.

Anonymous DNA markers can be used to help map human genes to a specific region of a specific chromosome, as shown in **Figure 11.17 a-b** and **Figure 11.18**. In order to map genes you must correlate the co-inheritance of an easily tracked DNA marker and the disease gene. The more such DNA markers there are the better the chance of finding one that is genetically linked to the gene of interest. The **advantages** of anonymous DNA markers include: (i) they are **abundant and easy to find**. SNPs (single nucleotide polymorphisms = anonymous markers) occur about once in every 1000 base pairs in the human genome so any region of the genomes of two individuals should contain such polymorphisms. (ii) You **do not have to rely on the special circumstance of finding a family with multiple afflicted individuals**. The **disadvantage of such markers is that they only provide signposts along the genome**. The linked SNP is only <u>near</u> the gene of interest, not <u>in</u> the gene.

11-3. When examining **anonymous DNA markers you are looking directly at the DNA sequence of an individual. The terms dominant and recessive can only be used when discussing the phenotype of an organism, so in one sense this question is meaningless**. Also, the majority of these anonymous DNA loci are not located in genes and so have no effect on the phenotype of the organism (problem 11-4). However, **geneticists often say that DNA markers are inherited in a codominant fashion to denote that the both alleles can be seen in the DNA sequence** and that the genotype of a heterozygote depends equally on both alleles.

11-4. SNPs in protein coding regions are more likely to have a deleterious effect on the organism as they are more likely to affect protein function, so they **are expected to be relatively rare**. Also **most SNPs are found in non-coding DNA simply because non-coding DNA in humans comprises a much greater proportion of the genome (98%)** than coding DNA.

11-5.

a. **Microsatellites are simple 1-3 bp sequences repeated in tandem 15-100** times. The **polymorphisms are different numbers of the simple sequence repeats**.

b. See <u>**Figure 11.9**</u> Changes in the number of repeats are caused by **slippage of DNA polymerase** during replication.

c. Minisatellites are 20-100 bp sequences that are repeated up to thousands of times/locus. Minisatellite polymorphisms occur by **unequal crossing-over if sequences on homologs align out of register during mitosis or meiosis**. The repeat sequence in a minisatellite is too long to cause slippage of DNA polymerase during replication.

11-6. The nucleotide sequences of polymorphic forms of the same gene should be 99.9% identical. In other words, the differences would be relatively rare, on the order of 1 difference in every 1000 bases. **Paralogous genes represent a gene duplication event that occurred before the species arose in evolutionary time**, so there has been substantial time for the two paralogous genes to diverge in sequence. Thus **the nucleotide sequence differences between paralogous genes in the same species would be more frequent, perhaps 1 in every 50 bases**. Furthermore f you are examining these DNAs in a species whose genome has been completely sequenced the answer will be immediately obvious. If the genes are paralogous, the genome sequence should contain the two different genes.

<u>Section 11.2 – Detecting Single Nucleotide and Small-Scale-Length Polymorphisms</u>
11-7.

a. **The SNP polymorphism must be within the sequence to which the oligonucleotide hybridizes** so the mismatch destabilizes the annealing of the oligonucleotide and the sample.

b. **The SNP polymorphism must be at the 3' end of the primer** so you can determine if the extension reaction is able to increases the length of the primer by the addition of the single nucleotide in the reaction mix.

c. In the case of the RFLP probe **the polymorphism must occur in a restriction site** (this is not necessary for parts a and b) and the probe simply has to be homologous to some part of the restriction fragment bounded by the polymorphic restriction site. The probe can actually be kilobases away from the actual polymorphism.

11-8. Amplification of the PAX-3 gene by PCR in a normal individual (homozygous for the wild-type allele) will give a single band. Most individuals with deafness due to the 18 bp deletion in the PAX-3 gene will be heterozygous for the mutant allele. Thus, **amplification of this gene in affected individuals will give two bands - a normal size band and a smaller band corresponding to the deletion allele.** The mutation in the β-globin gene that causes sickle cell anemia is a single nucleotide change (SNP). Thus **there is no size difference between the wild type and mutant alleles of the β-globin gene.**

11-9. For ASO analysis, **the potential mismatch should be in then center of the oligonucleotide.** Therefore a 19mer that extends 9 bp in either direction would be a good probe. You need two different oligonucleotides for the analysis - one that contains a sequence corresponding to the mutant and the other corresponding to the wild type sequence. The oligos would be complimentary to:

        5' CTATAAATGCGCTAGGCGT

and   5' CTATAAATGGGCTAGGCGT

11-10.

a. **The incubation temperature affects the accuracy of annealing between complimentary sequences.** At 100°C, no hybrids will form because the DNA remains denatured. At 80°C, the conditions are too permissive so that even oligonucleotide probes with a single base mismatch will hybridize. 90°C is a good temperature for differential hybridization using this allele specific oligonucleotide (ASO). Differentiation is made between DNAs having a complete match and those that have a single base difference.

b. **Individuals 1, 5, 6, 8 are homozygous for the A allele; individuals 2, 7, 9, 10 are AS heterozygotes; individuals 3, 4 are S homozygotes.**

11-11. See **Figure 11.5**. **Eggs are collected** from Angela. The eggs are **fertilized *in vitro*** with George's sperm. The resultant embryos are allowed to mitotically divide to the 8 cell stage. Researchers t**ake one cell from each embryo. The region of the β-globin gene is amplified from each cell using PCR. The PCR samples are digested with *Mst*II and electrophoresed on a gel.** Two bands on a gel means the β-globin gene has an *Mst*II site = AA genotype; one larger band on the gel means no *Mst*II restriction site = SS genotype; three bands on the gel = AS heterozygote.

11-12. A child should inherit 1 allele of each locus from each parent. Thus if the 10 bands in the daughter's pattern represent 5 heterozygous loci, she should share five in common with one parent and the other five in common with the other parent. The daughter shares 6/11 bands with male 3 and only 1/10 bands with male 4. Therefore, **male 3 is more likely to be the father**. The child may not have exactly the same 5 alleles as one parent - remember that microsatellite alleles change in length due to unequal crossing-over during meiosis, so the allele inherited by the child may be different than either allele in the parent.

11-13. The pattern seen in the sample from individual 3 is very similar to the pattern of the sperm taken from the crime scene. Thus **it is possible that individual 3 is the perpetrator of the crime**. If this minisatellite probe hybridizes to 10 unlinked lock then there is $0.375^{10} = 0.005\% = 1/20,000$ chance that this pattern will be found in another person. If you are looking at 24 separate loci the probability of finding the same pattern in another person is 1 in 17 billion.

11-14. Allele-specific PCR will distinguish between the two alleles of the β-globin gene. **One of the primers is β-globin gene-specific and must be complementary to a downstream, invariant sequence in the β-globin gene (primer 1)**. This downstream primer will hybridize to and direct PCR-based DNA synthesis from all DNA samples. **Two upstream primers are designed that are identical except for the 3' most nucleotide**. If the 3' end of a PCR primer is not an exact complement to the template DNA strand the DNA extension reaction will not occur. **One of these upstream primers is designed to amplify the wild-type DNA sequence, so it has the sequence 5' CACCTGACTCCTGA 3' (primer 2). The alternate upstream primer is designed to amplify the sickle cell allele sequence (primer 3) - its sequence is 5' CACCTGACTCCTGT 3'**. DNA from homozygous wild-type individuals will be amplified by primer pairs 1 and 2 but not 1 and 3. DNA from heterozygous individuals will be amplified by both sets of primers. DNA from homozygous individuals will be amplified by primer pairs 1 and 3 but not 2 and 3. Allele-specific PCR is basically the same as using ASOs to detect genotypes.

11-15. Huntington disease is late-onset dominant lethal disease. The disease is associated with the expansion of a CAG trinucleotide repeat in the HD gene. Normal individuals have up to 34 copies of the triplet. Individuals with repeat regions containing more than 42 repeats are susceptible to Huntington disease. In general, greater numbers of repeats correlate with younger age of onset of the disease (**Figure 11.11**).

a. **Individuals A, B, C and E** all have one PCR band (one allele) that is much larger (between 270 and 380 nucleotides long) than the second allele (between 200 and 220 nucleotides long). It isn't possible to correlate these band sizes with trinucleotide repeat numbers because we don't know where the PCR primers anneal with respect to the repeats. However these much larger bands must have more copies of the repeat than the smaller bands. The repeat region in **individual B** is the longest; so this person is likely to have the earliest onset of the disease.

b.   Both PCR bands for **individuals D and F** are smaller, so they appear to have received HD^+ alleles from both of their parents. Thus they should not have the disease.

c.   If we assume that the longest HD^+ allele shown in the figure (the smaller band in individual B) contains the maximum number of repeats for a non-disease-causing allele (34 repeats), then the 220 bp of this PCR product should include 34 x 3 = 102 bp of CAG repeats, plus 70 bp between the 5'-end of one PCR primer to the nearest CAG repeat. Therefore, 220 - 172 = **48 bp** will remain from the end of the CAG repeats to the 5'-end of the second PCR primer.

11-16.

a.   In the patient whose graph is shown **on the left**, the CAG repeat number in the HD^+ allele is **approximately 15**; the somatic cells of the patient **on the right** has **about 20** CAG repeats in the HD^+ allele.

b.   These results say a great deal about the mechanisms that give rise to mutant HD alleles. First, the **repeat number varies among different sperm** from the same individual, so these **processes take place in germline cells during spermatogenesis**. Second, it appears that **the larger the original number of repeats in any HD allele, the more likely it is that the number of repeats in the sperm will vary and the greater the degree of potential variation**. In the graph on the left, almost all of the sperm with the mutant HD allele have more than 62 repeats, some sperm have more than 120 repeats and the distribution is quite broad. The distribution of mutant sperm sizes in the graph on the right is tighter. In addition, note the fact that there is little if any variation in sperm size of the sperm containing an HD^+ allele. Third, it is interesting that the **sperm mostly seem to accumulate more CAG repeats rather than lose CAG repeats** - few if any sperm have less than the number of CAG repeats in the HD genes in somatic cells.

c.   There is some probability that the number of CAG repeats can expand during spermatogenesis, even with lower numbers of CAG repeats. A man with **an HD allele on the high side of normal (~30 CAG repeats) or in the grey area between 35 and 41 repeats** (who may not show any symptoms), **may produce sperm with more than 42 repeats**. Note that this data predicts that the unaffected father (but not the unaffected mother) of a Huntington disease patient should have an allele with between 30 and 42 repeats, since this particular repeat expansion occurs during spermatogenesis. Interestingly, in Fragile X syndrome, another trinucleotide repeat disease, the

expansion occurs during meiosis in the mother (see the <u>Genetics and Society box on</u> **pp. 208 -
209 of Chapter 7**).

d. You would expect that **each of the two blood cell samples would show a very tight
    distribution of CAG repeat numbers around the numbers seen in the normal and mutant
    HD alleles** (15 and 62 for the patient at the left, 20 and 48 for the patient at the right). There
    should be very little variation in repeat number between different blood cells from the same
    person since the expansion in repeat number seems to be restricted to the germline.

11-17. The first step is to identify which of the parents RFLP alleles are linked to the disease gene.
Begin by figuring out which allele the affected child inherited from each parent.

a. In the left hand pedigree **the affected child inherited the 10 kb band from the father which
    must be linked to his mutant FM allele**. Therefore, even though both parents have a 13 kb
    allele, the mother's 13 kb RFLP is associated with her mutant allele of FM while the dad's 13 kb
    allele is associated with his normal FM allele.

b. In the right pedigree the **affected child received the 10 kb band from his mother** which must
    contain her FM mutant allele. This child also must have inherited the 10 kb allele from the father
    which must also have his mutant FM allele.

c. The male child in the left pedigree inherited his mother's 7 kb allele with the normal FM allele
    and his father's 13 kb allele with the normal FM allele. Therefore this child is homozygous
    normal. The female child in the right pedigree inherited the mother's 7 kb allele with the mutant
    FM allele. Because this child is phenotypically normal, she must have inherited her father's 10 kb
    allele linked to his normal FM allele, so she is a carrier. There is a **0% probability** that this
    couple will have a diseased child.

d. The male from the left pedigree is homozygous normal and the female from the right pedigree is a
    carrier, so there is a **1/2 chance that their child will be a carrier**.

11-18. Detecting all 800 mutant alleles of CFTR is not possible, at least with current techniques.
However, it <u>is</u> possible to screen for some of the most common alleles, including the three nucleotide
deletion described in the problem. In fact, in 2001 **a committee from the National Institutes of
Health recommended that the broad population be screened for the 25 most common mutant
alleles. This would allow the detection of about 85% of potential CFTR mutations**. This still
means that 15% of the people diagnosed as "normal" by this screen could still have a child with cystic
fibrosis.

Screening of a large population for cystic fibrosis would almost certainly be worthwhile, as the
disease is relatively common, very destructive to health, and very expensive to treat. **The costs of a**

screening program would be more than balanced by potential savings in medical bills. If resources are limiting, it would make sense to target the screening program to populations in which the disease alleles are most prevalent.

It is important to realize that **broad screening of the populations must be done to detect prospective parents who are carriers**. This is inexpensive and risk free. Although it is possible to conduct prenatal screening via amniocentesis or chorionic villous sampling, these are much more expensive and present small but real risks to the fetus. Without a DNA-based screening program, the only way that two parents would realize that both are carriers is if they had an affected child. If the parents were carriers, then prenatal or preimplantation diagnosis of the fetus or embryo could be conducted as described in the text (**Figure 11.1**).

Section 11.4 – Positional Cloning

11-19. For a marker to be informative in a particular family the individuals must be **heterozygous for a polymorphism**. This **polymorphism must be closely linked to the disease gene**. The two different alleles of the marker allow the inheritance of each homolog and by extension the disease allele to be traced from one generation to the next.

11-20.

a.   The **SNP that is completely linked to the disease locus in this family is part of their haplotype** of this region of the chromosome. The SNP is a DNA sequence change that happened on the same chromosome <1 cM from the mutation in the disease gene. It is never found apart from the disease gene because a recombination event that would separate them is very rare.

b.   **Check the sequence of the SNP in other, unrelated families** - both those with the same disease and those without the disease. If the G allele is the cause of the mutation in the disease gene then the same change could occur in some of the other families. If the G allele is part of the haplotype of the original family but not associated with the disease then there will be no correlation between presence and absence of the G allele and the presence or absence of the disease allele. Also do positional cloning and look for candidates for the disease gene in this area of the chromosome. **Once you have identified the candidate gene you can determine if the G allele is within the disease gene.** For example, does the G allele affect the predicted amino acid sequence of the candidate gene? Does it occur in or near a proposed splice junction?

11-21. The affected first child shows which marker allele is linked to the disease allele in each parent. The affected child got Dad's large allele and Mom's small allele.

a. The fetus has Dad's small allele (the normal CFTR allele) and Mom's small allele (the CFTR mutant allele). Thus **the fetus is a carrier, so there is 0% chance the fetus will exhibit the disease**. If Dad's gamete is the result of a recombination event between the microsatellite marker and the CFTR mutation then the fetus could have inherited Dad's small microsatellite allele and the mutant CFTR allele. The probability of this happening is very small as the microsatellite marker is within an intron in the CFTR gene.

b. The child is a carrier, so 1/2 of the gametes produced will be mutant for CFTR. If 3% of the population are carriers for CFTR then the probability of the child marrying a carrier is 0.03. The probability of this couple having an affected child is: 1/2 probability of mutant allele from the carrier form part a x 0.03 probability that spouse is a carrier x 1/2 probability of mutant allele from spouse = **0.0075 probability of affected child**.

11-22. Assign d^+ as the normal allele of the disease gene and D^* as the mutant allele for the disease. Use the affected child in generation 2 in each case to determine which allele of the disease gene is linked to which allele of the marker in the affected person in the first generation. In the first pedigree the individual II-1 received marker allele 1 ($m1$) from the unaffected parent and marker allele 2 ($m2$) from the affected parent. Therefore, the genotype of II-1 is $m1d^+/m2D^*$. The genotype of II-2 is $m3d^+/m2d^+$. Child A inherited the $m2d^+$ haplotype from II-2 and the $m1$ allele from II-1. The $m1$ allele is 10 cM from the D^* allele. Individual II-1 can produce the following gametes: $m1d^+$ (0.45 probability), $m2D^*$ (0.45 probability), $m1D^*$ (0.05 probability), $m2d^+$ (0.05 probability). The **probability of disease expression in child A is 0.05 probability of $m1D^*$ / 0.5 probability of inheriting the $m1$ allele = 10%**. Child B inherited $m2$ from II-1, so there is a probability of 0.45 $m2D^*$ / 0.5 probability of $m2$ = **90% probability of Child B being diseased**. In the second pedigree the genotype of II-1 is $m3d^+/m2D^*$. **Child C inherited $m2$ from II-1, so 0.45 $m2D^*$ / 0.5 probability of inheriting $m2$ = 90% probability of being diseased**. In the third pedigree III-1 inherited $m1D^*$ from II-1 and $m2d^+$ from II-2. Individual II-1 could be either of two genotypes: genotype (i) is $m1D^*/m2d^+$, 90% probability; or genotype (ii) is $m1D^*/m2d^+$, 10% probability. Child D inherited the $m2$ allele from II-1. Therefore, **the probability that child D (III-2) is diseased** is the probability of inheriting $m1D^*$ if II-1 is genotype (i) + the probability of inheriting $m1D^*$ if II-1 is genotype (ii) = 0.9 probability of genotype (i) x 0.1 probability of inheriting $m2D^*$ + 0.1 probability of genotype (ii) x 0.9 probability of inheriting $m2D^*$ = 0.09 + 0.09 = **0.18**.

11-23. An informative mating is one in which all possible meiotic product (gametes) are distinguishable from each other. For example, the genotype *Aa Bb* allows you to differentiate the AB and ab reciprocal pair of gametes from the Ab and aB pair. Thus if you know that the linkage of the alleles of the A and B genes is A B / a b you can determine if any meiotic product is parental (A B and a b) or recombinant (a B and A b). Therefore, an informative genotype is one in which the individual is heterozygous for two different genes. (For a further discussion of this see p. 64 in the Student Companion / Study Guide Chapter 5 Problem Solving Tips - especially see the first bulleted point.) Thus, **mating W is not informative** for either parent. The male parent is heterozygous for the B polymorphism but homozygous for A locus. Therefore you can not determine whether an A1 B3 gamete in the child is parental or recombinant. The genotype of the female parent is likewise non-informative because it is homozygous for the A locus. **Mating X is informative – both parents are doubly heterozygous** so all possible permutations of gametes can be distinguished. **Mating Y is non-informative** because both parents are homozygous at the B locus. **Mating Z is non-informative** because both parents are homozygous for both the A and B loci.

11-24.

a.   The **Pinocchio syndrome is autosomal dominant**. Autosomal because there is male to male inheritance and dominant because an affected child always has an affected parent.

b.   **The Pinocchio locus is linked to the SNP1 locus**. In generation II it is seen that the syndrome (P) is inherited with the $m2$ allele of the SNP locus. Thus the affected haplotype is $m1P$ and the unaffected haplotype is $m1P^+$. In generation III 7/8 children inherit either one of these two haplotypes which suggests very strong linkage. Child III-8 inherits a recombinant haplotype from his affected father, $m2P$, so the genotype of III-8 is $m2P/m1P^+$. This individual passes one or the other of these haplotypes to all of his children except IV-7 who receives a recombinant gamete of genotype $m2P$. Of the sixteen children in generations III and IV, 14/16 inherit a parental haplotype. Thus the SNP1 locus and the locus for Pinocchio syndrome are closely linked, with a recombination frequency between them of 2/16 = 0.125 = 12.5 cM.

c.   **It is not likely that the coding region containing SNP1 is the Pinocchio gene.** There are 12.5 cM between these 2 markers. In humans, 1 cM = ~1,000 kb, thus there is roughly 12,500 kb between SNP1 and Pinnochio.

11-25. In order to find genes in a large cloned and sequenced region you can **use conserved sequence comparisons with evolutionarily divergent species,** for example compare mouse sequences to human sequences. Stretches of highly conserved sequence >25 bp are candidates for genes or regulatory regions. Do a **computational search for open reading frames (ORFs) and exon/intron boundaries. Look for homology to ESTs** (expressed sequence tags) which are cDNA fragments from the organism.

11-26.

a.   One strategy is to **transform the mouse with the human gene cloned behind a promoter and an inducible regulatory regio**n. Thus the expression of the human gene can be turned on at will. The effects of over-expression can then be examined in a model organism.

b.   Hemophilia is caused by the lack of the factor VIII protein. Therefore, you must knock-out the mouse gene for the factor VIII protein. This knock-out strain will be hemophiliac.

11-27.

a.   If a DNA fragment hybridizes to a band in the Northern blot then at least part of the DNA fragment was transcribed into the hybridizing mRNA. Therefore **A, C and E** contain sequences homologous to genes.

b.   Because three differently sized mRNAs are hybridizing, **three different genes have been identified**.

c.   **Yes**, it is possible that there are more genes in this region. There could be a gene which is transcribed in very low amounts and would not be detected by Northern blot analysis. Fragments A - E could all be in one large intron of a gene which spans this entire region. Thus these sequences would not be homologous to this mRNA but DNA sequences on either side would hybridize to the same size mRNA.

d.   The transcripts recognized by **fragments C and E** are both genes that are expressed in the heart tissue, so they are candidates for the gene causing the disease.

e.   The **gene recognized by fragment E** is found only in the heart, so this seems to be the more likely candidate as the disease affects only heart tissue.

f.   **If there is a mouse model of this disease you would transform the mice with the cDNA clone of the candidate gene and look for the normal human gene to rescue the mutant phenotype in the mice**. If there is no mouse model of the disease you would compare the DNA sequence of the alleles from unrelated diseased families with the DNA sequence from normal people. Look for obvious mutations in the sequences from the diseased families which would alter the protein - changes to the predicted amino acid sequence of the protein, mutations that would affect intron/exon splicing, etc. Do Northern blot analysis on the heart tissue from affected cadavers - are there differences in the size or the amount of the candidate mRNA?

Section 11.5 – Complex Traits

11-28.

a.   The disease is **autosomal dominant**. Dominant because all affected children have an affected parent and autosomal because affected male III-1 passes the trait on to both daughters and sons.

b.   **Yes**, there are 2 possibilities to account for the disease in III-1. Perhaps either **II-3 or II-4** had the mutant allele but didn't express the phenotype. Otherwise, III-1 inherited a mutant allele because there was **a spontaneous mutation of the normal allele in either the gamete from II-1 or II-2**.

11-29.

a.   **The disease is autosomal dominant**. The disease appears to skip generations which would suggest the disease is recessive. However if it is recessive then both II-1 and II-4 would have to be carriers. You are told the disease is <u>very rare</u>, so this is extremely unlikely. Therefore, the disease is dominant with incomplete penetrance. It is autosomal affected male I-1 passes the disease to male II-5.

b.   **Yes, II-2 and III-1** must have the mutant allele but do not express the disease. II-2 has an affected child and III-1 is an identical twin of III-2 so they must both share the same alleles for all genes!

11-30. See Table 11.2.

a.   The fact that many people develop heart disease later in life suggests that environmental factors induce the disease over time. Therefore, **choose families where the onset of the disease is early - families where people develop the disease in their 20's and 30's**. These families are likely to have a mutation in a gene that is important in the development of the disease. Look only at diseased individuals and divide complete sets of families on other criteria such as age of onset.

b.   **You must have a mouse model of heart disease. You then transform these mice with your candidate gene and examine the transformed mice for less severe symptoms and longer life-spans**.

Section 11.6 – Genome-Wide Association Studies

11-31.

a.   If there are 25 different alleles of A and 50 different alleles of B and 10 different alleles of C then there are 25 x 50 x 10 = **12,500 different combinations or haplotypes** possible.

b.   There are $12,500^2$ = **156,250,000 possible diplotypes** (or pairs of haplotypes or genotypes).

c.   The haplotypes in **the father** are A25 C4 B7 and A23 C2 B35. In other words his diplotype or **genotype is A25 C4 B7 / A23 C2 B35. The mother's genotype is A24 C5 B8 / A3 C9 B44.**

d. The probability of the next child getting the A24 C5 B8 haplotype from the mother is 1/2. The probability of this child receiving the A25 C4 B7 haplotype from the father is 1/2. Therefore the probability of the next child having the same genotype as child #1 is 1/2 x 1/2 = **1/4**.

11-32.

a. Alleles at separate loci – in this case haplotype SNP markers and the Canavan disease locus – that are associated with each other at a frequency significantly higher than that expected by chance, are said to be in linkage disequilibrium. In this small sample, **the disease causing mutation appears to be in strong linkage disequilibrium with the SNP alleles SNP3 (G), SNP4 (T), SNP5 (T), SNP6 (T), and SNP7 (C)**. In the affected group the disease causing mutation is associated with these particular SNP alleles at a much higher frequency than these SNP alleles are found in the general population.

b. The Canavan disease gene is most likely found in the region of the strongest linkage disequilibrium, or the region with a haplotype whose frequency in the disease group is higher than that of the control unaffected group. This is **the haplotype you defined in part a. above**. Since the SNPs are roughly 100 kb apart, **the total region of interest is about 400 kb**.

c. The data suggests **two independent Canavan-causing mutations**. The first was induced on a chromosome that had a haplotype identical to that found in the region between SNPs 3 and 7 in individuals #1-4. The second disease causing mutation was induced on a completely unrelated chromosome with the haplotype seen in individual #5.

d. Since **individuals #1-#4 share the same haplotype for this region, it is most likely that they all share the same Canavan-causing mutation by descent from a common ancestor. This suggests that the mutation was induced on a chromosome in a common ancestor before there was a separation between Ashkenzic and Sephardic Jews** - that is, the mutation occurred at some point in time prior to 70 A.D. This conclusion is not firm because it is possible that the two populations were not completely separate so that the Sepahardic patient had an Ashekenazic ancestor at some later point in time.

e. The **advantages of looking at subpopulations are: (1) it is easier to find certain mutations in certain subpopulations because the frequency of the disease is higher in the subpopulation than in the population at large, and (2) it is likely that many affected individuals in the subpopulation may inherit the same allele by descent**, making it easier to detect haplotype associations - this is called a *founder effect* and will be explained in more detail in Chapter 19. The **disadvantage is the flip side of the second advantage - if the subpopulation was formed only relatively recently in the past, there will not have been sufficient time for recombination to shuffle the chromosomes. As a result, the haplotype regions associated**

with the disease-causing mutation will be very large, making it harder to find the disease gene.

f. The only way the researchers could determine the haplotypes is to **examine the alleles of each SNP present in the parents and (if available) siblings of the Canavan patients** (<u>problem 11-31</u>). For example, consider just SNP2 and SNP3 in patient 1. For example, when sequencing the patient's DNA, you might discover that the patient is heterozygous for the SNP2 marker and homozygous for the SNP3 marker - T and C for SNP2; G for SNP3). The DNA sequence from the patient's mother shows the genotype A and T at SNP2 and G and C at SNP3. The DNA from father shows C and G at SNP2 and A and G at SNP3. Therefore the child (patient 1) must have inherited haplotype T and G from the mother and C and G from the father. These SNPs are so close together (only 100 kb between SNP2 and SNP3) that recombination is extremely unlikely. This same logic is applied to the other SNP alleles to generate the entire haplotype for this region for each patient.

Chapter 12   The Eukaryotic Chromosome

Synopsis:

This chapter describes the structure of eukaryotic chromosomes and how that structure affects function. The very long, linear DNA molecules are compacted with proteins in the chromosomes to fit into the nucleus. Several structures are essential for duplication, segregation, and stability. Replication origins are necessary for copying DNA during S phase; centromeres are necessary for attachment to spindle fibers and proper segregation during cell division; telomeres are necessary at the ends of the linear DNAs to maintain the integrity of the DNA molecule. Chromatin structure (packaging of DNA in the chromosomes) can have consequences for gene activity. Areas of normally packaged chromosome can become decompacted to allow expression. Some regions of chromosomes or entire chromosomes are packaged in a different way that decreases gene activity as, for example, in heterochromatin (**Figures 12.11, 12.12**) or Barr bodies.

Significant Elements:

After reading the chapter and thinking about the concepts you should be able to:

♦ Describe the essential elements of eukaryotic chromosomes.

♦ Predict the stability of artificially constructed chromosomes based on the components they contain.

♦ Analyze data on changes in chromatin compaction.

♦ Understand the role of chromosomal origins of replication.

♦ Explain why centromeres are necessary for proper segregation during mitosis and meiosis (**Figure 12.20**).

♦ Discuss the role of telomeres (**Figure 12.18**).

♦ Understand how chromatin packaging influences gene activity including nucleosomes, nonhistone scaffold proteins, euchromatin and heterochromatin.

♦ Explain PEV (position effect variegation) in *Drosophila* (**Figure 12.12**).

♦ Explain X chromosome inactivation.

Problem Solving Tips:

♦ Put yourself in the position of being the researcher. When designing experiments consider the aim of the experiment, the concepts that apply to the problem, and think through experimental methods you know to find a relevant methodology.

Solutions to Problems:

12-1. a. **4**; b. **9**; c. **7**; d. **8**; e. **2**; f. **3**; g. **5**; h. **1**; i. **6**.

Section 12.1 – Chromosomal DNA and Proteins

12-2. Non-histone proteins, which make up ~1/2 the mass of proteins associated with DNA, are a very heterogeneous group of proteins. There are hundreds or even thousands of different kinds of proteins. **Some of these proteins play a purely structural role, (e.g., scaffold proteins, see Figure 12.2) while others are active in replication (DNA polymerase) and the processing of recombination (proteins in the synaptonemal complex).** Still **others are necessary for chromosome segregation (the motor proteins of the kinetochores, see Figure 12.3).** The largest class of non-histone proteins are **those that foster or regulate transcription and RNA processing.** In mammals there are 5,000-10,000 of these tissue specific transcription factors that are found in different tissues at different times in the life cycle. The distribution of the non-histone proteins along the chromosome is uneven. They are found in different amounts and in different proportions in different tissues.

Section 12.2 – Chromosome Structure and Compaction

12-3. See **Table 12.1. In interphase the chromosomes are compacted 40-fold more than naked DNA and during metaphase the chromosomes are compacted 10,000 fold more than naked DNA.**

12-4. The core histones (H2A, H2B, H3 and H4) are the core of the most rudimentary DNA packaging unit, the nucleosome. The core is an octamer made up of 2 of each core histone. Roughly 160 bp of DNA wraps twice around the core, leading to a 7 fold compaction over naked DNA. About 40 bp forms the linker that connects one nucleosome to the next. Histone H1 lies outside the core, apparently associating with the DNA where it enters and leaves the core. Removal of H1 causes some DNA to unwrap from each nucleosome, but the core 140 bp of DNA stays intact. H1 is involved in the next level of compaction, formation of a 300 Å fiber.

12-5.

a. There are 3×10^9 bp in a haploid human genome. If each nucleosome has a spacing of 200 bp then 3×10^9 bp in a haploid human genome / 2×10^2 bp in a nucleosome = 1.5×10^7

nucleosomes to cover the DNA in a haploid genome. This estimate is high because not all parts of the genome are uniformly arranged into the most densely packed nucleosomes. The human genome is diploid, and after S phase each cell would have 2 chromatids for each chromosome, so $4 (1.5 \times 10^7$ nucleosomes$) = 6 \times 10^7$ nucleosomes would be required per cell. Finally, each nucleosome contains two molecules of H2A, so cells would need roughly **1.2×10^8 molecules of H2A protein**.

b. Histone proteins need to be synthesized **during or just after S phase**, when the chromosomes have just replicated. This is when there would be new "naked" DNA needing nucleosomes.

c. Each cell needs to make about 6×10^8 molecules of each type of histone during S phase of the cell cycle (see part a above). In human cells S phase generally lasts between 3 and 6 hours depending on the cell type. This is a lot of molecules of protein in a short period of time. **Multiple copies of the histone genes mean more templates that the cells can transcribe simultaneously, allowing the more rapid production of histone proteins**.

12-6.

a. **p represents the short arm; q represents the long arm**.

b. 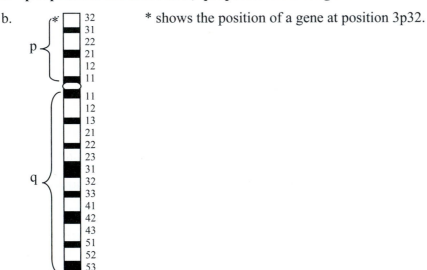   * shows the position of a gene at position 3p32.

12-7. Remember that the human genome contains about 3×10^9 bp. Therefore 3×10^9 bp / 2×10^3 G bands= 1.5×10^6 bp per G band = 1.5 Mb per G band. If there are about 30,000 genes in the human genome, then the average gene in the human genome is roughly 3×10^9 bp / 3×10^4 genes = 10^5 bp per gene or 100 kb per gene. On average one G band contains 15 genes (1.5 Mb per G band / 0.1 Mb per gene). **A deletion removing one G band would remove about 15 genes**. The number would be smaller if you could reliably discriminate the loss of a portion of a G band. There is considerable variation in the sizes of genes and G bands. For example the dystrophin gene is about 2.4 Mb long, so a deletion of this one gene would remove more than the amount of DNA present in a typical G band. Most small deletions that remove only a single gene or a part of a gene would not be detected in karyotype analysis.

12-8. Chromatin is the complex of DNA and proteins (histone and non-histone) which make up eukaryotic chromosomes. Evidently the nucleosome protects DNA from being digested with the micrococcal nuclease, so this enzyme preferentially attacks the DNA in chromatin somewhere in the linker DNA between nucleosomes. Thus, the pattern in lane A reflects the distribution of nucleosomes in chromatin - 200 bp is a single nucleosome, 400 bp is two nucleosomes, etc. **If the nuclease treatment is short enough cleavage of the double-stranded the DNA will occur in some linker regions but not others, producing chromatin fragments with one or more nucleosomes as in lane A. Longer periods of nuclease treatment result in more cleavage - the enzyme will attack all the linker regions, so that all the chromatin will be reduced to units of single nucleosomes (lane B). Finally, if the treatment of the chromatin and nuclease is sufficiently long, then all the linker DNA will be digested, leaving core nucleosomes, each with about 160 bp of DNA as seen in lane C.** These types of experiments were actually performed in the 1970s, and provided significant support for the emerging picture of the nucleosome.

12-9. Histone H1 is located on the outside of the complex and "locks" the DNA to the core. This protein is thus able to interact with the H1 proteins from other nucleosomes, forming the center of the coil that is thought to form the 300Å fiber. The histone core is made up of 2 subunits each of H2A, H2B, H3 and H4. These eight proteins are coated with DNA and thus unable to interact with each other, so they are unable to participate in forming the 300 Å fibers.

12-10.

a. Knowing the amino acid sequences of the proteins associated the human CAF-1 (chromosome assembly factor) complex means that you can **'reverse translate' the protein sequences and predict the degenerate nucleotide sequences of the proteins. The entire yeast genome has been sequenced, so you can search for orthologous yeast genes.** You expect the amino acid sequences, and therefore the DNA sequences, of these proteins to be very highly conserved. **Alternatively, you can clone the yeast genes by using the predicted DNA sequences of the human CAF-1 genes to make degenerate oligonucleotide probes and probe a yeast genomic DNA library.**

b. Why identify the yeast genes? **You can experimentally manipulate the yeast much more easily! It is possible to make mutations of the CAF-1 proteins and carefully examine s the affects on chromatin assembly in this model organism**. It may be possible to extrapolate your findings to human chromosome assembly.

12-11. Using the cloned gene, mutate the codon coding for the acetylated lysine so that it codes for a similar, non-acetylated amino acid. Then, transform this mutated allele into yeast cells on an autonomous plasmid so that both the mutant and wild type genes are present. If the acetylation is important for function then these transformed cells will grow more slowly.

Section 12.3 – Chromosomal Packaging and Function

12-12.

a. **300 Å fiber**;

b. **DNA loops attached to a scaffold**;

c. **heterochromatin**;

d. **metaphase chromosomes**.

12-13. The expression of **the Xist gene** is required for X inactivation. This gene produces is **a large, cis-acting mRNA** that stays in the nucleus and associates with the X chromosome that produced it, **causing inactivation of the X chromosome that produced it.**

12-14. Heterochromatin is the regions of darkly staining DNA which is much more condensed than the euchromatin (the rest of the chromosome). Constitutive heterochromatin are the areas of heterochromatin that remain condensed and heterochromatic most of the time in all cells (**Figure 12.11**). Facultative heterochromatin is that region of the chromosomes (or even entire chromosomes) that are heterochromatic in some cells and euchromatic in other cells.

a. **In *Drosophila* the centromeric regions of the chromosomes and the Y chromosome are examples of facultative heterochromatin. Constitutive heterochromatin is seen in cases of Position Effect Variegation (PEV).** The example discussed in the text is PEV of the white gene (Figure 12.12 and accompanying description of PEV) after a chromosomal rearrangement (an inversion that moves the gene next to the X chromosome heterochromatin). The mosaic of white and red patches seen in the eyes of these animals suggests that the decision about the heterchromatic spreading is the result of a random process which varies from cell to cell during development. Heterochromatization can spread over >1 Mb of previously euchromatic DNA. Autosomal genes can also show PEV as the result of either an inversion which moves the gene next to the centromeric DNA or a Y:autosome translocation which moves the gene next to Y chromosome DNA.

b. **In humans, the centromeric DNA and the great majority of the Y chromosome are also constitutive heterochromatin. The formation of Barr bodies due to the random inactivation of one of the X chromosomes in each cell early in female human fetal development is an example of facultative heterochromatin.**

12-15.

a. *Su(var)* mutations decrease the amount of PEV. **In the presence of a *Su(var)* mutant allele there will be fewer white patches in the eye and more red patches when the eyes are compared to a homozygous *Su(var)*$^+$ fly. The situation would be reversed with more white patches and fewer red (wild type) patches if the fly were heterozygous for the *E(var)* mutation (<u>Figure 12.12a</u>).**

b. The *Su(var)* and *E(var)* mutations both have phenotypes that lead you to think the proteins encoded by the genes are involved in chromatin condensation. Assuming the mutations are loss of function (null) alleles, then **the *Su(var)*$^+$ genes encode proteins that establish and assist spreading of heterochromatin.** Thus, loss of some of the gene product results in engulfment of neighboring genes by heterochromatin. **The *E(var)*$^+$ genes seem to encode proteins that restrict the spreading of heterochromatin**, since loss of one copy of the gene allows heterochromatin to spread into neighboring genes more often. The results also suggest that position effect variegation is very sensitive to amounts of either type of protein because a reduction of 50% of either type of protein causes the mutant phenotype.

12-16. a. **1**; b. **0**; c. **1**; d. **1**; e. **3**; f. **0**.

12-17. These twin sisters could still be monozygotic twins. They must both be carriers of the X-linked Duchenne muscular dystrophy (*Dmd*). **In the affected twin, the X^{Dmd+} homolog was inactivated in the cells that are affected by muscular dystrophy. In the unaffected twin, the other X chromosome (X^{Dmd}) was inactivated in those same cells.**

12-18. Girls of genotype of X^{CB}X^{cb} could have some patches of cells in the eye in which the X chromosome carrying the *CB* allele was inactivated and therefore those patches would be defective in color vision. Usually, enough cells have the *cb* allele inactivated and the *CB* allele

active that there is sufficient color vision and therefore no phenotypic effect of the *cb* mutant cells.

12-19.

a. Heterozygous *Oo* female cats have tortoiseshell coats. In some patches of cells the chromosome with the *O* allele is inactivated so the coat is black as determined by the *o* allele. In other patches of cells the chromosome with the *o* allele is inactivated so the fur is orange since the *O* allele is still functioning. Crosses that could yield *Oo* females include ***OO* x *o*Y** (orange females x black males), ***oo* x *O*Y** (black females x orange males), ***Oo* x *o*Y** (tortoiseshell females x black males), and ***Oo* x *O*Y** (tortoiseshell females x orange males).

b. Male tortoiseshell cats could **be XXY Klinefelter males who are heterozygous *Oo*.** One of their X chromosomes would be inactivated in some patches of cells; the other X chromosome would be inactivated in other patches of cells, just as for the tortoiseshell females in part a.

c. Coat color in calico cats requires the action of another gene to produce white fur, but the white cannot be epistatic to the *orange* gene since the black and orange patches would not be visible. One possibility is that the *Oo* or *Oo*Y cats are also heterozygous for an X-linked white coat color gene. However if this were the case, the white coat allele would have to be on either the X chromosome with *O* or that with *o*. These cats would be either white and black or white and orange. It turns out that **an autosomal gene called the *white-spotting* or *piebald* gene causes the white spotting - a dominant allele of this gene causes white fur, but in heterozygotes this allele has variable expressivity so some patches have a color dictated by the functional alleles of the *orange* gene.**

12-20.

a. All of the progeny will have **mutant coat color** since the wild-type allele is on the inactivated paternal X chromosome.

b. All of the progeny will have **wild-type coat color** since the mutant allele is on the inactivated paternal X chromosome.

c. To characterize alleles as recessive or dominant, you need to examine the phenotype of heterozygotes under conditions where both alleles are expressed. Here, **the heterozygotes are all females whose phenotype was determined by the allele received by the mother.**

d. In tortoiseshell cats the maternal X chromosome is inactivated in some embryonic cells and their descendants, whereas the paternal X chromosome is inactivated in other embryonic cells and their descendants. **In marsupials, the paternal X chromosome is always inactivated.**

12-21. DNase I can only digest "open" DNA. This is DNA that is not bound by proteins like histones. Such DNase hypersensitive (DH) sites are found in the promoter regions of genes that are being transcribed or that are being prepared for transcription in a later step of cellular differentiation (see Figure 12.14). Thus **a hypersensitive site would suggest choice b.**

12-22.

a. Studies of position effect variegation (PEV) show that **methylation of a lysine in histone H3 marks the DNA for assembly into heterochromatin** by signaling other proteins to interact with the DNA and further condense it into heterochromatin. In another example the RNA produced by the Xist gene on the X chromosome coats the DNA of the X chromosome that produced it. This leads to **the methylation and partial deacetylation of histones H3 and H4. These histone modifications in conjunction with the binding of other protein factors produce inactive, condensed heterochromatic DNA leading to the formation of a Barr body.**

b. **Synthesis of the four basic histone proteins increases during S phase of the cell cycle to incorporate histones onto the newly replicated DNA.** Special regulatory mechanisms tightly coordinate DNA and histone synthesis so that both occur at the appropriate time. The association of DNA with histones in the nucleosomes is the critical first step in packaging the 2 meters of DNA into the nucleus of a cell.

Section 12.4 – Replication and Segregation of Chromosomes

12-23.

a. The centromeric regions of human chromosomes are made up of **alpha satellite DNA**.

b. **Cohesin holds sister chromatids together until anaphase**, when it is cleaved and the sister chromatids are released. **Kinetochores attach chromosomes to the spindle poles and contain motor proteins that move the separated chromosomes to the poles.**

12-24. At 50 nucleotides/second, DNA Polymerase (DNAP) could synthesize about 270kb in 3 hours. Since DNAP can synthesize DNA in both directions from an origin of replication, there could be up to 540kb of DNA between origins. Thus the minimum number of origins expected for a 3 billion base pair genome would be about 3×10^9 base pairs per genome / 5.4×10^5 base pairs per origin = 0.55×10^4 origins = **5500 origins of replication.**

12-25.

a.   Proper mitotic chromosome segregation would be disrupted by mutations in **genes encoding cohesin proteins, genes encoding kinetochore proteins, genes encoding motor proteins that help chromosomes move on the spindle apparatus and genes encoding components of the spindle checkpoint that makes the beginning of anaphase dependent upon the proper connections of spindle fibers and kinetochores** (see **Figure 4.11**). **Mutations that alter the DNA comprising a centromere might also have similar effects.**

b.   In order to look for mutations that affect mitotic chromosome segregation you need cells containing a YAC with a marker conferring a visible phenotype like colony color. Treat these yeast cells with a mutagen and **look for colonies that contained many cells that had lost the YAC because of mitotic chromosome mis-segregation**. You could also try to **mutate the centromeric DNA of this YAC using in vitro mutagenesis. If the centromere were disrupted the YAC would not segregate properly and would be lost**. Again, you could follow this if the YAC carried a genetic marker that resulted in a visible phenotype like colony color.

12-26.

a.   In order to replicate the longest chromosome (66Mb) from one bidirectional origin of replication, 33 Mb would have to be copied along each replication fork during the 8 minute cycle (480 sec) = 33,000,000 bp replicated/480 sec = 68,750 bp/sec = ~69 kb/sec. Therefore, if a single origin of replication was used and replication took the entire 8 minutes of the cycle, **the rate of polymerization would be 0.069 Mb/sec or 69 kb per second**.

b.   If bidirectional origins of replication occur every 7 kb, then only 3.5 kb would have to be replicated during the 8 min cell division cycle. **The polymerization rate would be 3.5 kb/480 sec = 7.3 bp/sec, a _much_ more reasonable rate.**

12-27.

a.   In order to examine the end of one specific chromosome, **your DNA probe must contain unique DNA found next to the repeated 5' TTAGGG (telomere) sequenc**es.

b.   The sharpness of the band(s) seen after probing a genomic Southern with most DNA probes is due to the fact that the flanking restriction sites digest all the copies of the DNA into the same set of fragments. **The blurriness of the band seen when probing sequences found at the very ends of the chromosomes indicates that the hybridizing fragments from the end of the chromosome in a population of cells are not homogeneous in length**. In other words, the fragments at the ends of the chromosome are not the same size in all cells. **The number of repeat**

sequences at the telomere, and therefore the telomere length, varies from cell to cell, especially in actively dividing cells.

12-28. The new sequences that are added on to the end of this chromosome must be specific for the species which is adding them. **Because the YAC was transformed into yeast, telomerase in the yeast cell added on the sequence specific for yeast.**

12-29.

a. At high temperature the **CENP-A mutant dies while the CENP-B mutant is viable. Chromosome loss at elevated temperature cannot be measured in CENP-A because the cell dies. The CENP-B mutant, on the other hand, shows increased chromosome loss** at high temperature.

b. To measure chromosome loss in CENP-B mutants you need **cells with a marker** conferring a visible phenotype like colony color so that you can easily detect the loss of the marker. This marker must **be on a chromosome, or on an artificial linear chromosome (YAC), or on a circular plasmid containing a centromere.**

12-30.

a. **A plasmid containing only the URA^+ gene must integrate into the chromosome to be replicated and maintained because is has no origin of replication. Once it is integrated this gene will be stably maintained.**

b. **A URA^+, ARS plasmid can be maintained as a plasmid or it can integrate into the chromosome. If it remains as a plasmid, it will not be very stable and would be lost from many of the daughter cells during subsequent rounds of mitotic division. If this plasmid integrated, it would be very stable.**

c. **The URA^+, ARS, CEN plasmid could only be maintained as a separate plasmid in the cell.** If it did integrate into the chromosome, there would be two centromeres on that chromosome and during mitosis the chromosome would break. **The plasmid would be very stable from one generation to the next because the centromere sequence directs its segregation.**

12-31.

a. **Use the yeast CBF1 protein to make antibodies and then use these antibodies to probe the human cDNA expression library. Alternatively, you could use the cloned yeast gene as a probe to hybridize to clones in a human cDNA library.** To identify related genes in distantly related species, the stringency of the hybridization conditions is often lessened so you do not demand that every base be identical.

b. Use the human protein to make an antibody. This antibody will bind specifically to this protein in fixed cells. **Label or tag the antibody (with fluorescence for example). You can determine the location of the protein in the cell, for example the nucleus vs. cytoplasm or the centromere region of chromosomes vs. the telomeric region.**

12-32. The subcloned fragments that contain the centromeric DNA are those that show a high percentage of Trp$^+$ colonies after 20 generations without selection for the plasmid. These subclones include the 5.5 kb *Bam*HI, the 2.0 kb *Bam*HI-*Hind*III, and the 0.6 kb *Sau*3A. **Because the smallest of these has high mitotic stability and its ends are within the boundaries of the other fragments, the centromere sequence must be contained within the 0.6 kb *Sau*3A fragment**.

12-33. YAC clones can rearrange the insert DNA. BAC clones are not as likely to do this. The clones you have isolated from the BAC and YAC libraries have very different *Hind*III digestion patterns. In order to determine which of these restriction patterns most closely resembles pattern found in the human genome, **digest the BAC, the YAC, and the genomic DNA with several restriction enzymes and compare the restriction patterns of each when they are hybridized with a probe containing the BAC or YAC DNA**.

12-34. The Rec8 protein is found in the meiotic cohesion complex. Rec8 is degraded during anaphase II of meiosis allowing the sister chromatids to segregate to opposite poles during anaphase of meiosis II. During anaphase of mitosis the mitotic cohesion complex is degraded and the replicated chromosomes all split with one chromatid going to each daughter cell (see **Figure 12.20**). In mitotic cells expressing Rec8 the cleavage of the cohesion complex at the centromeric regions of the replicated chromosomes occurs normally at mitotic anaphase and the chromatids segregate normally. Shugosin protects Rec8 from degradation. If both proteins are expressed during mitosis then some of the centromeric regions will have both proteins. In these the Rec8 cannot be degraded and the sister chromatids will <u>not</u> segregate from each other. Therefore such replicated chromosome should undergo mitotic non-disjunction with both sister chromatids going to one daughter cell or the other. **This will result in an array of different aneuploid genotypes in the daughter cells. Some**

daughter cells will have 4 copies of some chromosomes (tetrasomic), 2 copies of some and no copies of others (nullosomic). Many of these genotypes will lead to cell death, so the phenotype would be slow growth of the yeast colonies.

Chapter 13   Chromosomal Rearrangements and Changes in Chromosome Number Reshape Eukaryotic Genomes

Synopsis:

Rearrangements of sections of chromosomes by duplication, insertion, deletion, inversion, or translocation (**Table 13.1**) can affect distances between genes and the function of genes in which they occur. Very large chromosomal rearrangements can be seen microscopically as changes in banding patterns. Many rearrangements are detectable by changes in linkage or their effects on meiotic products.

Transposable elements are segments of DNA that can move from one position to another in the genome. Different types of elements move using a transposase enzyme or by reverse transcription of RNA into a DNA copy.

Changes in chromosome number (**Table 13.1**) can be the result of loss or gain of one chromosome (aneuploidy) or changes in the numbers of sets of chromosomes (polyploidy). Aneuploid cells are generally inviable in humans, with the exception of those that involve sex chromosomes where there are still phenotypic consequences of extra or lost chromosomes.

The genetic imbalances and instabilities produced by rearrangements and changes in chromosome number usually place individual cells and organisms and their progeny at a selective disadvantage.

Significant Elements:

After reading the chapter and thinking about the concepts, you should be able to:

♦ Understand how the different types of chromosomal rearrangements are generated.

♦ Understand that meiotic pairing is the same pairing as seen in the polytene chromosomes in *Drosophila* (**Figures 13.6, 13.7** and **13.9**).

♦ Understand the implications of each rearrangement on viability, phenotype and linkage, both in the heterozygous and homozygous state.

   o Deletions remove DNA (**Figure 13.2a** and below) – small deletions affect only one gene while large deletions can remove tens or hundreds of genes. Deletions have an adverse effect on viability. At meiosis (or in the polytene chromosomes of *Drosophila*) the chromosomes of deletion heterozygotes form deletion loops. Heterozygosity for deletions affects map distances (**Figure 13.4**); heterozygous deletions "uncover" genes (pseudodominance, **Figure 13.5**) and this can be used to map genes (**Figure 13.8**); deletions can be used to locate genes (**Figure 13.8**). Homozygosity for a deletion is almost always lethal. There is no

<u>recombination</u> between genes within the deletion.

- o <u>Duplications</u> add DNA (**Figure 13.11**) – although some duplications affect phenotype (**Figures 13.12** and **13.13**), most have no effect. At meiosis (or in the polytene chromosomes of *Drosophila*) the chromosomes form duplication loops. Large duplications have an adverse effect on viability.

- o <u>Inversions</u> reorganize the DNA sequence of a chromosome (**Figure 13.14** and below). Most inversions have no affect on phenotype; if an inversion breaks within a gene the mutant gene can affect the phenotype. At meiosis (or in the polytene chromosomes of *Drosophila*) the chromosomes of an inversion heterozygote form an inversion loop (**Figure 13.15**); recombination within the inversion loop leads to genetically imbalanced chromatids while recombination outside of the loop is normal and gives balanced, recombinant gametes. Therefore heterozygosity for inversions reduces the total number of recombinant progeny. In paracentric inversions the centromere is outside of the inversion while in pericentric inversions the centromere is included in the inversion (**Figure 13.16**). Homozygosity for an inversion is viable and there is no inversion loop formed at meiosis.

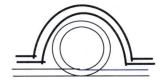

- o <u>Reciprocal translocations</u> attach part of one chromosome to another, non-homologous chromosome (**Figure 13.18a**). Heterozygosity for translocations reduces fertility and results in pseudolinkage (**Figure 13.21**). At meiosis (or in the polytene chromosomes of *Drosophila*) the chromosomes of a translocation heterozygote form a cruciform (**Figure 13.21b** and below). The components of the cruciform are labeled N1 and N2, representing the two normal non-homologous chromosomes, and T1 and T2 representing the two translocated chromosomes. During meiosis homologous centromeres segregate and non-homologous chromosomes show independent assortment – for example in an individual of the genotype *A/a*; *B/b* half of the meioses will yield 2 *AB* : 2 *ab* gametes and the other half will give 2 *Ab* : 2 *aB*. Independent assortment also occurs in a translocation heterozygote, where these segregations are called alternate (giving 2 N1 N2 : 2 T1 T2) and adjacent 1 (giving 2 T1 N1 :

2 T2 N2 gametes). A third type of segregation pattern, adjacent 2, is due to nondisjunction of homologous centromeres.

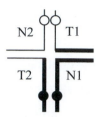

- ◆ Understand the role that transposable elements play in the formation of chromosomal rearrangements.

- ◆ Aneuploidy is the loss (monosomic) or gain (trisomic) of one or more chromosomes. Understand why aneuploidy of the sex chromosomes is tolerated in humans while aneuploidy of the autosomes is deleterious.

- ◆ Understand the role of meiotic nondisjunction (**Figure 13.32**), mitotic nondisjunction and chromosome loss in the formation of aneuploids.

- ◆ Polyploidy is the loss (monoploid) or gain (triploid, tetraploid, etc) of entire sets of chromosomes. Polyploidy is common in plants.

- ◆ The term x refers to the basic chromosome number. For diploid species x = n.

- ◆ Autopolyploids derive all sets of their chromosomes from the same species. Allopolyploids have chromosome sets from two or more distinct (though related) species (**Figure 13.38**).

Problem Solving Tips:

- ◆ When predicting the expected gametes from an individual that is heterozygous or homozygous for one of the chromosomal rearrangements you MUST diagram the pairing of the homologous chromosomes at meiosis! This will ensure that you are very clear about the chromosome composition going into and out of meiosis

- ◆ Remember to trace out meiotic products beginning from the centromere!

- ◆ Pay attention to whether the meiotic products are balanced (have one of everything – one allele of each gene and one centromere). If the gametes are balanced then they will give rise to viable progeny. If they are imbalanced (deleted or duplicated for large regions of a chromosome) then the progeny are usually inviable.

- ◆ Deletions, inversions, and translocations change the linkage of genes that surround or are within the rearrangement.

- ◆ Deletion:

- o   If the deletion of a gene on one homolog uncovers a mutation in the gene on the other homolog then the individual will show pseudodominance for the recessive mutant phenotype. If you see pseudodominance in a cross it indicates the presence of a deletion.

- o   Deletions of DNA can be analyzed using restriction analysis. In a diploid organism, a deletion on one chromosome will mean that a restriction fragment that comes from within the deleted region of the genome will be at half the concentration of that found in a normal cell that has the DNA on both chromosomes.

- ♦   Inversion:
  - o   Single crossovers within the inversion loop lead to recombinant gametes that are imbalanced for genetic material OUTSIDE the loop. One recombinant product is duplicated for the region at outside the loop at one end of the chromosome and simultaneously deleted for the material outside the loop at the other end of the chromosome. The reciprocal recombinant gamete has the reciprocal imbalance. All gametes have the expected amount of DNA for everything within the inversion loop. If the centromere is in the inversion (pericentric) then each meiotic product has a centromere. If the centromere is outside of the inversion loop then the meiotic products that result from a single cross over with in the loop are imbalanced for the centromere as well as the surrounding genes – one recombinant product will be dicentric and the other will be acentric (**Figure 13.16**).

- ♦   Therefore, severe reduction of recombination between genes within a rearrangement indicates the presence of a deletion (heterozygous or homozygous) or an inversion (heterozygous).

- ♦   Translocation:
  - o   More than half of the gametes formed at meiosis are imbalanced – the products of adjacent 1 and adjacent 2 segregations. Therefore the translocation heterozygote is semisterile. This can be detected in organisms such as corn where each kernel of corn on the ear is the result of an independent fertilization event. Because the alternate segregation is the only one that gives balanced gametes genes on the nonhomologous chromosomes involved in the translocation act as if they are linked. Imagine the genotype A N1/a T1; B N2/b T2. The only balanced gametes are AB and ab (the products of alternate segregation in the cruciform structure shown above). This is the result you would see from a double heterozygote (A/a; B/b) if the A and B genes were closely linked.

Solutions to Problems:

<u>Vocabulary</u>

13.1. a. **4**; b. **8**; c. **6**; d. **5**; e. **7**; f. **3**; g. **2**; h. **1**.

<u>Section 13.1 - Rearrangements of DNA Sequences</u>

13-2.

a. (i) **No**. (ii) **No**, (iii) **No**, (iv) **No**, (v) **Different chromosomes**.

b. (i) **Yes**, (ii) **Yes, if recombination occurs within the inversion loop**, (iii) **Yes, if recombination occurs within the inversion loop**, (iv) **Yes**, (v) **Same side of a single centromere**.

c. (i) **Yes**, (ii) **No**, (iii) **No**, (iv) **No**, (v) **The same side of the centromere**; if not, the chromosome would have two centromeres, and this is unstable.

d. (i) **No**, (ii) **No**, (iii) **No**, (iv) **No**, (v) **Different chromosomes**.

e. (i) **Yes**, (ii) **No**, (iii) **No**, (iv) **Yes**, (v) **Opposite sides of a single centromere**.

f. (i) **Yes**, (ii) **No**, (iii) **No**, (iv) **Yes**, (v) **Same side of a single centromere**.

13-3. In polytene chromosomes there are characteristic banding patterns for each chromosome. If there is a duplication of a region of DNA, **the bands in the duplication loop should be repeated elsewhere on the paired homologs**. That is, there should be three copies of the banding pattern in each genome: one in the duplication loop, one elsewhere on the chromosome carrying the duplication, and one on the wild-type homolog. The latter two copies should be paired with each other. If this is a tandem duplication, the loop should be adjacent to two paired copies. If the mutation is a deletion, **the looped out region in a deletion heterozygote contains the only copy of those bands found on the wild-type homolog**.

13-4.

a. Most deletions remove a large amount of very complicated DNA, as seen here. This DNA will never be restored, so **reversion of this deletion is impossible**. In theory a very small deletion of a few nucleotides could revert.

b. This rearrangement could revert because all of the original information still exists in the genome. **Reversion within the duplication could occur fairly frequently** because there is a mechanism that generates revertants - they can occur as a result of **unequal crossing over** in an individual homozygous for the duplication (**Figure 13.13**).

c. In theory a pericentric inversion could revert because all of the information still exists in the genome. However there is **no mechanism to ensure that the breaks will occur in the same** locations as the original breaks that gave rise to the pericentric inversion, so the rate of reversion should be extremely low. One exception to this are inversions that result from intrachromosomal crossing over between some sequence of DNA that is present at two locations on the same chromosome but in reversed order (**Figure 13.14b**). A similar cross over in the inverted chromosome could regenerate the original gene order.

d. If the organism with the Robertsonian translocation **has already lost the very small reciprocal chromosome generated in the process of translocation then the translocation can not be restored. If the small reciprocal chromosome still exists then it is possible that the translocation could revert. This would happen very rarely because there is no mechanism to ensure that the breaks will occur in the same locations as the first breaks.**

e. **For some types of transposable elements the mechanism that allows the transposable element to jump into the gene can also allow the element to jump out again**, often restoring the original DNA sequence of the gene (**Figure 13.27b right**). In these cases **the mutation will revert fairly frequently. If this jumping mechanism is lost then these types of mutations revert very rarely.**

13-5.

a. Fragments that are deleted on one homolog will be lighter in intensity than those that are not deleted (and are therefore present in two copies per cell). You can tell which fragments are contiguous by analyzing each deletion for the missing bands. Use this information about the deleted fragments in each deletion to order the fragments (as you did to order the genes in problem 13-6). New bands that appear in only one of the deletion strains represent new 'joining' fragments generated by the juxtaposition of the remaining bits of the two restriction fragments around the deletion breakpoints. For each of the strains, the following deleted fragments will be considered:

Strain	complete fragments deleted
Deletion 1	6.3, 5.6, 4.2
Deletion 2	6.3, 4.2, 3.0
Deletion 3	5.6, 0.9
Deletion 4	6.3, 3.0

Deletions 1 and 4 both delete the 6.3 kb fragment. The other deleted bands are not in common between deletions 1 and 4, so they must represent fragments that lie to either side of the 6.3 kb fragment. From strain 4, we know that the 3.0 kb fragment lies to one side of the 6.3 fragment, but since it is not lost in Deletion 1, the 5.6 and 4.2 kb fragments must lie to the other side (although we don't know the order yet). Deletion 2 also has a deletion of the 6.3 kb fragment as well as the 3.0 and 4.2 kb fragments. This tells us that the 4.2 kb fragment must be immediately adjacent to the 6.3 kb fragment and the order is 3.0, 6.3, 4.2, 5.6. Deletion 3, in which the 5.6 and 0.9 kb fragments are deleted, indicates that the 0.9 is adjacent to the 5.6 kb fragment but since it was not deleted in deletion 1 it must be on the side opposite the 4.2 kb fragment.

b. The approximate location of the genes can be determined based on which genes are pseudodominant in each deletion strain. Correlating the restriction map (part a) and the phenotypes of the strains, *rolled eyes* and *straw bristles* look like they are found somewhere within the 6.3, 5.6 and 4.2 kb fragments. But since deletion 3 also has the straw bristles phenotype, that gene can be placed at least partly within the 5.6 kb fragment. Deletion 2 has the rolled eyes phenotype as well, so the gene must be in one or both of the fragments that are common to deletions 1 and 2 (6.3 and 4.2 kb). However, since deletion 4 is missing fragment 6.3 but is not mutant for *rolled eyes*, the gene must lie within the 4.2 kb fragment. *Apterous wings* is mutant in deletions 2 and 4 which have in common deletion of the 6.3 and 3.0 bands, but since the mutant phenotype is not seen in deletion 1, the gene lies in 3.0 kb fragment. *Thick legs* is mutant in deletion 3 (5.6 and 0.9 kb fragments), but is not mutant in deletion 1 (5.6 kb fragment), so the gene lies in the 0.9 kb fragment.

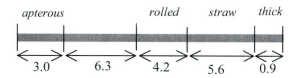

The map above provides only an approximation of gene position. For example, consider the location of the *apterous* gene. It is possible that *apterous* could actually lie in the left half of the 6.3 kb fragment (the part of the fragment removed by Deletion 4 but not by Deletion 1). In addition, because Deletions 2 and 4 (the two deletions that uncover *apterous*) both remove DNA to the left of the map, its also possible that *apterous* lies to the left of the sequences on the map. More accurate mapping would require examination of more deletions and more restriction enzyme sites.

13-6.

a. **Each of the strains shows pseudodominance for some of the mutant alleles**; that is, each strain is mutant for one or more of the marker genes. Furthermore, after the diploids undergo meiosis, **two of the spores die**. All of these are indications that the **X-rays induced deletion mutations**.

b. Two spores in each ascus die because they receive the deleted homolog. **The deletions remove some essential genes from the chromosomes and this is lethal in a haploid**.

c. There is only one X-ray induced mutation per strain, so **all genes that show pseudodominance are on the same chromosome**. Using this logic, all four genes: *w, x, y,* and *z,* are on the same chromosome.

d. **The order is *w y z x = x z y w***. Genes *w* and *y* are deleted in strain 1, uncovering the *w⁻* and *y⁻* alleles, so *w* and *y* are adjacent. Genes *x, y,* and *z* are deleted in strain 2 so they must be adjacent. Combined from the information from strain 1, this means the order must be *w y* [*x z*], with the brackets indicating that you don't know the relative order of *x* and *z*. Strain 3 is deleted for *w, y* and *z*, therefore the gene that follows *y* must be *z*. Note that two answers are given because you cannot determine the left-to-right orientation of this group of four genes.

13-7. Remember that sister chromatids are identical to each other. The probe will anneal to homologous DNA on the chromosome and produce a signal at that position.

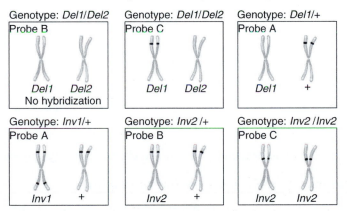

Note that in the paracentric Inversion 2, regions homologous to Probes A and B are not inverted and thus stay at the original position near the telomere, while regions homologous to Probe C in the bottom right figure are inverted and thus move closer to the centromere.

13-8. The deletion data allows you to narrow down the region in which the genes lie. All of these deletions remove portions of polytene chromosome region 65 (which turns out to be on chromosome 3, a *Drosophila* autosome). Deletion A shows pseudodominace for javelin and henna, so all or part of both genes must lie within the deleted region - between A2-3 and D2-3. Deletion B, pseudodominant for henna, indicates that *henna* lies between C2-3 and E4-F1. Combining the results for Deletions A and B, *henna* must be between C2-3 and D2-3. Because Deletion B is javelin$^+$, *javelin* must be located between A2-3 and C2-3 (the part of Deletion A that is not removed in Deletion B). Deletions C and D tell you that the *henna* gene cannot lie to the right of bands D2-3 on the figure in the text, delimiting *henna* to the interval between C2-3 and D2-3.

Inversions do not remove genes, they just relocate them. Therefore, if the inversion is made in a wild type chromosome, the inverted homolog will have the wild type alleles for all of the genes. Inversion B gives the expected result, and does not help locate either of the two genes. Inversion A, however, has a mutant javelin phenotype, indicating that there is a mutant allele of *javelin* on the inverted homolog. Thus, the javelin gene was broken by the inversion, so *javelin* is located in band 65A6. (This is consistent with the region containing *javelin* determined from the deletions above.) Very few *Drosophila* genes extend beyond one band, so we can assume that A6 is the location of *javelin*. In summary, **the *javelin* gene is in band 65A6 and the *henna* gene is between 65C2-3 and 65D2-3.**

13-9. The genotype of the female is:

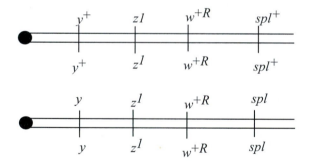

a. The remainder of the male progeny (76,671) are **the parental types, so they will be y^+ $z1$ w^{+R} spl^+ / Y (zeste) and y z^1 w^{+R} spl / Y (yellow zeste split).**

b. Classes A and B are a reciprocal pair of products. **They are the result of crossing over anywhere between the *y* and *spl* genes** resulting in the reciprocal classes: y^+ z^1 w^{+R} spl / Y (zeste split) and y z^1 w^{+R} spl^+ / Y (yellow zeste).

c. Remember that w^{+R} allele is really a tandem duplication of the w^+ gene and that the zeste eye color depends on having a mutant z^1 allele in a genome that also contains two or more copies of w^+. **Classes C and D, are the result of mispairing and unequal crossing over between the two copies of the w^+ gene.** The misalignment can occur in two different ways, see the figure below. Misalignment I gives y^+ z^1 [3 copies w^+] spl (zeste split = same phenotype as class B) as

one product and $y\,z^1$ [1 copy w^+] spl^+ (yellow wild type eye = class C) as the reciprocal product.
In misalignment II, the recombinant products are: $y^+\,z^1$ [1 copy of w^+] spl (wild type eyes split =
class D) and $y\,z^1$ [3 copies of w^+] spl^+ (yellow zeste = same phenotype as class A). Each of
these misalignments thus produces one class of recombinant products that results in a wild-type
eye color.

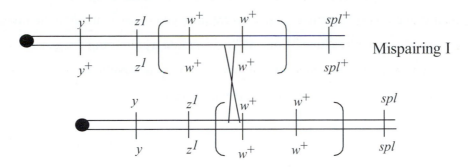

Mispairing I

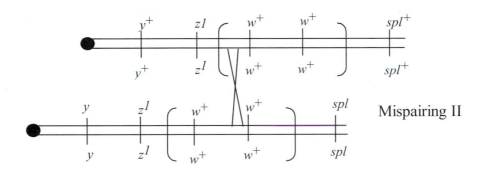

Mispairing II

d.   The **genetic distance between y and spl** = # recombinants between y and spl / total progeny =
2430 (class A) + 2394 (class B) + 23 (class C) + 22 (class D) / 81,540 = **5.9 mu**. This
recombination frequency includes all the recombinants, because classes A and B contain the
reciprocal events to classes C and D.

13-10. Each of the original strains is true breeding and shows the same recombination frequency of
21 mu between genes a and b. However, in the F_1 heterozygote genes a and b are only 1.5 mu apart.
The only rearrangements that affect recombination frequency in the heterozygote are deletions and
inversions. There are two reasons why this cross cannot involve a deletion: both parental strains are
homozygous (true breeding) and homozygous deletions are lethal; and in both parents genes a and b
are 21 mu apart but in a deletion homozygote the genes would be less than 21 mu apart. This
reduction of recombination between the 2 genes in the F_1 is therefore caused by an inversion. One
parental strain has normal chromosomes and the other parental strain is homozygous for an inversion.

There are 2 possibilities for the inverted region: (i) it includes almost all of the region between genes *a* and *b*, but does not include the genes themselves, or (ii) the inversion includes both genes and the DNA in between them.

There are 2 other possibilities for the location of the inversion with respect to the genes which can be ruled out. (iii) the inversion includes gene *a* and almost all of the DNA between the genes; and (iv) the inversion includes gene *b* and almost all of the DNA between the genes. These possibilities can be ruled out because in both cases the recombination frequency between *a* and *b* in the inversion homozygote would be much less than 21 mu, as the inversion would bring the genes closer together.

In any case, the F_1 progeny is an inversion heterozygote. The inversion loop occupies either (i) almost all of the region between genes *a* and *b*, although the genes are not included in the loop, or (ii) the loop includes both genes and the DNA in between them. Any single crossovers within the inversion loop (the huge majority of crossovers between the genes) will result in inviable gametes. The few recombination events that occur between gene *a* and the inversion loop or between the inversion loop and gene *b* will result in balanced, viable, recombinant gametes in scenario (i), as will some of the double cross over events within the inversion loop in both scenarios (i) and (ii).

13-11.

a. **2, 4**. Inversion loops are seen during MI of meiosis only if the cells are heterozygous for an inversion.

b. **2, 4**. Single crossovers within the inversion loop in inversion heterozygotes generate genetically imbalanced chromosomes. The genetic imbalance involves deletions and duplications of regions outside the inversion loop. If the inversion is paracentric, then the recombinant products are also dicentric and acentric. Crossovers in inversion homozygotes do not cause genetic imbalance.

c. **2**. An acentric (and the reciprocal dicentric) fragment is produced from a single crossover within a paracentric inversion in an inversion heterozygote.

d. **1, 3**. In an inversion homozygote, crossovers within the inversion yield 4 viable, balanced, spores all of which have the inverted gene order.

13-12. The data shows unexpectedly reduced recombination frequencies between certain pairs of genes in Bravo/X-ray and Bravo/Zorro heterozygotes. This reduction in recombination will be seen in both deletion heterozygotes and in inversion heterozygotes. You are told that the 3 strains have variant forms of the same chromosome, and that the number of bands in the polytene chromosomes are the same in all 3 strains. Thus, none of the strains is a deletion homozygote. The chromosomal rearrangements here must be inversions.

a. Recombination frequencies between genes in the Bravo/X-ray heterozygote are normal for *a-b* and *b-c* and *g-h* but are reduced for all other gene pairs. Thus, the inversion in the X-ray strain breaks between genes *c-d* and between genes *f-g* and inverts genes *d, e,* and *f.* The *c-d* and *f-g* intervals must still include some non-inverted DNA to allow recombination that produces viable gametes. Similarly, the genetic distance is reduced in the *b-c* and *f-g* intervals for Zorro, where the inversion end points are found and minimal in those intervals completely within the inversion. **The order of the genes in X-ray is: *a b c f e d g h*. The order of the genes in Zorro is: *a b f e d c g h*.**

b. **The physical distance in the X-ray homozygotes between *c* and *d* is greater** than that found in the original Bravo homozygotes. The inversion occurred in this portion of the chromosome, so *c* and *d* are now separated by many more genes (all the inverted DNA).

c. **The physical distance between *d* and *e* in the X-ray homozygotes is the same** as that found in the Bravo homozygotes because this interval is completely within the inverted segment. The relationship of *d* to *e* has not changed.

13-13. The diploid cell contains a pericentric inversion on one homolog. The pairing of the homologous chromosomes during metaphase I of meiosis is shown below. Use this drawing to trace the consequences of crossovers in different regions.

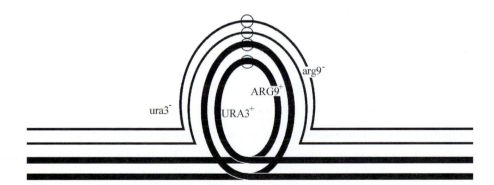

a. A single crossover outside the inversion produces **four viable spores: 2 *URA3 ARG9* (prototrophic for uracil and arginine) : 2 *ura3 arg9* (auxotrophic for uracil and arginine).**

b. A single crossover within the inversion loop, in this case between *URA3* and the centromere, results in imbalanced recombinant gametes. Both recombinant gametes have a duplication and a deletion of the material <u>outside</u> of the inversion loop. One recombinant is duplicated for the region outside the loop on the left and deleted for the information outside the loop on the right. The other recombinant is the reciprocal - deleted for the information on the DNA to the left of the loop and duplicated for the information outside the loop and to the right. This genetic imbalance is usually lethal, so the two spores containing the products of the recombination will die. The single cross over inside the loop gives rise to **2 parental spores and 2 recombinant, lethal spores as in the following ascus: 1 *URA3 ARG9* : 1 *ura3 arg9* : 1 *URA3 arg9* (lethal) : 1 *ura3 ARG9* (lethal)**.

c. This 2 strand DCO within the inversion loop produces four viable spores, **two parental spores and 2 recombinant spores: 2 *URA3 ARG9* (1 parental, 1 recombinant) : 2 *ura3 arg9* (1 parental, 1 recombinant)**.

13-14. In this problem, the diploid cell contains a paracentric inversion on one homolog. The pairing of the chromosomes in the inversion heterozygote are shown below. Use this drawing to trace the consequences of crossovers in different regions.

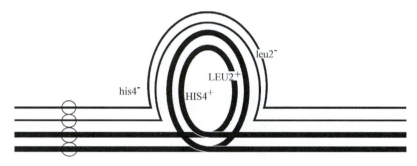

a. A single crossover within the inversion (between *HIS4* and *LEU2*) leads to 2 parental spores (one of each type) and 2 recombinant spores. The recombinants are duplicated and deleted for the regions outside the inversion loop (<u>problem 14-13</u>). In this case, one of the recombinant spores will be duplicated for the region containing the centromere and deleted for the DNA on the other side of the inversion loop (dicentric). The reciprocal recombinant spore will be deleted for the region containing the centromere and duplicated for the region to the right of the inversion loop (acentric). Neither type of chromosome segregates properly so the spores that receive these chromosomes will definitely die. **The ascus will contain: 1 *HIS4 LEU2* (prototrophic for histidine and leucine) : 1 *his4 leu2* (auxotrophic for histidine and leucine) : 1 dicentric (lethal) : 1 acentric (lethal)**.

b. This is a 2 strand double cross over within the inversion loop double. All four spores are viable: **2 *HIS4 LEU2* (1 parental and 1 recombinant) : 2 *his4 leu2* (1 parental and 1 recombinant)**.

c. A single cross over between the centromere and the inversion loop will give 2 parental spores and 2 balanced recombinant spores: **2 *HIS4 LEU2* (1 parental and 1 recombinant) : 2 *his4 leu2* (1 parental and 1 recombinant)**.

13-15. A tetratype ascus means there has been a single crossover between the 2 genes (**Figure 5.11**). Because both *HIS* and *LEU* are in an inversion loop a single crossover would give inviable spores. In this case the recombination event must be more complicated than a single cross over. **One possibility is a 2 strand double cross over. This could occur in 2 ways: one recombination between LEU and HIS and the second recombination between HIS and the end of the inversion loop; or one crossover between LEU and HIS and the second one between LEU and the end of the inversion loop. In both cases, the second recombination event must occur within the inversion loop.** Such a 2 strand double cross over will give the following tetratype ascus: 1 *HIS4 LEU2* : 1 *his4 leu2* : 1 *HIS4 leu2* : 1 *his4 LEU2*.

13-16. Any haploid spores with a deletion are dead (white). An octad has 8 spores.

a. **0 white spores.** The inversion has no effect if recombination does not occur.

b. **4 white spores.** Only two chromatids are involved in the crossover. The recombination gives 2 unbalanced gametes. The remaining two chromatids survive as haploid products and divide mitotically to form 4 viable spores in the ascus.

c. **0 white spores.** If a crossover occurs outside the inversion loop all the products are viable.

d. **8 white spores.** All of the resulting gametes would be genetically imbalanced and would die.

e. **0 white spores.** Alternate segregation produces balanced gametes.

f. **0 white spores.** The crossover in the translocated region would simply cause the reciprocal exchange of DNA between homologous portions of the chromosome. All spores live.

13-17.

a. **1, 3, 5 and 6.** Translocation heterozygotes can produce gametes with any pairwise combination of N1, N2, T1, and T2.

b. **2 and 4.** Translocation heterozygotes cannot produce gametes with two copies of the same chromatid.

c. **1 and 3.** These arise from alternate segregation, so they are balanced.

d. **5 and 6**; these arise from adjacent-1 segregations. **2 and 4**; these arise from adjacent-2 (nondisjunction) segregations.

13-18.

a.   Diagram the cross. Because the two genes are on different autosomes, they should assort independently:

$cn\ cn^+$; $st\ st^+$ x $cn\ cn$; $st\ st$ → 1/4 $cn\ st$ (white) : 1/4 $cn\ st^+$ (cinnabar) : 1/4 $cn^+\ st$ (scarlet) :

1/4 $cn^+\ st^+$ (wild type).

b.   The genes show pseudolinkage in this male. The $cn\ st$ and $cn^+\ st^+$ allele combinations seem to be linked. **This result suggests that the unusual male has a translocation between chromosome 2 and 3 with the mutant cn and st alleles either on the translocated chromosomes or on the normal chromosomes**. The figures below show two of the four possible genotypes for the unusual male fly. Both diagrams show the mutant alleles on the normal order chromosomes. Instead, the mutant alleles of the genes could both be on the translocated chromosomes. Although both the cn and st genes may actually be on the same chromosome after in the translocation (right hand panel), this is not necessary. The pseudolinkage will be seen even if the 2 genes are still on separate chromosomes (left hand panel). Also remember that male *Drosophila* do not recombine.

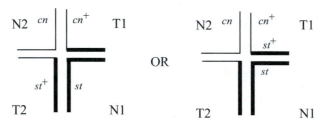

c.   The wild-type F_1 females are translocation heterozygotes containing the cn^+ and st^+ alleles on the translocated chromosomes (she obtained the cn and st alleles from her non-translocated mother). The pairing in meiosis would be the same as shown for the male in part b, except that recombination can occur in the female. **A crossover between cn and the translocation breakpoint or between st and the translocation breakpoint followed by alternate segregation produces gametes with the genotype $cn\ st^+$ (cinnabar) $cn\ st^+$ (scarlet)**. These classes allow you to calculate the map distance between the st and cn genes in the translocation. The rf = 10 mu.

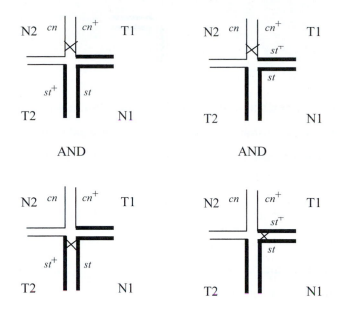

13-19. The semisterile F$_1$ is a translocation heterozygote and will produce 1/2 fertile : 1/2 semisterile progeny from alternate segregation. Products of adjacent-1 or adjacent-2 segregation are imbalanced and therefore inviable - this is the basis of the semisterility. Because the only viable gametes are the result of alternate segregation, genes that are on the chromosomes involved in the translocation will not show independent assortment. Instead, if the genes are located very close to the translocation breakpoints, they will show only the parental classes; that is, the genes will display pseudolinkage. Genes that are on any other chromosome will assort independently of the translocation (i.e., such genes will assort independently from fertility/semisterility).

a. If the *yg* gene is on a different chromosome than those involved in the translocation, the traits will assort independently. The product rule says you can cross multiply the 2 monohybrid ratios: 1/2 *yg*$^+$ (normal leaf color); 1/2 *yg* (yellow green) and 1/2 fertile : 1/2 semisterile to give: **1/4 fertile yg$^+$: 1/4 fertile yg : 1/4 semisterile yg$^+$: 1/4 semisterile yg**.

b. If the translocation involved chromosome 9, the fate of the fertility and leaf color phenotypes are connected – these genes will show pseudolinkage. The original cross was: semisterile *yg*$^+$ x fertile *yg* → F$_1$ semisterile x fertile *yg*. This means the normal, non-translocated chromosome 9 has the *yg* allele, while the translocated chromosome 9 has the *yg*$^+$ allele. Thus, the chromosomes of the heterozygous F$_1$ at meiosis I would look like:

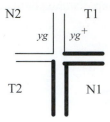

(For simplicity, this figure shows only one chromatid per chromosome.) From the translocation heterozygote, the products of an alternate segregation are balanced and viable. **The progeny (after crossing with a fertile *yg* homozygote) will be: 1/2 fertile *yg* (N1 + N2) : 1/2 semi-sterile *yg*$^+$ (T1 + T2).**

c. **The rare fertile, *yg*$^+$ and semisterile, *yg* gametes result from recombination between the translocation chromosome and the homologous region on the normal chromosome** with which it is paired. After the recombination event, the N1+ N2 fertile gamete will contain *yg*$^+$ and the T1 + T2 semisterile gamete will contain *yg*.

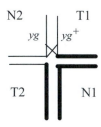

The frequency of crossing-over, as represented by the rare fertile green and semisterile yellow-green progeny, will give you the genetic distance between the translocation breakpoint and the *yg* gene.

13-20. Individuals that are homozygous for a translocation are fertile. However, **when insects that are translocation homozygotes mate with insects with normal chromosomes, the F$_1$ progeny will be translocation heterozygotes. The fertility of these F$_1$s should be reduced by about 50% and half of their progeny will also have reduced fertility.** If the released insects are homozygous for several different translocations, then the fertility of the F$_1$ individuals should be reduced by a further 50% for each translocation for which they are heterozygous. For example, in an F$_1$ insect that was heterozygous for 3 different translocations (T1 - T3), only 1/8 of its gametes (1/2 balanced for T1 x 1/2 balanced for T2 x 1/2 balanced for T3 = 1/8 balanced gametes) would be balanced and give rise to progeny.

13-21. Remember that the Y chromosome pairs with the X chromosome during meiosis, so the N1 + N2 chromosomes in the male translocation heterozygote will be the autosome on which *Lyra* is normally found and the X chromosome, as shown on the figure below.

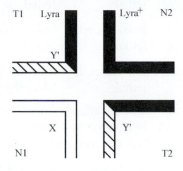

There are only two kinds of genetically balanced gametes produced by the male during alternate segregation. The T1 and T2 segregant has the Y chromosome and the autosome with the mutant *Lyra* allele. This gamete would fertilize an X bearing egg, producing Lyra males. N1 and N2 would yield a gamete with the X chromosome and the autosome with *Lyra*$^+$. This gamete would produce wild-type females when fertilized with *Lyra*$^+$ gametes from the wild-type mother. **The progeny will be 1/2 *Lyra* males : 1/2 *Lyra*$^+$ females**.

13-22.

a. A translocation heterozygote will make 2 types of gametes as a result of alternate segregation and 2 types of gametes as a result of adjacent-1 segregation in a 1:1:1:1 ratio. The 2 products of alternate segregation are N1 + N2 and T1 + T2, both of which are balanced gametes. The two products of adjacent-1 segregation are N1 + T2 and N2 + T1, both of which are imbalanced gametes. When a translocation heterozygote is crossed to a homozygous normal individual, the imbalanced gametes from the translocation parent never give rise to viable progeny. However, if both parents of a cross are translocation heterozygotes, then it is possible for an imbalanced gamete from one parent to be fertilized by the reciprocally imbalanced gamete from the other parent. For example, a N1 + T2 gamete from one parent can fertilize an N2 + T1 gamete from the other parent, creating a zygote that is a balanced translocation heterozygote!

	N1 + N2	T1 + T2	N1 + T2	N2 + T1
N1 + N2	homozygous normal	translocation heterozygote	imbalanced, lethal	imbalanced, lethal
T1 + T2	translocation heterozygote	translocation homozygote	imbalanced, lethal	imbalanced, lethal
N1 + T2	imbalanced, lethal	imbalanced, lethal	imbalanced, lethal	translocation heterozygote
N2 + T1	imbalanced, lethal	imbalanced, lethal	translocation heterozygote	imbalanced, lethal

Thus, among the viable progeny you would expect a ratio of 2/6 fertile (homozygous normal + translocation homozygote) : 4/6 semisterile (translocation heterozygote) = **1/3 fertile : 2/3 semisterile**.

b. This problem involves the self-fertilization of a particular translocation heterozygote. Instead of producing 2/6 fertile : 4/6 semisterile progeny as in part a, this plant produced a ratio of 1/5 fertile : 4/5 semisterile. These numbers suggest that one out of the two fertile and viable classes in part a did not survive in this cross. Thus, one possible explanation for these results is that **the translocation homozygotes die because the translocation breakpoint interrupts an essential gene**.

13-23. If the primers are to detect the presence of the inversion one of the primers must anneal to the inverted sequence. The other primer can be outside the inversion. The longest product will be produced from a primer within the inversion and a primer at the right end of the sequence shown in Solved Problem I of Chapter 6. PCR primers are extended at their 3' ends. Therefore the sequences of **the 11 base long primers must be 5' GTTCGCATACG 3' and 5' GTGTACGCACG 3'**. These primers will give a product that is 34 base pairs long. The primer pair 5' TAAGCGTAACC 3' and 5' CGTATGCGAAC 3' will also generate an inversion specific product but this product will only be 28 base pairs long.

13-24. A Robertsonian translocation (Figure 13.22) is a reciprocal translocation between two telocentric non-homologous chromosomes. One chromosome breaks in the short arm near the centromere. The other chromosome must break in the long arm near the centromere. One translocated product has the long arm and centromere of the first chromosome joined to the long arm of the second chromosome. The other product of the translocation is has the short arm of the first chromosome attached to the centromere and short arm of the second chromosome. This product is very small and does not contain many genes. This small product may eventually be lost during meiosis leaving just the large translocation product.

Based on the location and direction of primer A on chromosome 21 (shown in black below) one translocation breakpoint must occur just to the right of the primer A arrow. Thus the Robertsonian translocation between chromosome 21 and chromosome 14 (shown in gray below) discussed here must contain the entire long arm of chromosome 21 (21q), the chromosome 21 centromere and a small amount of the short arm of chromosome 21. The rest of this translocation product must be the long arm of chromosome 14 (14q) excluding the chromosome 14 centromere. Thus the chromosome 14 break must be just to the right of the centromere. This would generate the Robertsonian translocation shown below, and **primer pair A and 5** would be the best primer pair to produce a diagnostic PCR product. Primer 6 is too far away from the translocation breakpoint to produce a PCR product with primer A.

Section 13.2 – Transposable Genetic Elements

13-25. The homolog with the transposon forms a loop which contains the un-paired transposon sequences. The probe will hybridize to one spot at the base of the loop on the normal, non-inserted homolog. On the homolog with the transposon insertion, the probe will hybridize to the DNA on both sides of the base of the loop.

13-26. If two transposons are near each other on the chromosome and have normal gene-containing chromosomal DNA between them they could transpose the entire large segment of DNA when transposase acts on the ends the transposons (Figure 13.27).

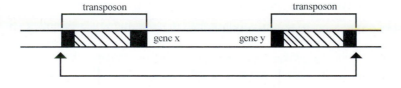

transposase acts on far ends and
transposes the whole section of
the chromosome

13-27. *Ds* **is a defective transposable element that does not encode transposase. The** *Ac* **element is a complete, autonomous copy of the same transposon that encodes transposase.** *Ds* **can thus transpose (move around the genome) only in the presence of** *Ac*. **Chromosomal breakage at** *Ds* **insertion sites could be due to errors in the transposition mechanism catalyzed by the** *Ac* **transposase. A new insertion of** *Ds* **into a gene might yield a mutant allele that is unstable in the presence of** *Ac* **transposase (that is, the transposase could subsequently move** *Ds* **out of the location where it caused the mutation). Because** *Ac* **is complete and autonomous, it contains ends that can be recognized by the transposase. Thus, in different strains** *Ac* **has transposed itself into different chromosomal locations**.

13-28. The original *ct* mutant allele was caused by the insertion of the gypsy transposable element. **The stable** *ct*⁺ **revertants are likely to be precise excisions in which the** *gypsy* **element has moved out of the gene, restoring the normal** *ct*⁺ **sequence. The unstable** *ct*⁺ **revertants are likely to be** cases in which the transposition process altered the *gypsy* element in the *ct* gene so that the gene could function normally. However, the bit of the transposon that remains may try to transpose when in the presence of the transposase. These attempts might cause deletions or rearrangements of the *ct* gene, thus creating new *ct* mutant alleles. The stronger *ct* mutant alleles could result from imprecise excision of the *gypsy* element, leading to the deletion or alteration of sequences within the *ct* gene that would strongly affect its functions (**Figure 13.29**). Alternatively, these stronger *ct* alleles could result from the movement of the *gypsy* transposon into other parts of the gene that would compromise gene function more seriously.

13-29. Create a probe composed of the DNA sequence preceding the 200 A residues. Hybridize the probe to a genomic Southern blot. If other copies of a retrotransposon are present, you would expect to see several hybridizing bands. You could also use the probe to do *in situ* hybridization to human chromosomes. If the probe is homologous to a retrotransposon (or if the sequence is repeated in the genome for any reason), there will be several bands of hybridization.

13-30. It is easiest to work this type of problem if you figure out the sizes of the genomic fragments and place these on the map. Remember that the probe extends from 2.6 kb to 14.5 kb.

a.   **5**. The base change exactly at coordinate 6.8 will alter the restriction recognition site at that position and therefore the 1.1 and 3.0 kb fragments will disappear to be replaced by one new 4.1 kb fragment.

b.   **3**. A point mutation at position 6.9 has no effect on these restriction fragments since there is no EcoRI site at that position.

c.   **2**. This deletion removes 0.3 kb of DNA within the 2.5 kb fragment, resulting in the loss of the 2.5 kb fragment and the appearance of a new 2.2 kb fragment.

d.   **6**. This deletion removes 0.3 kb of DNA including a restriction site. The 1.1 and 3.0 kb restriction fragments would disappear and be replaced by one new 3.8 kb fragment.

e.   **8**. The insertion of a transposable element at coordinate 6.2 will change the size of the 1.1 kb fragment. The 1.1 kb fragment will disappear and be replaced by and a new larger fragment. If the transposable element has a restriction site in it, the 1.1 kb fragment will be replaced by 2 new fragments, one a minimum of 0.5 kb and the second a minimum of 0.6 kb in size. Because the new fragments actually seen are 4.9 and 2.3 kb long, the results suggest that the transposable element is at least $(4.9 + 2.3 - 1.1) = 6.1$ kb long; it could be even larger if the transposable element has more than one site for this restriction enzyme.

f.   **10**. An inversion with breakpoints at 2.2 and 9.9 will alter the two fragments in which the breakpoints are located but will not affect the 1.1 and 3.0 kb fragments that occur in between. Thus the 5.7 kb and 2.5 kb fragments will disappear and be replaced by a 5.9 kb fragment and a 2.3 kb fragment. The total size of this region of the genome will not change.

g.   **7**. The reciprocal translocation will cause the 2.5 kb fragment to disappear. Two new fragments will be seen, each containing part of the 2.5 kb fragment. One of the new fragments will be a minimum of 0.3 kb and the other will be a minimum of 2.2 kb.

h.   **1**. A reciprocal translocation with a breakpoint at 2.4 will cause the 5.7 kb fragment to disappear. This will be replaced by 1 new fragment with a minimum size of 3.3 kb. The sequences between coordinates 0 and 2.4 will also be connected to DNA from the other chromosome; though this will make a new restriction fragment, it will not be visible in this Southern blot because the probe does not include any of this DNA.

i.   **4**. The duplication will increase the size of the 2.0 kb fragment by 3.0 kb, creating a new 5.0 kb fragment. The restriction map makes it clear that these duplicated sequences do not contain a site recognized by the restriction enzyme.

j.   **9**. The 2.5 kb fragment is maintained but the 4.6 increases to 6.6 kb. Because there is a restriction site within the region that is tandemly duplicated, a new fragment of 2.0 kb is now present.

Section 13.3 – Rearrangements and Evolution

13-31.

a. Note that almost all of the *K. waltii* genes are homologous to a *S. cerevisiae* gene. The *K. waltii* genes that are shown in black are homologous to two different genes in *S. cerevisiae*. For example the left most *K. waltii* gene is homologous to both *S. cerevisiae* gene 206 on chromosome 4 and gene 233 on chromosome 12. Likewise the other darkly shaded *K. waltii* gene is homologous to both *S. cerevisiae* gene 201 on chromosome 4 and gene 238 on chromosome 12. Therefore the Scer 4 gene 206 and Scer 12 gene 233 are the result of a duplication, so **the black *K. waltii* genes are duplicated in *S. cerevisiae*.**

b. Both *S. cerevisiae* and *K. waltii* are descendants of a common ancestor. **At some time after the evolutionary lines for these two species separated a portion of the *S. cerevisiae* genome was duplicated in a progenitor of *S. cerevisiae*. Over time one copy was lost of many of the duplicated genes. Occasionally both copies of a gene were retained**, probably because they had changed by mutation into genes with slightly different functions which were both valuable to the organism. This hypothesis explains both the presence of the duplicated genes in *S. cerevisiae* and the 'interleaving' pattern of genes from *K. waltii* that are found in the same order on the two different *S. cerevisiae* chromosomes. The figure below diagrams the processes described.

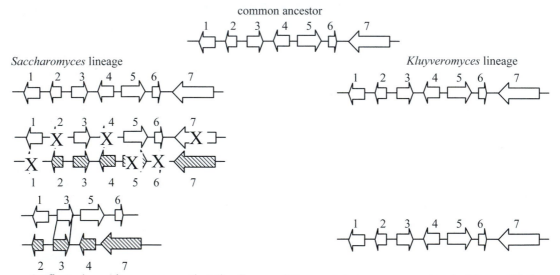

At first glance it may seem that the *S. cerevisiae* genome was reconstructed by multiple rearrangements of a single genome as shown in Figure 13.1. Multiple rearrangements can account for the differences between human and mouse chromosomes. However, such rearrangements do not explain why the same two *S. cerevisiae* chromosomes are always involved instead of any of the other sixteen chromosomes. Multiple rearrangements also cannot account for the conserved relative order of the genes between the three chromosomes in the two species nor for the duplicated genes.

In fact similar studies on the entire genome of both species of yeast suggest that the entire genome was duplicated in an ancestor of *S. cerevisiae* but not during the evolution of *K. waltii*. Subsequently one copy of most of the duplicated genes was lost in the evolution of *S. cerevisiae*. Only about 500 duplicated genes remain today in the *S. cerevisiae* genome. Presumably these duplicated genes provide an evolutionary advantage to this species, or they would have been lost as well.

13-32. Look specifically at the genes that are duplicated in *S. cerevisiae* but found only once in *K. waltii* and compare their DNA sequences. In the first model one of the S. *cerevisiae* genes should be more similar to the *K. waltii* gene than the other *S. cerevisiae* gene does to the *K. waltii* gene. The more similar genes retain the original function which must be conserved in *K. waltii* because it has only one copy, while the other *S. cerevisiae* gene would acquire mutations more rapidly. In the second model the two *S. cerevisiae* genes would have roughly an equal number of sequence changes relative to the gene in *K. waltii*. The data analyzed in these cases suggests that almost all of the duplicated genes in *S. cerevisiae* have evolved in a fashion consistent with the first model.

Section 13.4 - Changes in Chromosome Number

13-33.

a.   The **x number in *Avena* is 7**. This represents the number of different chromosomes that make up one complete set.

b.   **Sand oats are diploid ($2x = 14$); Slender wild oats are tetraploid ($4x = 28$); Cultivated wild oats are hexaploid ($6x = 42$).**

c.   The number of the chromosomes in the gametes must be half of the number of chromosomes in the somatic cells of that species. **Sand oats: 7; slender wild oats: 14; cultivated wild oats: 21**.

d.   **The *n* number for each species is the number of chromosomes in the gametes and therefore is the same as the answer in c.**

13-34.

a.   **15 ($= 2n + 1$)**

b.   **13 ($= 2n - 1$)**

c.   **21 ($= 3n = 3x$ because the original species was diploid)**

d.   **28 ($= 4n = 4x$)**

13-35.

a. (i) **aneuploid**, (ii) **monosomic for chromosome 5**, (iii) this will **probably be an embryonic lethal** because of genetic imbalance – this plant has probably evolved to require two alleles of a least some of the genes on chromosome 5.

b. (i) **aneuploid**, (ii) **trisomic for chromosomes 1 and 5**, (iii) this will **probably be an embryonic lethal** because of genetic imbalance – there are many genes where too much gene product is detrimental to the organism.

c. (i) **euploid**, (ii) **autotriploid**, (iii) **adults** should be **viable but they will essentially be infertile**.

d. (i) **euploid**, (ii) **autotetraploid**, (iii) **adults** should be **viable and fertile**, particularly if there is a mechanism that makes the chromosomes pair as bivalents or as quadrivalents (<u>problem 14-42</u>).

13-36.

a. (i) $5x$ = **45 chromosomes**, (ii) **allopentaploid**, (iii) should be **sterile** - there are an odd number of chromosomes so there is no way to get an even distribution of chromosomes to the gametes during meiosis.

b. (i) $4x$ = **36 chromosomes**, (ii) **autotetraploid**, (iii) should be **fertile if the chromosomes could pair as bivalents or as quadrivalents** (<u>problem 13-44</u>). The gametes should have half the number of chromosomes, so n = **18**.

c. (i) $3x$ = **27 chromosomes**, (ii) **autotriploid**, (iii) should be **sterile** as there is an odd number of chromosomes.

d. (i) $4x$ = **36 chromosomes**, (ii) **allotetraploid**, specifically an **amphidiploid**,(iii) should be **fertile** as the chromosomes in the two B genomes can pair with each other as bivalents and the chromosomes in the two D genomes can do the same, n = **18**.

e. (i) $3x$ = **27 chromosomes**, (ii) **allotriploid**, (iii) **infertile**; the chromosomes cannot pair at all.

f. (i) $6x$ = **54 chromosomes**, (ii) **allohexaploid**, (iii) should be **fertile** as each chromosome has a pairing partner of its own type, n = **27**.

13-37. Remember that nondisjunction in MI causes one copy of each homolog to be found in the gamete, while nondisjunction in MII causes both copies of a single homolog to be found in the gamete. Both nondisjunction in MI and nondisjunction in MII also produce gametes without the chromosome (nullo). See <u>Chapter 4 Synopsis</u> and <u>Problem 4-34</u>. **Possibility A** received 2 copies of the smaller band from Fred; this is **due to non-disjunction during meiosis II in the father. Possibility B is caused by non-disjunction in meiosis I in the mother. Possibility C arose by non-disjunction in meiosis I in the father. Possibility D arose by non-disjunction in the mother in meiosis II.**

13-38.

a. **Perhaps the most obvious mechanism for uniparental disomy is the fusion of a nullo gamete from one parent (the result of nondisjunction in either MI or MII in that parent, <u>Problem 4-34</u>) with an MII nondisjunction gamete from the other parent** (the mutant homolog must be the one that undergoes MII nondisjunction). **However, several other scenarios are also possible**, of which a few are described here. A second mechanism begins with the fusion of a nullo gamete from one parent (due to MI or MII nondisjunction) with a normal gamete carrying the mutant allele from the other parent. This generates a monosomic embryo, which then undergoes mitotic nondisjunction of the monosomic chromosome very early in development to create an individual homozygous for that chromosome. A third possible mechanism begins with the formation of a normal heterozygous embryo. Very early in development this embryo undergoes mitotic nondisjunction, leading to loss of the normal allele and retention of one homolog with the mutant allele. This loss of the normal homolog is then followed by a second mitotic nondisjunction to generate a homozygous mutant genotype. A fourth mechanism is fusion of a normal wild type gamete from one parent with a gamete carrying 2 copies of the affected homolog (the result of nondisjunction in MII). This would generate a trisomic embryo. Early in development a mitotic nondisjunction or chromosome loss event would cause the loss of the normal allele, leaving the 2 mutant alleles.

   There is no easy way to distinguish between these possibilities, because they all lead to the same result of uniparental disomy. In very rare cases, it might be possible to discriminate between these scenarios if the individual were a mosaic and you could find some cells with aneuploid chromosome complements predicted by one of the proposed mechanisms.

b. **Girls with unaffected fathers that are affected by rare X-linked diseases could be produced by any of the mechanisms described in part a**. The mother must be a heterozygous carrier, and the father's chromosome is the one that must be lost. **The transmission of rare X-linked disorders from father to son could only be explained by the first mechanism described in part a.** For the son to be a male and also affected, the son <u>must</u> inherit <u>both</u> his father's X and Y chromosomes. The maternal X chromosome must also be lost; this could occur by a meiotic non-disjunction event in the mother, or by a mitotic non-disjunction/chromosome loss event early in the development of an XXY zygote.

c. Another way in which a child could display a recessive trait if only one of the parents was a carrier involves mitotic recombination early in the development of a heterozygous embryo. The recombination event must occur between the mutant gene and its centromere **(Figure 5.24)**. The zygote then forms from the recombination product that is homozygous for the mutant allele. To detect the occurrence of mitotic recombination, **you must examine several DNA markers along the chromosome arm containing the gene involved in the syndrome**. In particular, you want to compare markers close to the centromere with those near the telomere. **If the presence of the disease were due to mitotic recombination, then the person would be heterozygous for markers near the centromere but homozygous for those near the telomere. If uniparental disomy caused by any of the mechanisms described in part a was instead involved, the individual would be homozygous for all of the DNA markers** (assuming that the individual is not a mosaic).

13-39. Meiotic nondisjunction should give roughly equal numbers of autosomal monosomies and autosomal trisomies. In fact, the total number of monosomies would be expected to be greater, because chromosome loss produces only monosomies. The actual results are the opposite of these expectations. The much higher frequency of trisomies seen in the karyotypes of spontaneous abortions suggests that human embryos tolerate the genetic imbalance for 3 copies of a gene much better than one copy. Also, one copy of a chromosome will be lethal if that copy carries any lethal mutations. Monosomies usually arrest zygotic development so early that a pregnancy is not recognized, and thus they are not seen in karyotypic analysis of spontaneous abortions.

13-40. Both types of Turner's mosaics could arise from chromosome loss or from mitotic nondisjunction early in zygotic development. Chromosome loss would involve the loss of one of the X chromosomes early in development in an XX embryo (producing a mosaic with both 46, XX and 45, XO cells) or the loss of the Y chromosome in the XY embryo (yielding a mosaic with 46, XY and 45, XO cells). Mitotic non-disjunction in a normal XX embryo should produce an XXX daughter cell in addition to an XO, while mitotic non-disjunction in an XY embryo yields an XO and an XYY daughter cell. If the XXX or XYY daughter cells did not expand into large clones of cells during development, karyotype analysis would not be able to detect their presence. Note that for mitotic non-disjunction to have given rise to the described mosaic individuals, the nondisjunction event must have occurred after the first mitotic division so that there would be some XX or XY cells.

13-41. You have 3 marked 4th chromosomes: ci^+ ey, ci ey^+ and ci ey. *Drosophila* can survive with 2 or 3 copies of the 4th chromosome, but not with 1 copy or 4 copies. You are looking for mutations which are defective in meiosis and cause an elevated level of nondisjunction.

a. **Mate potential meiotic mutants that are $ci^+ ey / ci ey^+$ with $ey ci / ey ci$ homozygotes. The normal segregants should be $ci^+ ey / ey ci$ (ey) and $ci ey^+ / ey ci$ (ci). Nondisjunction in MI will be seen as the rare $ci^+ ey / ci ey^+ / ey ci$ (wild type) progeny**. Nullo-4 gametes without any copy of chromosome 4 would produce zygotes with only 1 copy of this chromosome that would not survive.

b. **The cross in part a will detect nondisjunction in MI, but it will not distinguish MII nondisjunction**.

c. Diagram the test cross: $ci^+ ey / ci ey^+ / ey ci$ x $ey ci / ey ci$ → ?
Remember that in a trisomic individual, two of the three copies of the chromosome pair normally at metaphase I of meiosis, while the third copy assorts randomly to one pole or the other. There are 3 different ways to pair the 4th chromosomes in the trisomic individual. The first option is: 1/3 ($ci^+ ey$ segregating from $ci ey^+$ with $ey ci$ assorting independently) = 1/6 probability of (1/2 $ci^+ ey / ey ci$: 1/2 $ci ey^+$) and 1/6 probability of (1/2 $ci^+ ey$: 1/2 $ci ey^+ / ey ci$). The second option is: 1/3 ($ci^+ ey$ segregating from $ey ci$ with $ci ey^+$ assorting independently) = 1/6 probability of (1/2 $ci^+ ey / ci ey^+$: 1/2 $ey ci$) and 1/6 probability of (1/2 $ci^+ ey$: 1/2 $ey ci / ci ey^+$). The third option is: 1/3 ($ci ey^+$ segregating from $ey ci$ with $ci^+ ey$ assorting independently) = 1/6 probability of (1/2 $ci ey^+ / ci^+ ey$: 1/2 $ey ci$) and 1/6 probability of (1/2 $ci ey^+$: 1/2 $ey ci / ci^+ ey$). Each option gives the following progeny phenotypes (which is the same as the gamete genotypes, as this is a test cross) in the following frequencies: option 1 = 2/12 + ey : 2/12 ci + ; option 2 = 1/12 wild type : 1/12 ci ey : 1/12 ci + : 1/12 + ey; option 3 = same as option 2. The 3 options can be summed to give the final result: **1/3 ci + : 1/3 + ey : 1/6 wild type : 1/6 ci ey**.

d. These compound 4th chromosomes (att4) can be used in crosses to assay potential mutants. For instance, cross a potential mutant that is $ci^+ ey / ci ey^+$ (as in part a) to a fly with attached 4th chromosomes that are not marked (that is, both are $ci^+ ey^+$); that is, **potential meiotic mutants of genotype $ci^+ ey / ci ey^+$ x att4 $ci^+ ey^+ / ci^+ ey^+$** → ? In this cross all of the normal progeny would have 3 copies of the 4th chromosome and would be phenotypically wild type. **Nondisjunction in MII would be seen as unusual + ey progeny or ci + progeny**. This occurs because half the gametes in the att4-containing parent would have no copy of the 4th chromosome (nullo-4) since the att4 chromosome does have a partner. The products of nondisjunction in MII in the other parent would thus be $ci^+ ey / ci^+ ey$ (+ ey) or $ci ey^+ / ci ey^+$ (ey +). **In this case, nondisjunction in MI is not distinguishable** because the resulting progeny would be $ci^+ ey / ci ey^+$ (wild type like the normal progeny). **Another possible cross would be:**

potential meiotic mutants that are *ci⁺ ey / ci ey⁺* x att4 *ci ey / ci ey* → . In this cross the normal progeny will be + ey and ci +. **Nondisjunction in MI will give unusual wild type progeny** (*ci⁺ ey / ci ey⁺* gametes from the potential mutant; nullo-4 gametes from the att4 parent). Nondisjunction in MI or MII would yield nullo-4 gametes from the potential mutant, so the progeny (which could only be formed with att4 *ci ey / ci ey*) would be phenotypically ci ey. **With this second cross you can screen for nondisjunction both in MI and MII.**

13-42. The genotype of the diploid *Neurospora* cell is shown below. For a review of recombination in fungi and tetrad analysis see <u>Chapter 5 Synopsis and Problem Solving Tips</u>. For a review of meiotic segregation and non-disjunction see <u>Chapter 4</u>. The figure below shows prophase of meiosis I that occurs during this mating. In the answers to parts a-f, asci are written as though they have 4 (rather than 8 spores), and all spores are written in order relative to the equator of the ascus:

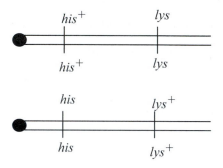

a.   A single crossover between the centromere and *his* will give a PD ascus that shows MII segregation for both *his* and *lys*: **1 *his⁺ lys* : 1 *his lys⁺* : 1 *his⁺ lys* : 1 *his lys⁺*.**

b.   A single crossover between *his* and *lys* will give a tetratype ascus showing MI segregation for *his* and MII segregation for *lys*: **1 *his⁺ lys* : 1 *his⁺ lys⁺* : 1 *his lys* : 1 *his lys⁺*.**

c.   Non-disjunction during MI causes one daughter cell to have both homologs (4 chromatids) and the other daughter cell to have none (nullo). At metaphase II of meiosis, the homologous chromosomes align independently of each other on the metaphase plate, and 1 chromatid from each segregates into the gametes. This leads to 2 gametes that are diploid for the chromosome that underwent non-disjunction. (The nullo cell will have two nullo daughters.) The gametes after MII segregation are: **2 *his⁺ lys / his lys⁺* (his⁺ lys⁺) : 2 nullo (aborted, white).**

d.   Non-disjunction during MII affects one of the 2 daughter cells formed after MI segregation. In the affected daughter cell, both chromatids go to one gamete and the other gamete is nullo. Two different results are possible, depending on which daughter cell undergoes MII non-disjunction: **2 *his⁺ lys* : 1 *his lys⁺ / his lys⁺* (his lys⁺) : 1 nullo (aborted) OR 1 *his⁺ lys / his⁺ lys* (his⁺ lys) : 1 nullo (aborted) : 2 *his lys⁺*.**

e. The single crossover between the centromere and *his* involving chromatids 2 and 3 gives the following meiotic structure:

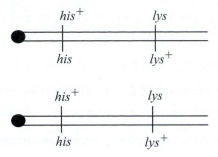

Non-disjunction in MI gives one daughter cell with both homologous centromeres and a second nullo daughter cell. At metaphase of MII the homologous centromeres line up on the metaphase plate independently of each other. MII segregation then causes one of each sister chromatid to segregate into the 2 resultant gametes. After the crossover the bivalents are no longer homozygous. As a result, there are 2 different segregation patterns that can happen, leading to the following types of asci: **1 *his⁺ lys* / *his⁺ lys* (his⁺ lys) : 1 *his lys⁺* / *his lys⁺* (his lys⁺) : 2 nullo (aborted) OR 2 *his⁺ lys* / *his lys⁺* (his⁺ lys⁺) : 2 nullo (aborted).**

f. The single crossover between *his* and *lys* gives the following meiotic structure at metaphase I of meiosis:

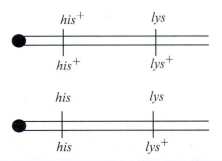

Non-disjunction at MI causes both centromeres to segregate to one daughter cell while the other daughter cell is nullo. Again, the homologous chromosomes line up independently of each other at metaphase II of meiosis, leading to 2 different possible segregation patterns: **1 *his⁺ lys* / *his lys* (his⁺ lys) : 1 *his⁺ lys⁺* / *his lys⁺* (his⁺ lys⁺) : 2 nullo (aborted) OR 1 *his⁺ lys* / *his lys⁺* (his⁺ lys⁺) : 1 *his⁺ lys⁺* / *his lys* (his⁺ lys⁺) : 2 nullo (aborted).**

13-43. See **Figure 13.36c**. Haploid plant cells in culture can be **treated with colchicine to block mitosis (segregation of chromosomes) during cell division and create a daughter cell having the**

diploid content of chromosomes. Once the diploid resistant cell is obtained, it is grown into an embryoid. Proper hormonal treatments of the embryoid will yield a diploid plant.

13-44. Only plants that are F^A- F^B- will be resistant to all 3 races of pathogen. Because the chromosomes of same ancestral origin still pair, i.e. F^a pairs with F^A and F^b pairs with F^B, the cross can be represented as a dihybrid cross between heterozygotes. This treats the resistance genes from the two ancestral species as independently assorting genes. You are doing a cross between parents that are heterozygous for 2 different genes. Thus, **9/16 of the progeny will have the F^A- F^B- genotypes and these plants will be resistant to all three pathogens**.

13-45. Karyotype analysis is done at metaphase of mitosis. The chromosomes are stained such that each different kind of chromosome has a characteristic banding pattern. **The banding patterns of the homologs in the autopolyploids should be the same (for example, in an autotetraploid you would see four copies of each type of banding pattern), but the chromosomes of the different species that formed the allopolyploids would probably have different banding patterns (so in an allotetraploid you would see two copies of each type of banding pattern).** Note that karyotype analysis allows you to determine the x number (the number of chromosomes in the basic set), because this will equal the number of different patterns you see. In turn, this allows you to determine the ploidy (the number of copies of the basic set), which is equal to the number of copies of each pattern.

13-46.

a.   In an autotetraploid species all four sets of chromosomes are from the same species. Normally these chromosomes pair two by two and form two bivalents (**Figure 13.36b**). However all four chromosomes are homologous, so this is not the only pairing option. Any one of the four chromosomes could pair with a second chromosome over part of its length, and with a third chromosome over the rest of its length. The remainder of the third chromosome would be available to pair with the fourth chromosome, making a quadrivalent. In the figure below, each solid circle is a centromere. **The figure shows one of the possible quadrivalent pairing configurations of the four chromosomes during meiosis I**. Each chromosome is drawn as a single line for simplicity, though of course in meiosis I each chromosome is actually composed of two sister chromatids.

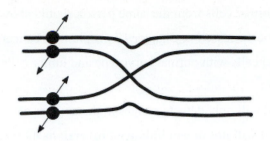

b. **To make euploid gametes, two of the chromosomes in the quadrivalent would have to go to one spindle pole during anaphase I, and the other two chromosomes to the other pole. The arrows in the diagram above show one way this can be done**. Paired centromeres (the two centromeres that are synapsed) will be connected to the spindle so that one centromere attaches to fibers from one spindle pole and the other centromere attaches to fibers from the other spindle pole. The pairing is usually ensured by molecular mechanisms at the centromeres that ensure centromere/spindle fiber connections are stabilized only when there is mechanical tension that pulls paired centromeres in opposite directions. In the two sets of paired centromeres shown above, two chromosomes will indeed go to one spindle pole and the other two chromosomes to the other spindle pole, thus giving euploid gametes.

c. An allopolyploid contains two different though related species. Allopolyploids are usually sterile because the chromosomes from the two species cannot pair with each other. Effectively the allopolyploid is haploid for two chromosome sets. Occasionally a fertile allopolyploid, or amphidiploid, arises when rare chromosomal doubling generates a homologue for each chromosome. **As long as the two genomes in the amphidiploid are sufficiently different from each other, none of the chromosomes of one genome should be able to pair and synapse with chromosomes of the other genome. Thus, quadrivalents should not normally form in amphidiploids**. However, if the two species that formed the hybrid amphidiploid had diverged from each other only in the very recent evolutionary past, it is possible that some of these chromosomes would have retained sufficient homology to form quadrivalents.

Section 13.5 – Beyond the Karyotype

13-47.

a. The data shows that 21 out of the 23 chromosomes are found in 2 copies (diploid). However **chromosomes 12 and the X are present in three copies (triploid)**. Note that a portion of chromosome 11 is found only in one copy, so this cell line is also **heterozygous for a large deletion of much of the long arm of chromosome 11 and for a shorter deletion in the long arm of chromosome 13.**

b. A virtual karyotype of normal cells from the same person should show **two copies of all regions in the genome**. As will be discussed in <u>Chapter 17</u> people with cancers are often genetic mosaics, consisting of both normal cells with normal karyotype and tumor cells having abnormal karyotypes.

c. This karyotyping method will only detect changes in the number of copies of a region or an entire chromosome. Therefore it **will not detect balanced inversions or translocations**. Neither heterozygosity nor homozygosity for these types of rearrangements changes the gene dosage. **It might also be difficult to detect small deletions or duplications**. The detection of such small rearrangements would depend on both the size of the individual DNA fragments being analyzed in each spot and on the density of these spots along the chromosome.

d. Any region of the genome with a change in dosage may contain a gene that contributes to leukemia. Thus **regions on chromosomes 11 and 13 might contain genes that cause cancer when dosage is decreased while chromosome 18 and the X chromosome might have genes that contribute to leukemia when there are extra copies**, (when the dosage is increased).

e. The **changes in gene dosage** (number) discussed in part d. **cannot affect the viability of cells** since tumor cells with these changes grow and divide perfectly well (too well in fact).

Chapter 14   Prokaryotic and Organelle Genetics

Synopsis:

This chapter describes characteristics of genetic analysis in bacteria, with a focus on gene transfer in *E. coli*. Two key features have made *E. coli* a powerful model organism for understanding basic cell processes. First, the ability to grow massive numbers of bacteria on defined media allows easy and quick isolation of mutants. Second, there are many naturally occurring ways to transfer DNA from one cell to another (**Figure 14.13**). The DNA transfer mechanisms - transformation, conjugation, and transduction (generalized and specialized) and the ways in which geneticists use these are described. In addition to their value for research, DNA transfer mechanisms are important for survival and evolution of bacterial species.

This chapter goes on to discuss mitochondria and chloroplasts. Mitochondria and chloroplasts are eukaryotic organelles that contain their own genomes. The genomes of both share similarities with prokaryotic genomes in the types of genes, the organization of the genes as well as chromosome structure (**Table 14.2 and Table 14.4**). The presence of several organelles per cell in most organisms leads to many different combinations of organelle types in the same cell. Heteroplasmy refers to the presence of more than one type of genome per cell; homoplasmy is the state in which there is uniformity of the genome copies. Mitotic segregation produces an uneven distribution of organelle genes in heteroplasmic cells. Inheritance of organelle genomes is not dependent on the same cell machinery as nuclear chromosomes use for mitosis and meiosis, so the inheritance of these organelles is non-Mendelian. In most organisms, organelles are inherited from one generation to the next from the mother. This leads to distinct patterns of inheritance as seen in pedigrees.

Mitochondria and chloroplast functions require cooperation between the organelle and nuclear genomes.

Significant Elements - Gene Transfer in Bacteria:

After reading the first part of the chapter and thinking about the concepts, you should be able to:

- Distinguish between selection (only one specific genotype can grow) and screening (more than one genotype can grow, but additional analysis is needed to establish the genotype of each cell).

- Describe how Hfr cells are formed (**Figure 14.17**) and how they are used for mapping genes (**Figure 14.21**) and complementation analysis.

- Understand the differences in DNA transfer between matings with F$^+$ cells (**Feature Figure 14.16**) and Hfr cells (**Figure 14.19**).

- Set up Hfr crosses to map genes (describe the genotypes of donor and recipient and the selective media used, **Figure 14.20**).

- Analyze time of entry data to map genes and origins of transfer for different Hfr strains (**Figure 14.21**).
- Map genes using cotransduction (**Figure 14.23**) or cotransformation frequencies.
- Describe the differences between transformation, transduction, and conjugation.
- Describe the similarities between transformation and generalized transduction.
- Map genes using specialized transduction (**Figure 14.26**).

Significant Elements - Genetics of Chloroplasts and Mitochondria:

After reading the chapter and thinking about the concepts, you should be able to:

- Differentiate between heteroplasmy and homoplasmy experimentally.
- Recognize organelle inheritance in human pedigrees.
- Suggest explanations for unusual phenomenon involving organelle genomes that may arise from heteroplasmy and different proportions of affected cells.
- Describe the endosymbiont theory and discuss the supporting evidence.
- Recognize a 4:0 segregation of a phenotype in the gametes as showing that the trait is controlled by a mitochondrial or chloroplast gene. Remember that nuclear genes always segregate 2:2.

Problem Solving Tips - Gene Transfer in Bacteria:

- The F plasmid can integrate at different locations around the *E. coli* chromosome to generate Hfr strains that have different origins of transfer. These locations for integration are determined by the presence of IS elements scattered around the genome. The same IS elements are found on the F plasmid.
- Genes closest to the origin of transfer are more likely to be transferred into a recipient than those further from the origin.
- The time at which genes are transferred into a recipient in an Hfr cross is a reflection of the distance from the origin of transfer.
- Hfr crosses are usually used to get a low resolution map; P1 transductions (generalized transduction) are useful for finer genetic mapping.
- Bacteriophage lambda (λ) can integrate into the chromosome (lysogeny) to generate a lysogen (**Figure 14.26**).
- Bacteriophage lambda can excise imprecisely from the bacterial chromosome, picking up adjacent bacterial genes.

Problem solving tips - Genetics of Chloroplasts and Mitochondria:

After reading the chapter and thinking about the concepts, you should be able to:

- Explain the characteristics of mitochondrial gene inheritance in humans, including maternal inheritance and variable levels of expression in affected individuals.
- Understand the implications of the fact that there can be several organelles in a cell and several copies of the genome in each organelle.

Solutions to Problems:

<u>Vocabulary</u>

14-1. a. **4**; b. **5**; c. **2**; d. **7**; e. **6**; f. **3**; g. **1**.

<u>Section 14.1 – General Overview of Prokaryotes</u>

14-2. It is not easy to discriminate unicellular organisms as eukaryotes, bacteria, or archaea. Eukaryotes have nuclei and mitochondria enclosed in their own membranes, while bacteria and archaea (being prokaryotes) do not. However, because many unicellular organisms have cell walls it is not always possible to look through the cell walls to visualize the presence or absence of nuclei, and mitochondria are often hard to see. Furthermore, the nature of the genome is not completely predictive. Most bacteria have a single circular chromosome while eukaryotes have multiple linear chromosomes, but some bacteria have linear chromosomes or multiple circular chromosomes. All known archaea have circular chromosomes, but there is no particular reason a currently uncharacterized species of archaea might have a linear chromosome. Neither cell shape nor cell size is diagnostic. There are many other interesting comparisons that could be made between these three groups: for example, eukaryotes and archaea have translational systems with methionine as the initiating amino acid, not formylmethionine as in bacteria; eukaryotes and bacteria have lipids of a particular type while those in archaea are of another type; all three groups have genes interrupted by introns although these are far more common in eukaryotes than the other two groups. **The most reliable measure for characterizing unicellular organisms is actually DNA sequence comparisons with known members of each group**. Historically bacteria and archaea were grouped together until DNA sequence comparisons made it clear that they comprised two groupings that were separated a very long time ago in evolution.

14-3. The initial tube of bacteria has 2×10^8 cells/ml. The first step of the dilution series is a 10^{-2} dilution (0.1 ml of the initial tube into 9.9 ml of diluent = 0.1 ml/10.0 ml = 1/100). The second step of the dilution is again 10^{-2}, so at this point the total dilution is 10^{-4}. The third step is a 10^{-1} dilution (1 ml/10.0 ml = 1/10) for a total dilution of 10^{-5}. You then put 0.1 ml (10^{-1} ml) of this dilution on the petri plate. **Therefore, you are putting 10^{-5} x 10^{-1} ml x (2x10^8 cells/ml) = 2x10^2 cells on the first petri plate, which should grow into 200 colonies.** The fourth step of the series is another 10^{-1} dilution, for a total dilution of 10^{-6}. You again plate 0.1 ml of this dilution, so you expect 10^{-6} x 10^{-1} ml x (2x10^8 cells/ml) = 2x10^1 cells = **20 colonies on this second plate.**

14-4.

a. In order to determine the number of nucleotides necessary to identify a gene, you must calculate how many bases represent a unique sequence in a DNA molecule of the size of the *E. coli* genome (4.6 Mb or 4,600,000 bases). There are 4 bases possible at each position in a sequence, so 4^n represents the number of combinations found in a sequence n bases long. For example, 4^2 is the number of unique sequence combinations that could be made with 2 positions. Therefore, if you were looking for a unique 2 nucleotide sequence, you might expect to find it, on average, every 1/16 nucleotides. A sequence of 11 nucleotides would appear $1/4^{11}$ or once in every 4×10^6 bases (4 Mb); a sequence of 12 nucleotides would appear uniquely $1/4^{12}$ or one in 16.8 Mb. Therefore **you need a sequence of about 12 nucleotides in order to define a unique position in the *E. coli* genome.** This problem can also be solved using the equation $4^n = 5 \times 10^6$ (5 Mb). To solve for n, rearrange the equation: $n\log4 = \log5 \times 10^6$; $n = 11.1$. Thus you need more than 11 nucleotides to find a unique nucleotide sequence.

b. This problem assumes that you have obtained a sequence of contiguous amino acids within a protein. The 12 nucleotides shown in part a define a unique position in the genome would encode 4 amino acids. However, because of the genetic code's degeneracy, you actually only know the identity of about 8 of these nucleotides. (For amino acids specified by 6 codons, you would potentially know one less nucleotide per codon; for tryptophan and methionine, which are specified by only a single codon, you would know all three nucleotides in the codon.) **If you had a sequence of six amino acids, you would probably know at least 12 unique nucleotides.** Because many genes evolved through a pattern of duplication followed by divergence, some protein domains of 6 amino acids might appear in more than one protein. As a result, knowing a few more than 6 amino acids would make the case that you have identified the correct gene even stronger.

14-5. In this problem, "minimal media" contain no carbon sources, so they require supplementation with the indicated sugars. In most cases, if there is no carbon source specified (e.g. rich media + X-

Gal) then the sugar is glucose. Otherwise, all sugars in the media must be specified. Note that X-Gal is a substrate for the ß-galactosidase enzyme, but it cannot serve as a carbon source.

a.   **(iv)** Lac$^+$ cells are able to use lactose as the sole carbon source for growth, while Lac$^-$ cells would not be able to grow if the only sugar in the media were lactose. Medium iv is thus selective, because Lac$^+$ cells can grow on it but Lac$^-$ cells cannot.

b.   **(iii)** A screen is different than a selection. For a genetic screen, you need to be able to examine the phenotype of each individual cell or colony. To screen for Lac$^+$ cells, you therefore need a medium on which both Lac$^+$ and Lac$^-$ cells can grow but on which they have different visible phenotypes. In a selection, only certain genotype(s) of cells can grow (min + lac selects for Lac$^+$ cells and against Lac$^-$ cells). The rich medium (iii) allows both Lac$^+$ and Lac$^-$ to grow, so it is <u>not</u> selective. However, the X-Gal in these plates distinguishes between the two phenotypes (Lac$^+$ cells are blue; Lac$^-$ cells are white).

c.   **(ii)** To select for Met$^+$ cells, the medium should lack methionine, demanding that the bacteria must be able to synthesize methionine in order to grow. Medium ii is the only choice that lacks methionine.

14-6.

a.   To find the original linezolid-resistant strains, you would **take a wild-type (linezolid-sensitive) isolate of pneumococuccus and place it on petri plates containing linezolid. This is a direct selection** for the desired linezolid-resistant bacteria. You will find very rare resistant colonies amidst a very large number of originally sensitive bacteria without using replica plating or enrichment or treatment with mutagens.

   Finding linezolid-sensitive derivatives of the linezolid-resistant mutants is much harder. You must somehow screen to find the rare desired bacteria, since they will not grow in the presence of the antibiotic as the resistant bacteria do. You must use replica plating to identify the desired bacterial colonies that grow the absence of the drug and not in its presence. You may use mutagenesis to increase the frequency of mutations, and you want to do an enrichment procedure if this is possible with pneumococci.

b.   In general, mechanisms which affect the bacterial toxicity of an antibiotic include disrupting the import of the antibiotic into the cell or increasing its removal from the cell; altering the rate of chemical modification of the antibiotic and thus activation or detoxification; alterations to the cellular targets of the antibiotic.

   In terms of linezolid-resistant mutants, you could imagine either loss-of-function or gain-of-function mutations that would affect antibiotic import. For example, **a loss-of-function mutation** could destroy a receptor or pump that imports the antibiotic into the cell or, less likely, a gain-of-function mutation that would increase the efficiency of a system that might export the antibiotic out of the cell. Gain-of-function mutations in a system that normally detoxifies the antibiotic or, less likely, loss-of-function mutations in a system that normally alters the antibiotic to increase its toxicity are also possible. Finally, it is possible that a mutation might disrupt an RNA or protein component of the 50S ribosomal subunit so that it could no longer be bound by linezolid. Such mutations could not be loss-of-function mutations in terms of the function of the component in translation since the cell would die if it could not synthesize proteins. These are only some of the possibilities.

In this example the linezolid-sensitive derivatives could be due to a reversion of the original mutation or they could be caused by the opposite of any of the types of mutations described above, for example, a gain-of-function mutation in a system that is responsible for importing the antibiotic into the cell.

Section 14.2 – Bacterial Genomes

14-7. The *purE* and *pepN* genes will be cotransformed at a lower frequency if the *H. influenzae b* pathogenic strain was used as a host donor strain. There is a lower likelihood that the two genes will be on the same piece of DNA because they are separated by 8 more genes-worth of DNA (~ 8 kb of DNA) than in the *H. influenzae Rd* non-pathogenic strain.

14-8.

a. **(1) Pass an extract from *E. coli* cells which have been induced for β-galactosidase expression through an APTG-agarose resin.** β-galactosidase will bind to the resin along with any proteins that in turn bind to β-galactosidase. **You remove these proteins from the column using a detergent. (2) Digest the mixture of proteins with trypsin to generate a large number of smaller peptides. (3) Subject this mixture of peptides to mass spectrometry to get the molecular weight of the peptide fragments.** The molecular weights of each of the peptide fragments will be unique because of the range of molecular weights for individual amino acids. **(4) Take this profile of peptide molecular weights and compare it to a list of all the expected molecular weights for the tryptic fragments of all proteins in *E. coli*.** This list has already been determined using the genomic DNA sequence information. The comparison will tell you which proteins are present in the mixture, and therefore which genes encode proteins binding to β-galactosidase.

b. **Make a fusion between the gene for your protein of interest and the *lacZ* gene. Clone the gene fusion into an expression plasmid that will produce the fusion protein in bacterial cells.**

Then repeat the steps you performed in part a. When an extract from these cells is passed over an APTG-agarose column, you will isolate proteins that bind to your protein of interest as well as those that bind β-galactosidase. Mass spectrometry of the fragments resulting from tryptic digests will identify the proteins. Ignore the proteins that were also seen in the experiment discussed in part a, as this will be those that bind to β-galactosidase, rather than to the protein of interest.

14-9. A partial list of organisms and the genes introduced include: **plasmid transformation into** *Shigella dysenteriae*, **bacteriophage infection of** *Staphylococcus*, *Streptococcus* or *E. coli* **species and transposition of DNA (pathogenicity island) into** *Vibrio cholerae*.

14-10. Isolate genomic DNA separately from *E. coli B* **and** *E. coli K*, **digest these DNAs with the** *Eco*RI **restriction enzyme, electrophorese the DNAs on a gel, and transfer the DNAs to a nitrocellulose filter for a genomic Southern hybridization using an IS1 DNA as a probe.** The numbers of bands that appear correspond to the number of IS1 elements in the genome because *Eco*RI does not cut within the element. Instead, the enzyme will cut at the nearest recognition sites that flank each IS1 element. If a band is twice as intense as other bands, this would indicate there are two different fragments of the same size that contain an IS1, and such a band represents two different IS1 elements.

14-11. Do a mating between the mutant cell with 3-4 copies of F and a wild-type F⁻ recipient. If the mutation were in the F plasmid, you expect the recipient strain into which the plasmid is transferred (the exconjugant) to have the higher copy number. If the mutation is in a chromosomal gene, the higher copy number phenotype would not be transferred into the recipient. There is however a potential complication with this experiment. When you do an F⁺ x F⁻ mating, some of the F⁺ cells will have converted to Hfr cells that can transfer bacterial DNA. An alternative experiment could avoid this complication. **You could isolate the F plasmid DNA from the mutant cell, and then transform this plasmid into new recipient cells. By examining the number of copies of the F factor in the transformed cells, you could tell whether the trait was carried by the plasmid.**

14-12. An **imprecise excision of a bacteriophage from a bacterial strain with virulence genes produced a specialized transducing bacteriophage containing both bacterial DNA (virulence genes in this case) and bacteriophage genes.** This specialized transducing bacteriophage then infected the ancestral pathogen and integrated into the genome.

Section 14.3 – Gene Transfer in Bacteria

14-13.

a. The general mechanisms of gene transfer in bacteria **are (i) transformation, (ii) conjugation and (iii) transduction**.

b. Natural transformation usually involves recipient cells taking up small fragments of single stranded DNA from the surroundings. **However if the donor DNA used in the transformation includes plasmids recipient cells may take up the entire plasmid and acquire the characteristics conferred by the plasmid genes. Conjugation requires the presence of a conjugative plasmid in the donor cell.** Such plasmids carry the genes necessary for transfer of the plasmid DNA into a recipient cell. A few of these conjugative plasmids also allow transfer of donor bacterial genes to the recipient cell. An example of such a plasmid is the F plasmid.

c. **Bacteriophages are required for transduction**. There are two different sorts of phage-mediated bacterial DNA transfer - generalized transduction and specialized transduction.

d. Bacteria degrade linear DNA. Therefore any mechanism that transfers linear DNA to the recipient cells requires recombination in the recipient cell to produce genetically stable cells. There must be two recombination events between the circular bacterial chromosome and the homologous linear DNA. This results in the exchange of the alleles on the linear DNA with the alleles on the chromosome. The linear DNA is subsequently degraded leaving the transferred alleles stably integrated into the chromosome. Such mechanisms include **natural transformation with DNA fragments, conjugation with an Hfr and generalized transduction.**

14-14. The donor DNA is fragmented into small pieces of about 20 kb during transformation. Furthermore, the donor DNA replaces only a small percentage of the recipient's chromosome. Thus only genes that are close together on the chromosome can be cotransformed. The entire *E. coli* chromosome is about 4.2 Mb long. **If *purC* and *pyrB* are located half way around the chromosome from each other they are roughly 2.1 Mb apart making it impossible for them to be cotransformed.**

14-15. Transfer the plasmid (by transformation) into a non-toxin producing recipient strain. If the gene is encoded on the plasmid, the transformed cells will produce the toxin.

14-16.

a. The partner strain should be **F⁻, Str^R and mutant (negative) for all the markers to be transferred from the Hfr strain.**

b. Selecting for Pyr$^+$ exconjugants selects for an early marker transferred from the donor into the recipient. The frequency with which genes beyond *pyrE* are transferred decreases with distance from this early marker. This presumably occurs because of the fragility of the contact between the cells. The further a gene is from *pyrE,* the more likely it is that the connection between the donor and recipient will be broken before that gene can be transferred. This is very similar to what is depicted in **Figure 14.21**. **The order of genes is: *pyrE xyl mal arg met tyr his*.**

14-17. This is a three factor cross in which the selected marker, *ilv,* is the farthest from the origin of transfer. This ensures that all 3 genes were transferred into all exconjugants. The largest class of exconjugants is the one in which all of the genes recombined into the donor chromosome, which suggests that the genes are close together. The smallest class is the one in which quadruple crossovers occurred. In this case, Ilv$^+$ Bgl$^-$ Mtl$^+$ is the smallest class, so the *bgl* gene must be between *ilv* and *mtl.* The Ilv$^+$ Mtl$^-$ Bgl$^-$ class results from a crossover between *ilv* and *mtl* and another crossover on the other side of *ilv.* The Ilv$^+$ Mtl$^-$ Bgl$^+$ class comes from a crossover between *mtl* and *bgl* and another crossover on the other side of *ilv.* **The order is *ilv bgl mtl*.** Note that you can also obtain an estimate of the distances separating the three genes by looking at the frequencies of the latter two classes. The distance between *ilv* and *bgl* is about 3.3 times the distance between *bgl* and *mtl* (60:18).

14-18.

a. **The order of the genes is either *pab ilv met arg nic (trp pyr cys) his lys* or *pab ilv met arg nic (cys pyr trp) his lys*.**

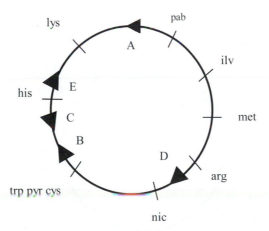

b. See problem 14-17. After mating with HfrB selecting for Nic$^+$ meant the *trp$^+$ pyr$^+$ cys$^+$* alleles had all been transferred to the F$^-$ cell. Trp$^+$ was selected and most of the exconjugants

(790/1,000) also recombined in the *pyr*$^+$ *cys*$^+$ alleles from the donor. This shows that the 3 genes are closely linked. Usually the smallest class of recombinants requires four crossovers instead of two. After the crossovers have occurred, the two outside genes will have the genotype of the Hfr parent and the gene in the middle will have the genotype of the F$^-$ parent. In this problem, however, the smallest class of exconjugants is represented by Trp$^+$ Pyr$^-$ Cys$^-$, so only the <u>middle</u> gene has recombined in the allele from the prototrophic Hfr parent. This happens when the selected gene is the gene in the middle, so **the order is *cys trp pyr*.** That is, the quadruple recombinant would have been Trp$^-$ Pyr$^+$ Cys$^+$, but you can't see this class because you selected for Trp$^+$. The 3 genes are very closely linked, so the double crossover giving Pyr$^-$ Trp$^+$ Cys$^-$ is rare. The two other classes represent double crossovers between *cys* and *trp* or *pyr* and *trp*. The recombination frequency between the genes = the number of recombinants between the two genes / by the total number of exconjugants. **Rf between *cys* and *trp* = (145+5)/1,000 = 0.15 = 15%. Rf between *pyr* and *trp* = (60+5)/1,000 = 6.5%.**

14-19.

a. Arbitrarily place the first Hfr insertion site (HfrA) on the bacterial chromosome and then order the genes that are transferred by that Hfr. When you place HfrB on the same map notice that the first gene transferred by this Hfr is lys. HfrB could be placed on either side of this gene. However, the second gene transferred determines on which side of the first gene the F factor is inserted and the directionality of transfer. The time of transfer for four of the markers *(gly, phe, tyr, ura)* is indistinguishable, so we cannot put them in an order on the map, but *lys, nic* and the cluster of four genes can be placed.

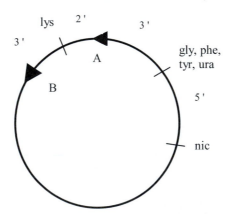

b. See **Figure 14.23**. *Phe* was cotransduced with *ura* more frequently than with *tyr*. Also note that all of the Tyr$^+$ transductants are also Ura$^+$. These two facts both indicate that the order is *phe-ura-tyr*. None of the cotransduction classes is very rare, so none of these result from quadruple crossovers. The relationship of these three genes to other markers and the order of *gly* gene is still unknown, and this is represented by placing *gly* in parenthesis and the flanking marker *lys* and *nic* in brackets.

c.   **To map the *gly* gene with respect to other markers, select for Gly$^+$ transductants on min + lys + phe + tyr + ura+ nic. Then score the other markers to determine which genes are cotransduced with Gly$^+$ at the highest frequency**. The Gly$^+$ transductants would be selected on min + lys + phe + tyr + ura, and then individual transductants would be tested for growth on media lacking one or more of the unselected amino acids.

14-20. See **Figure 14.22**. Transduction is the transfer of bacterial genes mediated by phage. The DNA is protected inside the protein head of the phage. Transformation is gene transfer using naked DNA, which is not enclosed in any protective structure. The DNA being transferred by transformation will therefore be susceptible to degradation by DNase. **If the transfer of the *amp^r* allele still occurs after DNase treatment, transduction must be occurring**.

14-21. To get a stable exconjugant cell from an Hfr mating requires *recA*-mediated recombination of some of the donor genes in the F$^-$ recipient cell. Therefore, **this assay would detect *recA$^-$* mutants in the F$^-$ cell based on the inability to form stable exconjugants** on the selective media.

14-22. In generalized transduction, the bacterial chromosomal DNA is packaged randomly into transducing phage particles that contain no phage DNA, only bacterial DNA. Thus it is possible for any region of the bacterial chromosome to be transferred. In specialized transduction, the phage in the lysate contains bacterial DNA covalently attached to phage DNA. The bacterial DNA that is transduced from any single lysogenic strain is always from the same region of the chromosome, which is adjacent to the insertion site of the integrated phage.

Section 14.4 – Bacterial Genetic Analysis

14-23. Recombineering is a method to identify the function of a gene of unknown function. A mutant allele of the gene is made *in vitro*. PCR is used to insert sequences from the gene on either end of a selectable marker, for instance an antibiotic resistance gene. This linear DNA is transformed into *E.*

coli cells, so the rest of the process occurs *in vivo*. A double cross over between the gene sequences on either side of the antibiotic marker lead to the replacement of the wild type allele of the gene in the bacterial chromosome by an allele with the antibiotic gene inserted into the middle of the gene (see Figure 14.28). The presence of the mutant allele is selected by growing the cells on media containing the antibiotic. The phenotype of these cells can now be determined allowing a determination of the function of the unknown gene.

a. **The primer DNA is a step in the *in vitro* portion** of the process.

b. **Antibiotics are used in the selective media to select cells with the mutant allele of the gene. This is part of the *in vivo* process.**

c. **The recombination enzymes are provided by the transformed *E. coli* cell in the *in vivo* portion of the experiment.**

d. **The PCR amplification is done at the beginning when you are generating the mutant allele *in vitro*.**

14-24.

a. The easiest way to get the plasmid into *S. parasanguis* cells is by transformation. *Streptococci* take up DNA naturally, as seen in the early experiments of Griffiths with *Streptococcus pneumoniae*, but **artificial transformation should be used to increase the efficiency of cells that uptake the plasmid. It is critical that the transformation be done at low temperature, so that the plasmid can replicate in transformed cells. Select for transformed cells containing the plasmid by plating on media containing kanamycin and erythromycin. Next raise the temperature to the restrictive conditions and allow the cells to grow.** The plasmid is unable to replicate at the high temperature, but the transposon will insert into random positions in the bacterial genome. These insertions will contain the Erm^R gene. **To identify the bacteria with insertions plate them at high temperature on medium containing erythromycin but not kanamycin. Use replica plating to make sure that individual colonies are susceptible to kanamycin. Each individual Ery^R Kan^S colony should have an *IS256* insertion in a different genomic location.**

b. Using a transposon-based mutagen has two advantages. First, as explained in part a, **you can select for strains that have a potentially mutagenic insertion of the transposon into the bacterial chromosome.** Second, **the presence of the transposon allows the researchers to rapidly identify new mutations that might disrupt the ability of the bacteria to cause dental plaque.** They could use cloning or PCR to purify DNA adjacent to the *IS256* in the key mutant strains, and then use DNA sequence analysis to determine the particular open reading frame disrupted by the insertion. This will simultaneously localize the site of insertion and tell you the nature of the gene product of the mutated gene.

Section 14.5 – Genetics of Chloroplasts and Mitochondria

14-25. a. **both**; b. **both**; c. **neither**; d. **both**.

14-26.

a. **No**. The hybridization of the plant chloroplast gene to the nuclear DNA sample from the red alga **could be due to the contaminating chloroplast DNA in the nuclear fraction**.

b. The first step would be to **hybridize the plant probe to the chloroplast DNA from the red alga**. If this hybridization is negative it provides strong evidence that the gene has moved to the nucleus in red alga. A second step would be to **hybridize the probe to DNA separated by different methods, for example Southern blots of nuclear DNA and Southern blots of chloroplast DNA**.

c. Do reciprocal crosses of two strains of the red alga with different alleles of this gene. **If the transmission of each allele follows a uniparental pattern, the gene is probably located in the chloroplast genome**. In other words, all of the offspring of one cross should have the same allele of the gene and would not contain the other allele while the progeny of the reciprocal cross would only have the second allele. **If the gene is nuclear then all progeny of both crosses should have both alleles of the gene** (they would show a biparental pattern of inheritance).

14-27.

a. See **Figure 8.3**. The 'universal' code is always true for nuclearly (n) encoded genes.

		Trp	His	Ile	Met
mRNA	5'	UGG	CAU/C	AUU/C/A	AUG
nDNA mRNA-like	5'	TGG	CAT/C	ATT/C/A	ATG
nDNA template	3'	ACC	GTA/G	TAA/G/T	TAC

b. See **Table 14.3**. The mitochondrial (mt) genomes of some organisms have differences from the universal code.

		Trp	His	Ile	Met
mRNA	5'	UGA/G	CAU/C	AUC/U	AUG/A
mtDNA mRNA-like	5'	TGA/G	CAT/C	ATC/T	ATG/A
mtDNA template	3'	ACT/C	GTA/G	TAG/A	TAC/T

14-28.

a.   **The large subunit of Rubisco is encoded in the chloroplast genomes of all three species. The small subunit is nuclearly encoded in green alga and chloroplast encoded in the red and brown alga.**

b.   **In red and brown algae the size of the transcripts** recognized by the large and small subunit probes is the same, so they appear to be **cotranscribed**. This is consistent with the fact that both genes are chloroplast encoded. The transcripts for the green algal Rubisco subunits are different sizes and therefore represent different transcripts. **In the green alga the genes are not cotranscribed** – cotranscription is impossible for genes located in different genomes.

14-29. a and d are characteristics of chloroplasts and mitochondria that are like those found in bacteria. While alternate codons (choice b) are used in mitochondria such variations from the 'universal' code are also found in eukaryotes like yeast as well as in bacterial. Introns (choice c) are found very rarely in bacterial genomes.

14-30. a. 2; b. 1; c. 4; d. 3.

14-31.

a.   Several changes must be made to a nuclear gene like *ARG8* in order for it to be transcribed and its mRNA translated in the mitochondria. First, any **introns would have to be removed**. This could be done by starting with cDNA, rather than genomic sequences. Next **some of the codons in the nuclear gene would have to be changed since the genetic code in the nucleus and in mitochondria is not identical**. The list in Table 14.3 shows the changes in the genetic code specific to vertebrate mitochondria, not yeast mitochondria. The actual changes in the yeast mitochondrial genetic code are: ATA encodes Met, not Ile; TGA encodes Trp, rather than signaling "stop"; and any codon whose first two letters are CT encodes Thr, not Leu. Thus, an ATA codon in the nuclear gene would have to be changed to a different Ile codon; TGA could not be used as a stop codon, but rather would cause Trp to be inserted into the protein. Using TGA for Trp codons in the *ARG8* gene has an additional experimental advantage: this would prevent expression of the novel gene in the event it somehow escaped from the mitochondria back into the nucleus. Any codon starting with CT would have to be changed to an alternative Leu codon. Third, **the open reading frame of the altered nuclear gene would have to be placed under the control of a promoter, a translational start site, and a transcriptional termination site that work in mitochondria.**

The researchers used a DNA synthesizer to make an *ARG8* open reading frame without introns and with the proper genetic code alterations. This was possible because the *ARG8* gene is quite small, but it would be possible to do this by altering an *ARG8* cDNA using *in vitro* mutagenesis. They then replaced the open reading frame of an actual cloned mitochondrial gene with the open reading of the altered *ARG8* that they made. This put the *ARG8* sequences in the proper position with respect to mitochondrial regulatory sequences. Next **they introduced the cloned gene into yeast mitochondria using microprojectile bombardment (the biolistic gun).**

b.   A yeast strain with *ARG8*-expressing mitochondria has at least two advantages. First, it provides a novel phenotype that is dependent upon mitochondrial gene expression. **Such a yeast strain allows one to select for function of the mitochondrial genetic system in mutants that are unable to respire** (grow on glycerol). If the nuclear *ARG8* gene was deleted or otherwise mutated, then the yeast cells could survive on medium lacking arginine only if they had mitochondria making this arginine biosynthetic enzyme. Second, since this *ARG8* gene was actually controlled by authentic mitochondrial regulatory regions like promoters, they could use this strain to study mitochondrial regulatory sequences. That is, they could try to **find arginine auxotrophs that could no longer make arginine because there was a DNA change that obliterated the function of the promoter**, for example. By sequencing such mutations, **they could figure out a lot about the function of regulatory elements in the mitochondrial genome**.

Section 14.6 – Non-Mendelian Inheritance of Chloroplasts and Mitochondria

14-32. a. **3**; b. **1**; c. **2**.

14-33. There are many mechanisms that cause the mitochondrial DNA from one parent to be absent from the zygotes. **The small size of the sperm can mean that organelles are excluded; cells can degrade organelles or organellar DNA from the male parent; early zygotic mitoses distribute the male organelles to cells that will not become part of the embryo; the details of the fertilization process may prevent the paternal cell from contributing any organelles (only the sperm nucleus is allowed into the egg); and in some species that zygote destroys the paternal organelle after fertilization.**

14-34. Offspring will resemble the mother because the sperm does not contribute significantly to the cytoplasm of the zygote.

14-35. Heteroplasmic cells have a mixture of the 2 genotypes in their organelles. Homoplasmic cells have only one type of genome. Thus, homoplasmic cells can either be totally normal or totally mutant. **If the mutation is very debilitating to the cell, either because of the loss of energy metabolism in the case of mitochondria or of photosynthetic capability in the case of chloroplasts, a cell that is homoplasmic for the mutant genome will die.** Therefore, you will only find the mutant plastid genome in heteroplasmic cells.

14-36. In order to determine if individual organelles are heteroplasmic, you **use method a.** Differentially label the probes (different fluorescent tags, for example) that are specific to each of the two genomes. Then do a double hybridization to cells *in situ* and see if both probes hybridize to the same organelles in individual cells. PCR amplification of DNA from a population of cells will <u>not</u> indicate the genotypes of <u>individual</u> organelles. For example, there could be cells containing all a^+ mitochondria and other cells in the population containing all a^- mitochondria. When DNA from such a population is used for PCR amplification, both mitochondrial genomes would be represented, suggesting heteroplasmy although individual cells are homoplasmic.

14-37.

a. **The zygote that formed this plant was heteroplasmic, containing both wild type chloroplast genomes and mutant chloroplast genomes.** The mutant chloroplast genomes are deficient in a gene needed to make chlorophyll. **These two types of genomes can segregate as tissue is propagated mitotically.** The type of leaf on a branch or the color of a portion of a leaf reflects the kind(s) of chloroplast DNAs present in the cells that gave rise to the branch or leaf region.

b. The variegated branch was formed from heteroplasmic cells so **many of the ovules generated on this branch are heteroplasmic and give rise to variegated plants.** However, **some of the cells on the variegated branch gave rise to ovules that segregated one or the other chloroplast genomes. The phenotype of the progeny will reflect the type(s) of chloroplast genome(s) in the ovules** – ovules with chloroplasts with only normal DNA give rise to green leaved plants while ovules with chloroplasts with only mutant DNA give rise to white leaved plants. These **white leaved plants cannot make chlorophyll and since they cannot conduct photosynthesis they will die.**

c. The main purpose of chloroplasts is to use sunlight to generate energy-rich carbohydrates. If **part of the plant can conduct photosynthesis then these carbohydrates can be made and transported to tissues that are unable to conduct photosynthesis.**

14-38.

a. For simplicity, the mitochondrial (double circle) and chloroplast (single) genomes are represented in circular form, and one pair of nuclear chromosomes are indicated, represented as vertical lines. Remember that mitochondria and chloroplasts are inherited maternally in plants. Therefore in this cross **the female parent, which is the male sterile strain, gives the organelle genomes and a haploid set of chromosomes while the male parent, which is the male fertile strain, only gives a haploid set of chromosomes. This cross is diagrammed below**:

female gametes x male gametes →   F$_1$ offspring

b. When **the F$_1$ in part a undergoes meiosis to make the female gametes recombination will occur between the homologous nuclear chromosomes,** as represented by the left hand linear chromosome below. This F1 is backcrossed to the same male fertile plant shown in part a. Therefore **the male gamete will be the same as the male gamete in part a**. The genotype of the resultants first backcross progeny is shown below on the left:

This backcross progeny now provides the female gamete in yet another backcross to the same male fertile strain. Again, nuclear recombination will occur. The resulting genotype is shown above on the right. Remember that the figures above only show one homologous pair of nuclear chromosomes, but the same replacement due to recombination is happening to all other nuclear chromosomes as well.

c. **Each generation of backcrossing increases the percentage of the nuclear genome from the male fertile strain.** The fact that sterility still occurred with each generation indicates that sterility was not a nuclear gene effect. Remember that in each generation the female parent is passing on the cytoplasmic elements (mitochondria and chloroplasts). **These backcrosses allow the researchers to study interactions of chloroplast and nuclear gene(s) that cause male sterility**.

14-39.

a. Remember that this type of CMS is caused by mutant mitochondrial genomes that prevent pollen formation. **If one of the parental inbred lines is male sterile, then this line can not self-fertilize and the seed companies would not have to do anything more to prevent self-**

fertilization. Therefore the sterile inbred line is used as the "female" parent, and the line being used as the "male" parent should be fertile. If the female parent was fertile, then much of the corn produced by the cross would not be hybrid corn, but instead simply more of the inbred line. Prior to the use of the CMS technique, seed companies would prevent self-fertilization by hiring high school and college students to go through the fields and manually remove the tassels before they produced pollen. This is very labor-intensive and expensive.

b. Mitochondrial inheritance in corn is uniparental from the female parent. Thus the F_1 corn plants in part a inherit mutant CMS mitochondrial DNA. These F_1 corn plants must be fertile so the kernels form in the F_1 ears of corn. If the Rf allele of the nuclear Restorer gene suppresses the mitochondrial sterility mutation, then **the sterile inbred line in part a must also be homozygous for the recessive for the *rf* allele of *Restorer*.** These plants are then male sterile and would thus be the female parent in the hybrid-generating cross, as discussed above. **The other inbred line, the male parent supplying the pollen for the cross, would have to have at least one (and preferable two) dominant *Rf* alleles of *Restorer*;** if the male parent was *Rf/-* it would not matter if these plants had normal or CMS mitochondria since the nuclear Rf allele suppresses the sterile phenotype. The F_1 hybrid corn would thus have CMS mitochondria but also carry *Rf*, so these corn plants would be fertile. If the inbred male line was *Rf/rf*, then half the hybrid plants would be sterile and the other half fertile; this would not be a practical difficulty since if these seeds were planted in the same field, the male fertile plants would produce enough pollen to fertilize the ovules from the male sterile plants.

c. One method to produce CMS plants that are also *rf/rf* is to **make a fertile "Maintainer" line that has mitochondria with a normal (non-CMS) genome but whose nuclear genomes are the same as the CMS plants and also *rf/rf*. The *rf/rf* CMS plants are used as the female parent** (they can not produce pollen). **When pollen from Maintainer plants fertilizes the CMS plants, the progeny will have CMS mitochondria and will also be *rf/rf*; in other words, these progeny will be identical to their maternal parents.** Problem 16-21 demonstrates one way to make such Maintainer lines by repeated backcrosses. A second, more complicated method uses CMS *R/ rf* plants of the same inbred background. One fourth of the progeny would be CMS *rf/rf* plants that could be used to make hybrid seed, but you would need some way to identify these plants and isolate them from their male fertile CMS *Rf/rf* siblings.

d. There are two potential issues with using hybrid corn, both of which revolve around the idea of *monoculture*. Since hybrid seed is made by seed companies, **farmers are now dependent upon the seed companies to provide their seed.** In the past farmers saved part of their harvest and used this for corn seed the next season. The fact that most farmers now pay to buy hybrid seed shows the economic advantages of the increased yield due to hybrid corn. Second, farmers now grow only the few varieties of hybrid corn sold by the seed companies. **The genetic variation of the corn crop overall is reduced**. If the hybrid corn is susceptible to a newly-emerging disease, this reduced genetic variation could have disastrous consequences. In 1970, a new mutant strain of a fungus causing leaf blight emerged that was particularly lethal to corn with a CMS cytoplasm called *Texas* (*T*). This fungus decimated much of the hybrid corn crop in the southern U.S. until new fungus-resistant forms of hybrid corn could be developed.

Section 14.7 – mtDNA Mutations and Human Health

14-40. One characteristic of a mitochondrial mutation in many organisms is **maternal inheritance**. Most of the offspring of an affected female are affected. None of the offspring of affected males are affected. Another indication of mitochondrial inheritance is **differing levels of expression of the mutant phenotype in different progeny** due to differing amounts of heteroplasmy in either the egg or the cells of the embryo.

14-41.

a. **The mother (I-1) may have had very low levels of mutant mitochondrial chromosomes**, while the proportion of mutant genomes in the daughter was much higher. Alternatively, there **may have been a spontaneous mutation either in the mitochondrial genome of the egg that gave rise to individual II-2 or in the early zygote of individual II-2**.

b. **You could look at the mitochondrial DNA from somatic cells from various tissues in the mother. If the mutation occurred in her germline and was inherited by II-2, the mother's somatic cells would not show any defective DNA.** Remember that it is impossible to make a firm conclusion based on negative results. Therefore, if no mutant mitochondrial genomes are found in I-1 it is possible that you haven't looked at enough different tissues. Taking the tissue samples is very unpleasant for the donor!

14-42. The zygote that formed these twins was heteroplasmic, with both wild type and mutant mitochondrial genomes. Early in embryonic development two cell masses separated to become the two identical twins. The ratio of wild type to mutant mitochondrial DNAs in the two cell masses may have differed. Furthermore, during the individual development of the twins the ratios of wild type to mutant mitochondria may vary in different tissues. The more affected twin probably had a higher proportion of mutant mitochondria in tissues such as muscles and brain that particularly depend upon energy supplied by mitochondria.

14-43. Because Kearns-Sayre disease arises as a new mutation in an individual, **the variation in affected tissues is due to differences in where and when during development the mutation occurred**. In different individuals, the mutation may occur in cells that give rise to different sets of tissues. If the mutation occurs early in development then more tissues will be affected. **Variation in the severity of the disease can be due to the proportion of mutant genomes (the degree of heteroplasmy) in the cells of different tissues.**

14-44. If the patient is male then you could reassure him that none of his children would be affected by the disease (assuming his mate is unaffected by MERRF). **If the patient is female then she is most likely heteroplasmic. There is a strong chance that her child might be affected by MERRF, but there is no way to determine the probability that the child would be affected or the severity of the disease**, because of the random events in mitochondrial distribution through many rounds of cell division that will influence the ratio of wild type and mutant mitochondria in the egg and in various tissues in the fetus. It is relatively easy by **amniocentesis and PCR to check for the mutant mitochondrial DNA in the fetus**. However, **these results are <u>not</u> diagnostic** because the distribution of wild type and mutant mitochondria can vary considerably in various tissues.

14-45. Gel electrophoresis is better suited for an overview of the differences between mitochondrial genomes. Deletions can be very large and might not be amplified by PCR. In addition, sequences to which primers bind might be deleted in some mutations so no information will be obtained about the sizes of these deletions.

Chapter 15   Gene Regulation in Prokaryotes

Synopsis:

This chapter describes gene regulation in bacteria including the genetic analysis that led up to the postulation of the operon theory - the paradigm of gene regulation. A basic principle derived from experiments on the *lac* operon is that proteins bind to DNA to regulate transcription (**Feature Figure 15.5**). How mutations in the components of the regulatory system proved Jacob and Monod's theory is an instructive lesson in the power of the genetic approach for understanding basic cellular processes. The development of molecular biology techniques and increased study of protein structure allowed the operon theory of gene regulation to be refined so we understand more about DNA binding proteins and their interactions with regulatory regions. Experiments on the lactose operon led to the development of fusion technology in which the *lacZ* gene is placed next to a regulatory region of another gene. Expression of the other gene could then be monitored by measuring expression of β-galactosidase. Another type of fusion was developed in which the regulatory region of the *lac* operon was fused to a gene whose expression was then controlled using the induction of the *lac* genes.

In addition to the negative (**Feature Figure 15.5**) and positive (**Figure 15.15**) control described for the *lac* operon, global transcriptional regulation based on changes in RNA polymerase and its subunits is described as is attenuation- a mechanism for fine-tuning transcription (**Figure 15.30**).

Significant Elements:

After reading the chapter and thinking about the concepts, you should be able to:

♦ Predict overall cellular expression of proteins when there are site or gene mutations on a chromosome and other mutations on the F' plasmid (merodiploid analysis).

♦ Distinguish between positive (inducers) and negative regulators based on the behavior of mutants.

♦ Propose models for regulation of a set of genes based on mutant and molecular analyses.

♦ Describe the process of attenuation and how mutations in the component parts might affect expression of the *trp* genes or other amino acid operons that are regulated by attenuation.

♦ Design molecular experiments to test predictions of models based on mutational analysis.

♦ Describe use of and distinguish between *lacZ* fusions (geneX-*lacZ*, **Figure 15.27**) and lac regulatory fusions (*lac*-geneX, **Figure 15.28**).

Problem Solving Tips:

♦ Negative regulation blocks transcription.

♦ Positive regulation increases transcription.

♦ Sites and proteins that bind to sites form the control system of gene regulation.

- Regulatory sites in the DNA (P, O) only affect DNA and structural genes adjacent to them (**Figure 15.14)** – they are cis-regulatory.

- Regulatory proteins diffuse in the cytoplasm and therefore can act on any copy of their binding site in a cell (**Figures 15.12** and **15.13**) – they are trans-regulatory.

- Regulatory proteins that bind to several molecules (e.g., DNA, another protein, and inducers) have distinct regions or domains in the protein for these interactions.

- Learn the nomenclature for the different sorts of mutations - P^-, O^C, I^-, I^S, etc.

- When asked to determine regulation in a partial diploid (merozygote or merodiploid or a cell with an F' plasmid) examine the promoters first. If one of the promoters is non-functional, then you can ignore all of the cis-regulated structural genes. Next, consider the O/repressor interactions - are they wild type or mutant (constitutive or permanently repressed)? Then examine the structural genes for their ability to give a functional enzyme.

Solutions to Problems:

Vocabulary

15-1. a. **4**; b. **8**; c. **5**; d. **2**; e. **7**; f. **1**; g. **3**; h. **6**.

Section 15.1 – An Overview of Prokaryotic Gene Regulation

15-2. The first step of gene expression is the binding of RNA polymerase to the promoter. At this point the system is simple and regulation of gene expression involves adjusting the level of RNA polymerase and/or the level of genomic DNA. If the main controls of gene expression were later in transcription the cells would waste energy making RNA that would not be utilized.

Regulation of gene expression at the level of translation might be more difficult. There are many more molecules are involved and it might thus be harder to evolve mechanisms that specifically regulate particular mRNAs. You should remember that evolution does not always develop the most efficient solution to a problem, although that is likely true in this case. Evolution selects for the fitness of random mutational events, so any moderately successful mechanism may have evolved and been conserved.

15-3.

a. **i, ii, iii**. The first two, i and ii, reflect the rate at which the transcription of different operons is initiated. This will depend on the intrinsic affinity of the promoter of each operon for RNA polymerases - some promoters are stronger than others. The regulation of transcriptional initiation is also affected by the binding of factors like repressors to DNA sequences near promoters. In addition, one can imagine that some promoters are recognized by different types of RNA polymerase. For example, some sigma factors may be needed for RNA polymerase to recognize some promoters and not others. Point iii indicates that the level to which an mRNA accumulates in the cell depends not only on its rate of transcription but also on its rate degradation. Different mRNAs can be degraded at different rates due to differences in their structures that are recognized by various ribonucleases in the cell.

b. **iv, v, vi**. Point iv is quite interesting. In operons the "distal" genes that are further from the promoter may be less efficiently transcribed or transcribed later than the "proximal" genes that are closer to the promoter. If the mRNA of the operon is long it will take RNA polymerase some minutes to reach the distal genes. The RNA polymerase might sometimes fall off the DNA or reach transcription termination signals before transcribing the distal genes. Since transcription and translation are coupled in bacteria the transcription of distal genes might actually be regulated by the rate at which the proximal genes in the same mRNA are translated. Point v indicates that the efficiency of translational initiation for genes in the same operon can vary. Point vi is important because the amount of protein that accumulates in a cell is not only due to the rate it is translated, but also to the rate at which that protein is degraded.

15-4. The lack of *rho* function must be lethal for the cell, so **the *rho* gene function is essential**. Conditional mutations are the only sort of mutation that can be isolated for essential genes.

<u>Section 15.2 – Regulation of Transcription</u>

17-5. Mutations in the promoter region can only act in cis to the structural genes immediately adjacent to this regulatory sequence. This promoter mutation will not affect the expression of a second, normal operon.

15-6.

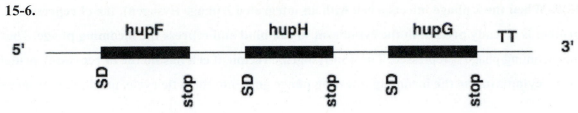

Transcription (**Feature Figure 8.11**) begins within the promoter region for this operon. Therefore, the 5' end of the mRNA shown above corresponds to the 3' end of the promoter region. The RNA polymerase continues transcribing the template DNA strand until it reaches the transcription termination signal (TT in the figure above). The transcription termination signal (**Figure 15.2**) could be a hairpin loop (an intrinsic signal) near but not directly at the 3'-end of the mRNA or it could be a sequence near the 3' end of the mRNA which is bound by a protein factor (an extrinsic sequence). Translation (**Feature Figure 8.25**) begins when a ribosome binds to the ribosome binding site in the mRNA which is made up of the Shine-Dalgarno sequence (SD in the figure above) and, a short distance downstream, the initiation codon. The solid rectangles represent the translated regions (open reading frame or ORF) for each gene. Note that there are 3 separate locations at which translation starts, one for each gene in the operon. The initiation codons for each ORF are not shown in the figure above, but they are a few nucleotides downstream from the SD sequences. In prokaryotes the initiating amino acid is fMet. Translation terminates at the nonsense codon (stop in the figure above) which leads to the release of the polypeptide and the dissociation of the ribosomal subunits. **Note that there are several regions in the mRNA which are NOT translated. These include the sequences upstream of the first SD sequence, known as the 5' UTR (untranslated region). Other non-translated regions of the mRNA include the sequences between each ORF (the intergenic regions) and the sequences downstream of the last ORF (the 3' UTR).**

15-7. Statement (b) will be true. When a positive regulator is inactivated there is no expression from the operon.

15-8.

a.   When the DNA binding protein is mutant you have constitutive expression. Therefore, **the wild type regulatory protein blocks transcription = negative regulator.**

b.   **Strain (i) will have inducible *emu1* and constitutive *emu2* expression, while strain (ii) will have inducible *emu1* and *emu2* expression.**

c.   **Strain (i) will have inducible *emu1* and *emu2* expression while strain (ii) will have inducible *emu1* and constitutive *emu2* expression.**

15-9. When the λ phage infects a cell with an integrated λ phage (lysogen), the *cI* repressor protein is already present in the cytoplasm so can bind and repress the incoming phage. Thus the incoming phage is repressed. **The non-lysogenic recipient cell has no the *cI* repressor protein in the cytoplasm, so the incoming infecting phage goes into the lytic cycle, producing progeny phage**.

15-10. These revertants of the constitutive expression of the lac operon could be changes in the base sequence of the operator site that compensate for the mutation in *lacI*. For example, if an amino acid necessary for recognition of operator DNA is changed in the *lacI* mutant, a compensating mutation changing a recognition base in the operator could now cause the mutant LacI protein to recognize and bind to the operator.

15-11. The *lacZ* gene codes for the β-galactosidase enzyme. The *lacY* gene codes for a permease. For an overview of the different types of mutations, see 'How to Begin Solving Problems' at the beginning of this chapter.

a.   *lacZ* **is constitutive;** *lacY* **is constitutive.**

b.   *lacZ* **is constitutive;** *lacY* **is inducible.**

c.   *lacZ* **is inducible;** *lacY* **is inducible.**

d.   **no expression of** *lacZ*; *lacY* **is constitutive.**

e.   **no expression of** *lacZ*; **no expression of** *lacY*.

15-12.

a.   In the presence of glucose, **the repressor protein (LacI) is bound** to the operator.

b.   in the presence of glucose + lactose there will be **no proteins bound to the regulatory region** of the lac operon.

c.   in the presence of just lactose, **the CAP-cAMP complex will bind to the promoter region** of the lac operon.

15-13. See 'How to Begin Solving Problems' at the beginning of this chapter.

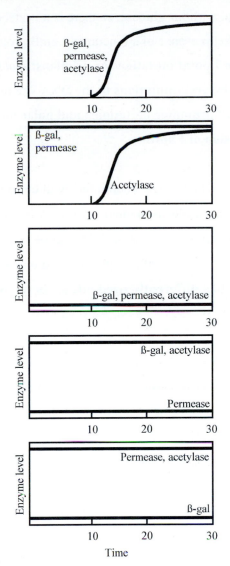

15-14.

a. The *malT* gene codes for the positive regulator that increases the expression of all three operons. A loss of function mutant in *malT* would be unable to transcribe the genes in the three operons, so the bacterial strain would be **unable to utilize maltose as a carbon source (mal⁻).**

b. **No,** the other operons would not be expected to have CRP binding sites. **The expression of the MalT positive regulatory protein is catabolite sensitive,** so this allows catabolite sensitive expression of all the other maltose structural genes.

c. Bacteriophage lambda binds to the LamB protein in order to initiate phage infection. Therefore the *E. coli* will only be sensitive to infection with lambda if the LamB protein is expressed. This will happen when the cells are grown in media **with maltose**, to activate *malT* which regulates expression of *lamB*, and **without glucose**, so there is no catabolite repression.

d. Any mutations that affect presence of a functional LamB protein would be resistant to lambda. These will include: ***lamB* point mutations or deletions; *malT* mutations that prevent positive regulation (superrepressors); mutations to the DNA site to which *malT* binds (operator mutations); *malK lamB* promoter mutations; and polar nonsense mutations in *malK* which will also block expression of LamB.**

15-15. If the three genes make up a single operon they will be cotranscribed as one polycistronic mRNA. Thus, **when a Northern blot is probed with DNA from each of these genes you will always see one band of the same size. This mRNA will be large enough to include the RNA of all three genes. If the genes are not part of an operon, each one will be transcribed separately so each probe will hybridize to a differently sized mRNA in the Northern blot.**

15-16.

a. The evidence that arabinose induces expression of the *araBAD* genes is seen in the genotypes and media conditions 1 and 2. **When all the genes are wild-type there is no expression of the structural genes in the absence of arabinose and the structural genes are expressed in the presence of arabinose.**

b. The conclusion that *araC* encodes a positive regulator of the operon is based on the data from the strains and conditions in 3 and 4. **The mutant version of *araC* blocks synthesis of the three structural gene products in the presence of arabinose.**

15-17. This problem can be approached by starting with the expression data and assessing what types of mutations could produce that particular phenotype or by starting with each mutant and matching it with an expression pattern. Using the latter approach, the superrepressor mutant (a) will show no expression under any conditions and therefore is either mutant 3 or 4. The operator deletion (b) will give constitutively high levels of expression with glycerol or lactose but will give low expression with lactose + glucose because it will be catabolite repressed. Therefore the operator deletion is mutant 5 or 6. The amber suppressor tRNA would have no effect on its own and is mutant 7. The defective CAP-cAMP binding site (d) produces the same low levels of expression with lactose or lactose + glucose because the CAP-cAMP complex cannot bind to the promoter to increase expression. Thus mutant 1 or 2 contains a mutation in the CAP-cAMP binding site. The nonsense mutation in β-galactosidase (e) will give no expression, so is mutant 3 or 4. The nonsense mutation in the repressor gene (f) leads to constitutive expression of β-galactosidase, as in mutants 5 and 6. A defective *crp* gene (g) means the CAP-cAMP complex cannot form so it cannot bind to the promoter region as seen in mutants 1 and 2. To summarize, mutations 1 and 2 are the mutant CAP-cAMP binding site and the defective *crp* gene; mutations 3 and 4 are a superrepressor and a nonsense mutation in the β-galactosidase gene; mutations 5 and 6 are an operator deletion and the nonsense mutation in the repressor gene; and mutation 7 is an amber suppressor tRNA.

The next stage of the analysis involves understanding the results of the double mutant and merodiploid genotypes. An amber nonsense mutation of the *lacZ* gene would be suppressed by mutation 7 (amber suppressor tRNA) in the same cell. Therefore, mutant 3 contains the nonsense mutation in β-galactosidase (*lacZ*) gene and mutant 4 must be a superrepressor mutation. Likewise, a nonsense mutation in the repressor gene would be suppressed by mutation 7 in the same cell, so mutant 5 is the mutant repressor gene and mutant 6 is the operator deletion. The defective *crp* gene can be distinguished from the defective CAP-cAMP binding site by the merodiploid genotypes presented. In both merodiploids the bacterial chromosome has a mutant gene for β-galactosidase, but all other parts of the lac operon expression system are wild type. The F' element has either a mutant CAP-cAMP binding site or a defective *crp* gene. In the latter case the presence of the trans-acting wild type crp gene on the bacterial chromosome will give normal regulation of the lac operon. However if the mutation on the F' element is in the cis-acting CAP-cAMP binding site the phenotype of the merodiploid will still be mutant. Thus, mutant 1 is the *crp* mutation and mutant 2 is the altered binding site. Therefore, **a. 4; b. 6; c. 7; d. 2; e. 3; f. 5; g. 1.**

15-18.

a. **If you only screen for Lac$^+$ revertants, some of the revertants will be compensating mutations in just the lac operon CAP-cAMP binding site. Demanding that the suppressor mutations affect both Mal$^+$ and Lac$^+$, will give more general revertants, which affect <u>all</u> CAP-cAMP binding or activity.**

b. **The α subunit of RNA polymerase interacts directly with the CAP protein.**

15-19.

a. Mutations **i, iii, v and vi** would all prevent the strain from utilizing lactose.

b. The *lacY* deletion will be complemented by any other mutation that expresses the *lacY* protein. This will include **mutations ii, iii and iv**.

c. First look for striking patterns that might indicate a specific mutation. For example, mutant 6 shows expression of *lacZ* when combined with any of the other mutations. Of the possible choices, this could only be an O^c mutation. Strains containing mutation 5 and any of the mutations other than mutation 6 also have a very consistent pattern. *lacZ* is never expressed, except when combined with mutation 6 (the O^c mutation). Mutation 5 therefore shuts down the other copy of the operon in addition to its own copy, which can be explained by a superrepressor (I^S) mutation. Looking at the remaining mutations, the inversion of the *lac* operon (iv) should not have an effect on expression and should not influence expression from other copies of the operon. Mutation 4 leads to inducible *lac* expression except when it is combined with mutation 5, the superrepressor. The inversion that does not include *lac I, P,* and *O* should not show expression because the regulatory region is now in the opposite orientation from the genes. It also should not influence expression from the other mutant copy. Looking at the patterns for mutations 2 and 3, mutation 3 does not show expression except when combined with 4, so fits with inversion. Mutation 2 is a *lacZ* mutation. Therefore, **mutation 1 is i; mutation 6 is ii; mutation 2 is iii; mutation 4 is iv; mutation 5 is v; and mutation 3 is vi**.

15-20.

15-21.

a. **Mutations in O$_2$ or O$_3$ alone have only small effects on synthesis levels** and would therefore be difficult to detect in screens for mutations that affect the regulation of the lac operon.

b. The functional lac repressor is a tetrameric protein. A functional lac operator must have two binding sites for the repressor protein on the same face of the DNA. Two subunits of the tetramer bind to the two recognition sequences at the O$_1$ site and the other two subunits of the tetramer bind to either the O$_2$ or O$_3$ site, bending the DNA in between the two operators. These two operators must be on the same face of the DNA. **Small DNA insertions between O$_1$ and O$_2$ may change the face and either change the ability of the repressor to bind one of the sites or change the ability of the bound repressor to bend the DNA leading to an O^c mutant phenotype.**

c. The O^c mutation described in part b will probably be **insensitive to a I^S repressor protein**. The degree of insensitivity depends on the extent to which repression is due to the binding of the repressor to the individual O sites vs. the looping of the DNA after repressor binding. Superrepressor mutations (I^S) may still bind individually to each of the three operator sites but once bound they may not be able to generate looping of the DNA. Previous studies have also found that repressor binding to O_2 or O_3 does not repress expression in O^c mutations containing a defect in O_1.

15-22.

a. **The operator begins at the left end of the sequence give**n, after the endpoint of deletion 1 and before the endpoint of deletion 5. The right endpoint of the operator cannot be determined by this data.

b. **The deletion may have removed bases within the promoter that are necessary for the initiation of transcription**.

15-23. The protein coding region of your gene must be in the same reading frame as the *lacZ* gene.

Section 15.3 – Attenuation of Gene Expression

15-24.

a. **At least three ribosomes** are required for the translation of *trpE* and *trpC* from one mRNA molecule. The fact that there is a full length mRNA means that early in transcription one ribosome must have initiated translation of the leader (attenuator). This ribosome must have stalled at the tryptophan codons in the leader. Therefore the remainder of the operon is transcribed giving a full length mRNA. A second ribosome must have bound to the ribosome binding site (problem 7-6) at the beginning of the *trpE* open reading frame, and a third ribosome must bind to the ribosome binding site at he beginning of the *trpC* open reading frame.

b. If the two tryptophan codons in the leader were deleted the 3-4 stem-loop would form (**Figure 15.30**) causing transcription termination. Thus, **the transcription of the *trpE* and *trpC* genes would be rare** regardless of tryptophan concentration, since the level of full length mRNA from the *trp* operon would be very low.

15-25. Analyze this mRNA sequence. There is one open reading frame beginning with the first nucleotide. The predicted amino acid sequence is:

N Met Thr Arg Val Gln Phe Lys His His His His His His His Pro Asp C

There are 7 histidines in a row out of a total of 16 amino acids! Thus, if the cell is starving for histidine, the ribosomes will pause at the His codons (CAC or CAU) in the sequence because the tRNA^{His} molecules will not be completely charged with histidine. This ribosome pausing causes the RNA polymerase to complete transcription of the operon 100% of the time, giving maximal production of the polycistronic mRNA and maximal expression of the structural gene proteins which synthesize histidine.

15-26. In order to determine the effect of the various genotypes on the expression of the operon(s) focus on the mutant portions of the operons.

a. If the promoter is deleted **there is no expression of either *trpE* or *trpC*** because transcription cannot occur.

b. The repressor is defective so transcription of the structural genes is constitutive. However, the attenuator is normal and this results in partially constitutive expression of the structural genes. There will be a **lower level of expression of *trpE* and *trpC* when tryptophan is present and high level of expression when there is no tryptophan**.

c. The repressor cannot bind tryptophan and thus cannot bind to the operator so transcription of the structural genes is constitutive. However, the attenuator is normal and this results in partially constitutive expression of the structural genes. Thus there will be a **lower level of expression of *trpE* and *trpC* when tryptophan is present and high level of expression when there is no tryptophan**.

d. The repressor cannot bind to the operator, so expression of *trpC* and *trpE* is constitutive. The attenuator is also mutant so the operon will always be transcribed at the maximal level. Thus, ***trpC* and *trpE* will show completely constitutive expression**.

e. There is **inducible expression of *trpC*** (remember that r^+ is a <u>trans</u> regulatory protein), and **partially constitutive expression of *trpE*** because the cis acting operator is defective but the attenuator is still functional.

f. There is **inducible expression of *trpC*** (remember that r^+ is a <u>trans</u> regulatory protein), but **no expression of *trpE*** because the cis acting promoter is defective

g. Expression of ***trpE* is fully constitutive** because the cis acting operator is defective and the cis acting attenuator site is defective. The ***trpC* expression is partially constitutive** because the $trpC^+$ gene is cis to a defective operator and a functional attenuator.

15-27.

a. The operon **seems to be a biosynthetic operon** because in a wild type operon the addition of Z decreases the expression of the A, C and D genes. **The operon is repressible.**

b.

Condition	Gene A	Gene B	Gene C	Gene D
wildtype	completely repressible	constitutive	completely repressible	completely repressible

Nonsense in A	not expressed	constitutive	completely repressible	not expressed
Nonsense in B	partially repressible	not expressed	partially repressible	partially repressible
Nonsense in C	not expressed	constitutive	not expressed	not expressed
Nonsense in D	completely repressible	constitutive	completely repressible	not expressed
Deletion of region incl. E	partially repressible	constitutive	partially repressible	partially repressible
Deletion of F	partially repressible	constitutive	partially repressible	partially repressible
Deletion of G	not expressed	constitutive	not expressed	not expressed

c. A, C, and D are structural genes needed for the biosynthesis of compound Z. Compound B is constitutively produced so it is not part of the transcriptional unit that includes the structural genes. A mutation in B leads to a partial turn-off of the structural genes A, C and D. B could be a repressor that shuts off synthesis of the structural genes when compound Z is present. Obviously, the total inhibition of transcription has a second component - an attenuator. This explains why there is still some transcription even when B is mutant. The nonsense mutation in C must be a polar nonsense mutation, as it affects not only the presence of the C protein, but also the A and D proteins, so C must be the first structural gene in the operon. A nonsense mutation in gene A blocks production of A and D while a nonsense mutation in D only affects the synthesis of D. The order of structural genes is C A D. Deletion of site G blocks all expression of the three structural genes, so site G is probably the promoter of the operon. The removal of either E or F lead to reduced repression which occurs if there is a repressor binding to an operator and an attenuator site in the operon, each being responsible for a 10-fold repression when compound Z is available. Sequence E codes for a small peptide and since small peptides are part of attenuator sites in amino acid biosynthetic operons, we can assume that deletion of E and adjacent DNA would delete an attenuator site. Site F then is the operator site. The operator and promoter are first, then the attenuation region including the leader peptide, then the structural genes. B is not cotranscribed with the other genes, so we do not know its location and it is shown on the map below in brackets {}.

15-28.

a. There are several possible explanations for polarity, but all of them are dependent on three basic ideas: first, transcription and translation in bacteria are coupled; second, the entire operon is transcribed into a single mRNA; and third, polarity must be associated with the failure to translate one gene since its open reading frame is disrupted by the polar nonsense mutation.

   One possible model is that ribosomes might protect mRNA from degradation by ribonucleases. If the ribosomes get "stalled" at the stop codon then the mRNA distal to the stop codon would be destroyed, thus preventing translation of downstream genes. **A second possible model is that recognition of the ribosome binding sequences (Shine-Dalgarno sequences) for the distal genes depends on complete translation of the open reading frame of the proximal gene** – perhaps ribosomes might be needed to bend these regions of the operon mRNA into the correct configuration. **A third possible model is that the stalling of the ribosomes at the nonsense codon in the mRNA might expose cryptic transcription termination sequences in the mRNA**. Such cryptic transcription termination sequences are usually non-functional because they are hidden by the ribosomes. This is by no means a comprehensive list of possibilities.

b. Note that **the presence of a nonsense suppressing tRNA restores expression of both the gene with the polar nonsense mutation <u>and</u> the downstream genes. This shows that the polarity of the nonsense mutation is due to the termination of translation and not simply the presence of a stop codon at that position**, but it does not support any of the models presented in part a above over any other. However the **rest of the data strongly supports the third model** above. Remember that the Rho factor is necessary for Rho-dependent transcription termination (**Figure 15.2**). When the Rho factor is missing the expression of the genes distal to the polar nonsense mutation is restored. Therefore, when cryptic Rho-dependent termination signals are exposed they will not act as transcription termination signals so the expression of distal genes is restored. Obviously this has no effect on the nonsense mutation, so the expression of that gene is still defective.

Section 15.4 – Global Regulatory Mechanisms

15-29. The loss of LexA function leads to the new expression of many genes. Therefore, the wild type LexA protein binds to the operators of these genes to shut them off. This is **negative regulation**.

15-30. Fuse the *lacZ* gene to the 3' end of one of the motility genes that was increased in expression during growth under poor carbon sources and *lacZ*. This fusion gene is under the control of the promoter and regulatory elements of the motility gene. **Then introduce this fusion**

into a *lacZ⁻ E. coli* **strain.** This bacterial strain will now make high levels of ß-galactosidase in media containing poor carbon sources but lower levels in media with rich carbon sources. Next, **mutagenize this strain and look for mutants in which** *lacZ* **expression does not increase under poor growth conditions.** Such mutants can be assayed with a ß-galactosidase substrate like X-Gal that changes color when cleaved. Mutant cells would not produce ß-galactosidase and so could not cleave X-Gal and would give rise to white colonies. The cells which do produce ß-galactosidase will cleave the X-Gal and give rise to blue colonies.

It is important to determine if the mutation affects expression of other motility genes under the growth condition of poor carbon sources. If the mutation <u>does</u> affect other motility genes then may be a mutation of a regulatory gene that globally influences the expression of several target genes, either directly or indirectly through other proteins.

15-31.

a. You would use **two probes. One consists of labeled cDNA corresponding to the mRNA extracted from the culture grown at the higher temperature, and the other consists of cDNA corresponding to the mRNA in the culture grown at the lower temperature**.

b. **Each spot on the microarray would have a DNA sequence representing a single** *E. coli* **gene.** There are 5000 genes in the *E. coli* genome so the microarray should have **a minimum of 5000 spots**. The DNA in each spot could be either a PCR product or a synthetic oligonucleotide specific to a particular gene.

c. **Use microarrays to compare the gene expression changes in cells grown under different osmotic conditions and those that are heat-shocked**. The genes that change expression under both conditions are likely to be involved in a general response.

15-32.

a. The data suggests **a minimum of eight promoters**, shown as ↓ in the figure below. Remember that genes shown above the line are transcribed from right to left and all use the same DNA strand as template, while genes shown below the line are transcribed in the opposite direction (left to right) using the other DNA strand as a template. One promoter is an early promoter and transcribes **an operon containing genes** *36* **and** *12* (open boxes); another is a middle promoter and transcribes only **gene** *33* (lightly stippled box). The remaining 6 promoters are late promoters (striped boxes); **there are 4 operons (genes *26* and *25*; genes *6* and *17*, genes *44* and *48*, genes *29* and *50*), and two "singleton" genes *15* and *21*.**

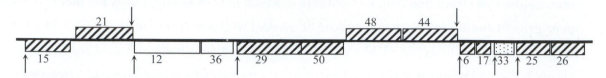

b. **You would expect *e* to be a late gene.** The bacteriophage must ensure that no endolysin is made until progeny bacteriophages are ready to be released from the host cell, or very late in the infection. It would be fatal for the bacteriophage if endolysin is made too early since there would be no progeny bacteriophage if the host lyses before the progeny are assembled. The *e* gene may actually be transcribed during the early or middle periods but the mRNA is not translated into the endolysin protein until late in the infection. It turns out that *e* is indeed mostly transcribed late, but there is a small amount of *e* mRNA that is also transcribed from an early promoter. The early *e* mRNA forms a hairpin structure that prevents its translation; the late *e* mRNA is transcribed from a different promoter and thus has a different 5' end that does not form the hairpin, thus allowing translation.

15-33.

a. **All of these turn out to be early genes.** The *motA* and *asiA* genes must be transcribed and translated early to enable the transcription of the middle genes. The *regA* gene product ensures that the transcripts of the early genes will be rapidly degraded. Figure A shows that the early mRNAs are degraded before there is much transcription of the middle genes. This makes the most sense if *regA* is an early gene, although it is also possible that *regA* could be a middle gene if the expression of just a small amount of this protein were sufficient for degrading the early mRNAs. The *55* gene is transcribed early, though it actually makes more sense for it to be a middle gene since it is only needed for transcribing late genes. Presumably the translation of the *55* protein is delayed until it is needed.

b. Only the **early genes** have promoters recognized by the *E. coli* σ^{70} RNA polymerase holoenzyme. At the beginning of the infection T4 has not yet made any proteins that could alter transcription. In fact, the early genes have promoters that attract the *E. coli* σ^{70} RNA polymerase holoenzyme more strongly than most genes in the host chromosome. This is because the ~40 T4 early promoters must compete with ~900 *E. coli* promoters for the same holoenzymes.

c. **Transcription of the large majority of *E. coli* genes would be drastically decreased** within the first 10 minutes of T4 infection. This is because most of the host genes are transcribed by the *E. coli* σ^{70} RNA polymerase holoenzyme, but the anti-σ factor encoded by the T4 gene *asiA* prevents the functioning of this holoenzyme. In other words, this anti-σ factor "hijacks" the host transcriptional machinery so that it can now only transcribe T4 genes rather than host genes.

d. Loss-of-function mutations in **the *motA* gene prevents transcription of the middle genes**; this will presumably also shut off transcription of the late genes if middle gene-encoded proteins are needed for late gene transcription (for example, factors that might enable the translation of the gene *55* mRNA as explained in part e). There should be no effect on the transcription of host mRNAs. Loss of function mutants in *asiA* **should lower the transcription of middle and late T4 genes**, since most RNA polymerase molecules will be in the form of the σ^{70} holoenzyme that cannot transcribe these genes. For the same reason the mRNAs of *E. coli* host genes and T4 early genes will accumulate to higher levels than in a normal infection. Finally, loss-of-function mutations in **the *55* gene should prevent the transcription of late transcripts but have little effect on the transcription of host genes** (which will be shut off) or of the early or middle T4 genes.

e. **The *reg-A*-encoded ribonuclease is specifically required for the rapid destruction of T4 early mRNAs.** Figure (a) in problem 15-33 shows that the accumulation of these mRNAs peaks within the first 3 minutes of the infection. These mRNAs disappear within 1-2 minutes. The reasons for the disappearance are (i) after the AsiA protein is made, the *E. coli* σ^{70} RNA polymerase holoenzyme can no longer transcribe from early promoters, and (ii) the RegA ribonuclease rapidly degrades the early mRNAs. Notice that the middle and late T4 mRNAs are more stable - it takes about 5-7 minutes for the middle mRNAs to disappear after their peak has been reached.

Chapter 16   Gene Regulation in Eukaryotes

Synopsis:

This chapter describes different ways in which gene expression is regulated in eukaryotes. Regulation of transcription is a major means of controlling expression. *Trans*-acting regulatory factors (activator and repressor proteins and miRNAs) bind to *cis*-acting sites in the DNA near the gene to turn up or down expression. The low level of basal transcription of a gene requires a set of transcription factors binding at the promoter. Gene expression that occurs in only certain cells or is specific for a stage in development or occurs in response to external factors is mediated by additional proteins binding to *cis*-acting sites (enhancers and LCRs) near the coding region of a gene. Characteristic DNA sequence motifs that are important for DNA binding or interaction with RNA polymerase or with other activator proteins can be recognized by regulatory proteins that bind to DNA.

Other controls of gene expression include chromatin structure (nucleosome remodeling and heterochromatization), imprinting (epigenetic methylation), posttranscriptional events such as alternate splicing events (producing different mRNAs from the same gene) and RNAi in which miRNAs affect mRNA stability, and posttranslational modifications such as ubiquitination, phosphorylation and dephosphorylation.

Significant Elements:

After reading the chapter and thinking about the concepts you should be able to:

♦ Distinguish between positive and negative regulators based on the effects of mutations.

♦ Design experiments to determine the sequence(s) of a regulatory region of a gene that are important for basal transcription and which sequences are important for tissue specific expression.

♦ Interpret expression of a gene using a reporter gene fused to a regulatory region.

♦ Identify interacting proteins in regulation of a gene.

♦ Interpret expression of a paternally or maternally imprinted gene (**Feature Figure 16.23**).

♦ Describe the importance of chromatin structure for eukaryotic gene regulation (**Figures 16.21** and **16.22**)

♦ Understand how miRNAs affect gene expression (**Figures 16.25, 16.26** and **16.27**)

♦ Understand the sex determination in *Drosophila* (**Figures 16.29-16.33** and **Tables 16.2** and **16.3**).

Problem Solving Tips:

♦   When a regulatory region of gene "X" is cloned next to the *lacZ* gene and reintroduced into eukaryotic cells, the expression of β-galactosidase will reflect expression patterns of gene "X".

♦   Deletions of sites to which activators bind will lead to lower expression (**Figures 16.7 and 16.8**); deletions of sites to which repressors bind lead to increased expression (**Figure 16.12**).

♦   Several regulatory proteins can interact to regulate one gene.

♦   Mutations in regions of proteins that interact with another protein or a site will only affect that one specific function.

♦   DNase hypersensitive (DH) sites indicate that the DNA is in a less compacted form and is more available for transcription.

♦   For maternally imprinted genes, the copy of the gene inherited from the mother will not be expressed in either her sons or daughters. For paternally imprinted genes, the copy of the gene inherited from the father will not be expressed in either his sons or daughters.

Solutions to Problems:

Vocabulary

16-1. a. **7**; b. **4**; c. **6**; d. **2**; e. **9**; f. **8**; g. **5**; h. **3**; i. **1**.

Section 16.1 – Overview of Eukaryotic Gene Regulation

16-2. See **Table 16.1**; a. **eukarotyes**; b. **both prokaryotes and eukaryotes**; c. **eukaryotes**; d. **prokaryotes**; e. **both prokaryotes and eukaryotes**.

16-3. Events that can affect the type or amount of protein produced in a cell **include transcript processing (including alternate splicing of the RNA), export of mRNA from the nucleus, changes in the efficiency of translation (including miRNAs), chemical modification of the gene products and localization of the protein product in specific organelles.**

Section 16.2 – Control of Transcription Initiation

16-4. a. **pol III**; b. **pol II**; c. **pol I**; d. **pol II**.

16-5. a. **i**; b. **ii**.

16-6. Maximal gene expression occurs when there is both a promoter and an enhancer upstream of the gene. There is <u>no</u> expression of a gene without a promoter (P) sequence. In the presence of a promoter sequence, there will be a low (basal) level of β-galactosidase reporter gene expression. An enhancer sequence alone shows no expression, while promoter + enhancer will give high levels of expression. In order to discuss the fragments shown in the restriction map, number them from left to right. Fragment #1 is M to M, fragment #2 is M to H, fragment #3 is H to M, fragment #4 is M to M and fragment # 5 is M to H. **The promoter must be located in M-H fragment #5** since each of the clones showing basal levels of transcription contain this fragment and clones without this fragment produce no β-galactosidase. **The enhancer is present in H-M fragment #3** since that sequence is found in the two clones (sixth and eighth) showing high expression. Clone #2 also contains this fragment but does not show any expression because there is no promoter upstream of the reporter gene.

16-7. The constitutive expression of the galactose genes in yeast will be seen if there is no way to prevent the binding of the GAL4 protein to the DNA. Therefore, **a *GAL80* mutation in which the protein is not made or is made but cannot bind to the GAL4 protein will prevent repression and lead to constitutive synthesis. A *GAL4* mutation which inhibits binding to the GAL80 protein will also be constitutive. A mutation of the DNA at the binding site for the GAL4 protein will also give constitutive synthesis.**

16-8. Isolate mRNA from cells expressing the 3 genes. Do a Northern blot analysis with this mRNA using probes from each of the three genes. If the 3 genes are co- transcribed as one polycistronic mRNA, you will see the same large band when the Northern blot is probed with DNA from any of the 3 *gal* genes. However, **you will see a differently sized mRNA band for each gene when the blot is probed with each gene separately**. None of these bands will be large enough to contain the coding information for all three genes.

16-9.

a. **DNA binding, <u>Figure 16.8</u>;**

b. **DNA binding;**

c. protein-protein **dimer formation, see <u>Figure 16.10</u>;**

d. **transcription activation** - protein-protein interactions between activating factors and basal proteins bound to the DNA at the transcriptional activation region, see **<u>Figures 16.6</u> and <u>16.7</u>;**

e. **DNA binding.**

16-10. Isolate mRNA from sporulating cells and make cDNA copies. The cDNAs then could be cloned into a vector to generate a cDNA library representing genes expressed during sporulation. The cDNAs can also be used to probe a genomic library to clone the genomic versions of the expressed genes. These methods will isolate all genes that are expressed in sporulating cells, both the sporulation-specific ones and the genes that are generally expressed (the house-keeping genes).

An alternate method that allows isolation of the genes which are specifically expressed during sporulation as well as those whose level of expression increases or decreases during sporulation involves **microarrays**, discussed in Chapter 10 (see **Figures 10.16** and **10.17**). The DNA sequence of the entire genome of yeast is known and all potential genes have been identified by computer alogorithm. Oligonucleotides from the 3' ends of all the putative genes can be arranged in a microarray. These arrays can be **probed with two types of differentially labeled mRNA** - that isolated from sporulating cells (labeled with a red fluorescent dye, for instance) and mRNA isolated from normal, vegetatively growing diploid cells (labeled with a green fluorescent dye). When a mixture of the 2 groups of mRNA are mixed and probed to the arrays, you will be able to distinguish genes that are expressed only in the vegetatively growing cells (red spots) from those that are only expressed in sporulating cells (green spots) from those expressed in both types of cells (**Figure 10.17**). Those genes expressed in both states but whose level of expression changes will be seen as intermediate shades. It is possible to measure the levels of the two dyes for each spot on the microarray to determine patterns of gene expression.

16-11. If the Id protein binds the DNA enhancer sequence then antibodies against the Id protein should bind to the chromatin in *in situ* hybridizations. Also, there should be a DNA binding domain (for e.g. a zinc finger or helix-turn-helix or helix-loop-helix sequence) in the *Id* gene. If the Id protein acts by quenching, it interacts with the MyoD protein, so there is no need for a DNA binding sequence in the *Id* gene. However the *Id* gene should have a protein-protein binding sequence, like a leucine zipper (Figure 16.10). In this case, the Id protein would bind to the MyoD protein but not to the chromatin.

16-12.

a.   Gene *A* = transcription factor 1; gene *B* = transcription factor 2. If both transcription factors are required for expression, only the *A- B-* genotype will produce blue flowers. Diagram the cross: *Aa Bb* x *Aa Bb* → 9/16*A- B-* (blue) : 3/16 *A- bb* (white) : 3/16 *aa B-* (white) : 1/16 *aa bb*(white) = **9/16 blue :7/16 white flowered plants**.

b. If either transcriptional factor is sufficient to get blue color, then three of the four genotypic classes from a dihybrid cross will produce blue flowers: 9/16 *A- B-* (blue) : 3/16 *A bb* (blue) : 3/16 *aa B-*(blue) : 1/16 *aa bb* (white) = **15/16 blue : 1/16 white flowers.**

Section 16.3 – Chromatin Structure and Epigenetic Effects

16-13.

a. You <u>must</u> have **a *Drosophila* promoter sequence** to allow transcription of the *lacZ* coding sequence. The **promoter must be added somewhere upstream of the DNA encoding the initiating AUG**. Other elements are optional but helpful. It would be advantageous to **place a *Drosophila* poly-A addition sequence and a transcription termination signal downstream of the *lacZ* coding sequence**. These are not absolutely required since transcription of *lacZ* would eventually end by chance, but they would increase the amount of ß-galactosidase produced.

b. **The type of construct you made is called an *enhancer trap*.** As the name implies, **these different insertions signal a position in the genome adjacent to a tissue-specific enhancer**. Remember that enhancers work in concert with the basal promoter to activate transcription, but the enhancers can be located some distance from the promoter. **In strains in which *lacZ* is expressed in the head, your construct must have integrated into the genome very near to an enhancer that helps activate transcription in the head. In other strains, your construct integrated into the genome near enhancers that are specific for other tissues like the thorax. Since the density of enhancer elements in the genome is low, most of the time new integrations of your construct would be located too far from an enhancer, so there would be no *lacZ* expression and no blue color.**

   This technique allows you to find genes that are normally expressed in a tissue-specific manner. For example, if you wanted to study *Drosophila* genes involved in the development of the head, you would collect many different enhancer trap lines with blue staining in the head. In each of these lines there is likely to be a gene specifically expressed in the head that would be controlled by the nearby head-specific enhancer.

16-14. An enhancer is an expression-promoting DNA element that helps activate transcription from a nearby minimal (weak) promoter. Function of the enhancer is usually independent of its orientation and its distance from the minimal promoter. To test that the orientation of the presumptive enhancer does not matter, you **create a construct in which this region is inverted with respect to the *lacZ* coding sequences.** You then introduce this construct into neurons in tissue culture and determine if *lacZ* expression remains high. Alternatively **you could add some extra DNA sequences between the minimal promoter and the presumptive enhancer.** You expect the *lacZ* expression to remain high even if the enhancer is moved a few kilobases distant from the minimal promoter.

16-15. The conclusion that chromatin structure has a major effect on the level of gene expression is based on many observations. The earliest involved the **differing levels of gene expression depending on their association with highly compacted, heterochromatic DNA vs. euchromatic DNA** (see Chapter 12). One example is **Position Effect Variegation in *Drosophila*** where the expression of a gene can be abolished when it is placed next to heterochromatin. Another is **Barr body formation in human females**. In this case almost all of the genes on the inactive, heterochromatic X chromosome are permanently inactivated in the mitotic descendants of that cell. Many genes in euchromatin must be decompacted in order to be transcribed (**Figure 16.21**). **Decompaction affects the location of the nucleosomes, and gives rise to DNase I hypersensitive sites where nucleosomes have been removed and the DNA is available for binding by RNA polymerase or regulatory proteins.** Such work lead to the discovery of boundary elements which are DNA sequences that help define the limits of the localized decompaction of the DNA. **Transcriptional silencing, on the other hand, involves methylation of the DNA (Figure 16.22).** A more recent example is imprinting (**Feature Figure 16.23**), which is a sex-specific inactivation of genes that occurs during meiosis and is passed on the zygote. In imprinting, the inactivation of genes is associated with increased methylation of the DNA. Studies on imprinted genes lead to the discovery of insulators, which are DNA sequences that seem to determine the limits of such methylation. The evidence is now compelling that chromatin structure plays a major role in gene expression in eukaryotes.

16-16. Option b. is suggested by the presence of a DNase I hypersensitive site.

16-17. The gene is more likely to be transcribed in liver cells, based on the DNaseI profile, than in muscle cells. **Liver cell DNA has a DNaseI hypersensitive (DH) site 4 kb from 1 end of the *Eag*I fragment. This site is probably the promoter region for your gene.** This site is not be exposed in the muscle cells, indicating lower (or no) transcriptional activity in muscle cells.

16-18.

a.   The deletion to -85 upstream from the transcription start site has full activity indicating that the binding site of the transcription factor (enhancer) is downstream (to the right of) -85. The region between -85 and -75 contains at least part of the enhancer that helps activate transcription of the adjacent genes in the glial cells. Within the 19 bp region that is protected in the DNase-I footprinting experiment, there is a short sequence that is almost repeated (between nucleotides -83 and -65). It is possible that this is the monomer binding site. This site must be repeated to get active dimer binding, as seen in the **region including base pairs -83 to -65**.

b. The binding site(s) for the glial-specific transcription factor are at least partially contained in the region from -85 to -67 which is protected from DNase-I digestion. This sequence, **beginning at -85, is:**

$$\text{5' CTTCTGCGGACCGCAAACTG}$$
$$\text{3' GAAGACGCCTGGCGTTTGAC}$$

c. The probable binding site for the glial cell transcription factor is **5' TGCGG 3'** (read off a single strand); its position in the DNase-I protected region is underlined in part b above. Note that the binding sites for the two monomers are in opposite orientations along the DNA. Protein dimers often display radial symmetry of their subunits so the subunits are oriented in opposite fashion and must recognize oppositely-oriented binding sites. The reason that the DNase-I protected region is longer than the binding sites because the transcription factor only needs to recognize the DNA nucleotides that are underlined. However once it is bound the protein dimer occupies more space on the DNA and thus protects a longer region from DNase-I digestion.

16-19.

a. **1, 2** and **4 (Feature Figure 16.23** and **Figure 16.22)**;

b. **1** and **4**;

c. **1** and **4**;

d. **3 (Figure 16.22)**.

16-20. Epigenetic changes are those caused by altered chromatin structure causing changes in gene expression. These changes in gene expression can cause phenotypic changes in a cell or an organism. These changes are not due to changes in the DNA sequence but are modifications of the genomic blueprint.

a. The methylation of the N-terminal tails of histones H3 and H4 have been closely linked to transcriptional regulation. Such modifications of the histones affect many processes including the differentiation of pluripotent stem cells into specific tissue lineages. Thus histone covalent modifications seem to have crucial roles for the establishment of genetic programs during development. Once these modifications are made they will be **inherited by the mitotic daughter cells. However such modifications will not be inherited by the subsequent meiotic generations of the organism**. Such developmental changes must be re-initiated anew each generation.

b. **DNA Methylation occurs at the fifth carbon of the cytosine base in a CpG dinucleotide pair** (see Figure 6.9a). Such methylation is associated with transcriptional silencing. Again **this sort of methylation will be inherited by the mitotic descendants of the cell.** This is accomplished by the activity of DNA methylases that recognize methyl groups on one strand of a double helix and respond by methylating the opposite strand. This sort of methylation of the DNA is called imprinting.

    In mammals like mice and humans a small subset of genes are imprinted and therefore transcriptionally inactivated by the male (paternally imprinted). A different subset of genes is imprinted by the female (maternally imprinted). **The meiotic offspring will inherit an inactive allele of a paternally imprinted allele from the male parent and a non-imprinted, functional allele of the same gene from the female parent. The inverse will happen for a maternally imprinted gene - the meiotic offspring inherits a functional allele from the father and an inactivated allele from the mother.** This DNA methylation is erased and regenerated during meiosis. If the meiosis is happening in a female then all alleles of the maternally imprinted genes will be methylated and inactivated while none of the paternally imprinted genes will be inactivated. In meiosis in a male all imprinting is erased and the reverse imprinting happens - all alleles of the paternally imprinted genes are methylated and none of the alleles of the maternally imprinted genes.

c. **A change in the DNA sequence is NOT an epigenetic change.** Such a change to the nucleotide sequence is permanent (barring direct reversion) and will be inherited by both the next mitotic generation of cells and by the meiotic progeny of the organism.

Thus the only epigenetic change that can be inherited from one meiotic generation to the next is imprinting (b.).

16-21. Remember that Prader-Willi syndrome is caused by a mutation in a maternally imprinted gene. In order to answer this question, you must be able to figure out the genotype of the original affected individual. Remember that all individuals have only one transcriptionally active allele of this gene, and that is the allele they inherited from their father.

a. An affected male has a mutant allele of the gene that he inherited from his father and an imprinted (transcriptionally silenced) allele of the gene that he inherited from his mother. This imprinted allele is normal, even though the affected male can not express it. When the affected male makes gametes, he removes all imprinting. As this gene is maternally imprinted, he does NOT imprint either allele of this gene. Therefore, half of his sperm will have the non-imprinted normal allele and the other half will have the non-imprinted mutant allele. All of his children will receive an inactivated (presumably wild type) allele of this maternally imprinted gene from their mother and so **half of his sons** and half of his daughters **will be affected**.

b. **True**, see part a.

c. **False.** An affected female has a mutant allele of the gene that she inherited from her father and an imprinted (transcriptionally silenced) allele of the gene that she inherited from her mother. This imprinted allele is normal, even though the affected female can not express it. When the affected female makes gametes, she removes all imprinting. As this gene is maternally imprinted, she DOES imprint BOTH alleles of this gene. Therefore, half of her eggs will have the imprinted normal allele and the other half will have the imprinted mutant allele. All of her children will receive an activated wild type allele of this maternally imprinted gene from their father, so none of her children will be affected.

d. **False**, see part c.

16-22. See **<u>Feature Figure 16.23</u>**. The boy received an imprinted (transcriptionally inactivated by methylation) allele of this gene from his father. The boy should be transcribing and translating the supposedly normal, active allele from his mother. If the boy has a mutant phenotype, it must be due to a **mutant allele of the gene that he received from his mother**.

16-23. Draw out the pedigree, including the information given in the problem. {} represents an allele that is transcriptionally inactivated (imprinted) and the alleles of the maternally imprinted *IGF-2R* gene are *60K* and *50K*.

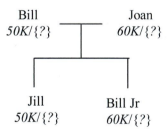

a. Both of Joan's alleles will be imprinted in her gametes, so Jill and Bill Jr each express the allele they received from Bill. Therefore, **Bill Sr's genotype is 50K/{60K}**. With this information, nothing further can be said about **Joan's genotype (*60K/{?}*)**, and there is no way of telling which of Joan's alleles Jill and Bill Jr received.

b. Bill Sr's complete genotype and the further information about Pat and Tim are presented in the pedigree below.

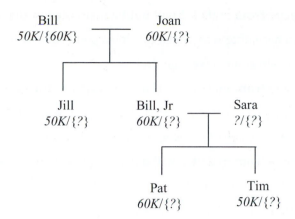

As in part a for Joan, both of Sara's alleles will be imprinted in her gametes. Therefore, Pat and Tim express the transcriptionally active alleles they received from Bill Jr. Because Tim's phenotype is 50K, Bill Jr must be heterozygous *60K/{50K}*, with his imprinted *{50K}* allele having come from Joan. Therefore, **Joan's genotype is *60*K/*{50K}* and Bill Sr's genotype is *50K/{60K}*.**

16-24. This is a paternally imprinted gene. Neither parent in generation I expresses the disease, but 2 out of 3 children do. At first glance, either parent could have the mutant allele of the gene. Because neither expresses the disease, they must have received an inactivated mutant allele from their father and a normal, transcriptionally active allele from their mother. However, further consideration shows that it can not be the male parent (I-2) who is the source of the mutant allele. He can only provide transcriptionally inactive alleles. If the children show the disease phenotype, then they must have received an inactive normal allele of the gene from him and an activated, mutant allele of the gene from their mother, I-1. In the genotypes below, an allele in {} brackets is imprinted (transcriptionally inactivated) and *a* is the mutant allele of the gene.

a.   The genotype of I-1 is **A/{a}**; see explanation above.

b.   The genotype of II-1 is **A (from I-1)/{A} (from I-2)**.

c.   The genotype of III-2 is **A (from II-2)/{A} (from II-3)**.

16-25. See **Feature Figure 16.23**. This pedigree tracks a paternally imprinted gene.

a.   **The alleles of the gene are not expressed in the germ cells of male I-2 (Feature Figure 16.23).** The gene is paternally imprinted, so the imprinting is removed during gametogenesis. By the end of meiosis both alleles of this gene will be imprinted, so male I-1 will not express the gene in germ cells.

b.   **The allele of the gene from male I-2 will not be expressed in the somatic cells of II-2.** The gene inherited from her father was turned off in his germ cells, so the somatic cells arising after fertilization will contain an inactive copy from I-2.

c.   **The allele of the gene from male I-2 will be expressed in the germ cells of II-2.** The imprinting from the I-2 is erased during gametogenesis and a female will not inactivate this gene, so both alleles of this gene will be active in the germ cells of II-2.

d.   **The allele of the gene from male I-2 will not be expressed in the somatic cells of II-3.** The allele inherited from I-2 is imprinted, so the son's somatic cells will not express that allele.

e.   **The allele of the gene from male I-2 will not be expressed in the germ cells of II-3.** The imprinting from I-2 is erased during gametogenesis but the male will re-imprint both alleles of the gene in his germ cells.

f.   **The allele of the gene from male I-2 will be expressed in the germ cells of III-1.** The III-1 inherited the non-imprinted allele of the gene from his mother (II-2), so the gene is expressed in the somatic cells of III-1.

g.   **The allele of the gene from male I-2 will not be expressed in the germ cells of III-1.** Male III-1 will have imprinted both alleles of this gene in his germ cells.

Section 16.4 – Regulation After Transcription

16-26. One gene can give rise to different (or variant) proteins through the action of alternative splicing. Alternative splicing allows transcripts to be combined in different ways and increases the variety of proteins produced by a genome. Splicing is also used to regulate gene expression by restricting or allowing certain combinations of transcripts. One example of this is the alternative splicing of the Sxl gene in *Drosophila melanogaster* (see Figure 16.24).

16-27. Introns are spliced out of a primary transcript (Figures 8.15 and 8.16), a ribonuclease cleaves the primary transcript near the 3' end to form a new 3' end to which a poly-A tail is added (Figure 8.14), and a methyl-CAP is added to the 5' end of the transcript (Figure 8.13).

16-28. The mRNA for this gene is made in all tissues tested, while the protein is made in only muscle cells. **Expression of this gene is translationally regulated.**

16-29. The 5' and 3' untranslated regions (UTRs) from the *hunchback* gene could be cloned at the 5' or 3' ends of a reporter gene (for example, *lacZ*) that is transformed back into *Drosophila* early embryos to see if either sequences cause translation of the reporter protein at the appropriate times during development.

16-30. The 1.2 kb mRNA in the skin cells and the 1.3 kb mRNA in the nerve cells could be due to **either a different start site for transcription in the two cell types or alternative splicing of the primary transcript** in the two cell types.

16-31. The protein in the fat cells may be post-translationally modified (for example, phosphorylated or de-phosphorylated) so that it is only active in fat cells. Alternatively, the protein may need a cofactor to be activated, and this cofactor is only transcribed in fat cells.

16-32.

a.   The basic idea of an RNA interference (RNAi) screen is simple: if you expose cells of an organism to short interfering RNAs (siRNAs) that complementary to the mRNA of a specific gene then the expression of that gene may be severely reduced or abolished. You can then ask whether the biological process you are investigating, for example mitosis, is affected or impaired because of the lower expression of the gene.

To perform the RNAi screen you need a way to make the siRNA and then you must transport the siRNA into cells. There are at least two ways to **make the siRNA in the test tube.** Since siRNAs are only 21-23 bases long you can **make the two strands of the siRNA with an automated oligonucleotide synthesizer. Alternatively you can make long transcripts of both the sense and antisense strands of the gene by transcribing the gene sequences** *in vitro* **with RNA polymerase.** For example you can PCR amplify the gene with primers at both ends that include DNA sequences for a strong promoter that can be used to transcribe the gene in both directions.

Either the short siRNAs or the longer gene length double-stranded transcripts can then be introduced into cells. If you are studying mitosis then you will **use tissue culture cells.** In such cases **simply incubating the cells with the RNAs is often sufficient.** Usually the longer double-stranded RNAs are converted into siRNAs within the cell by the activity of enzymes like Dicer that are already in the cells. **Getting the RNAs into an animal to look for genes involved in the development of specific body parts is more complicated.** The <u>Tools of Genetics</u> in Chapter 16 of the text outlines two **possible strategies - attaching cholesterol to the RNAs or incorporating them into viral particles.**

Other strategies for RNAi screens include synthesizing the double-stranded RNAs in the cells. This strategy begins by transforming cells or organisms with a transgene containing two copies of the sequence of the gene of interest. These copies are inverted repeats. Thus, when the transgene is transcribed in the cell the mRNA will be able to loop back on itself and make a double-stranded RNA that can be converted in the cell to siRNA by enzymes like Dicer.

Researchers studying the nematode *C. elegans* have developed a bizarre but powerful dietary variant of this last strategy. They make *E. coli* cells that contain transgenes that can make such a snap-back RNA. Since the nematodes eat *E. coli*, they simply feed bacteria expressing the transgenes to the worms; amazingly the double-stranded RNA gets into most of the worm cells.

b. The main advantage is that **using RNAi-based screens researchers can systematically look at every gene catalogued through genome projects**. For example *C. elegans* researchers have made about 19,000 different *E. coli* strains that express snap-back RNAs corresponding to every known gene in the nematode genome. By contrast, traditional genetic screens affect DNA sequences more-or-less at random. Because some genes are small and thus poor targets for mutagens it is very unlikely that mutations will be generated in every gene.

c. There are several shortcomings of RNAi-based screens. **It is a lot of work to make all the constructs needed to target every gene** (19,000 in *C. elegans*) **and get these double-stranded RNAs into the cells of the organism. RNAi screens only generate loss-of-function phenocopies** - researchers cannot identify rare informative gain-of-function mutations that might be found in a classical genetic screen. Furthermore, **if the RNAi treatment has no phenotypic effect the researchers have no way of knowing whether the siRNA has actually affected the expression of a gene sufficiently**. Finally, both types of screens (classical and RNAi) suffer from the possibility that **there might be redundant gene functions**, so that you would never see a phenotype if you only knocked out one of the genes that could perform this function.

16-33.

a. It is true that transcription, processing of the primary transcript and export of the mature mRNA to the cytoplasm take place before translation, but these events should only take a few minutes rather than 6 hours (1/4 of a day). Therefore **the difference in first detection of the mRNA of the interesting gene and the protein ß-galactosidase protein probably results from the different sensitivity in detecting mRNA versus protein**.

The difference between the disappearance times of the mRNA and the ß-galactosidase protein can be explained in at least three ways. First, because no data is presented for the stability of the normal protein from the interesting gene, **it is possible that the normal protein is present until day 12 also and the proteins are more stable than the mRNAs so they remain in the cells for several days longer.**

If the normal protein disappears at day 10.5 when the mRNA disappears, then other possibilities are that **the *lacZ* mRNA in mouse embryos is more stable** (less easily degraded) than the mRNA for the interesting gene, allowing the *lacZ* mRNA to stay in the cell and be translated for a longer period of time. A more likely alternate possibility is that both the mRNAs (interesting gene and *lacZ*) get degraded at about the same time. Thus they both disappear after 10.5 days of gestation, but **the ß-galactosidase protein is more stable** and remains in the mouse embryonic cells for several days. Both of these latter two possibilities assume that the transcription of the mRNAs for the interesting gene and ß-galactosidase stop at the same time. It is possible that **the transgene is transcribed until day 12** thus accounting for the presence of the ß-galactosidase protein.

b.   The **ß-galactosidase protein expression should accurately indicate the onset of gene activation**. Presumably **the *cis*-controlling DNA elements controlling the transcription of the *lacZ* reporter gene allow the transcription of the reporter gene to be turned "on" and "off" with the same schedule as the transcription of the interesting gene**. However, as explained in part a above, if mRNA or protein stability for the two genes differs then the activity of ß-galactosidase may cease sooner or continue longer than the activity of the protein product of the gene of interest.

Section 16.5 – Sex Determination in *Drosophila*

16-34.

a.   **Null alleles of *Sxl* will have no effect in males but are lethal in XX females.** This is because *Sxl* is needed for normal female development but not normal male development. Therefore the XX animals with null *Sxl* mutations would "think" they were males. These animals die for reasons discussed part b below. The XY animals do not make the Sxl protein (**Figure 16.30**) so they would not be affected by its absence. **A constitutively active *Sxl* allele would be expressed both in XX and XY animals and both would "think" they would were females. The lethality of XY animals that start developing as females is discussed in part b below.**

b.   **The reason for the lethality of XX animals with null alleles of *Sxl* and of XY animals with constitutively active alleles of *Sxl* is that dosage compensation is aberrant.** Dosage compensation depends on the proper amount of Sxl protein: in a wild type XY animal the absence of Sxl doubles the level of transcription of most X-linked genes ensuring that the XY animal makes the same amount of X-linked gene products as a normal XX animal. **If an XX animal has no Sxl protein then it will make twice the normal level of X-linked gene products as a normal XX animal.** This amount of gene product is lethal. In contrast, **an XY animal with constitutively active Sxl protein will now only make half the amount of the proteins from the X chromosome that are required for survival.**

c. **The Tra protein is needed for female development while its absence results in male development. Therefore, null alleles of *tra* convert XX animals into sterile males** but do not affect the sexual development of XY animals. In contrast, **constitutively active *tra* alleles would cause XY animals to develop as females** while having no affect on the development of XX animals.

d. XX animals with null *tra* alleles develop as males but they are sterile since **they do not have a Y chromosome which carries genes needed for male fertility in *Drosophila*.**

e. The implications of this question are important: **dosage compensation involves the Sxl protein, but not the Tra protein. Therefore XX animals with null *tra* alleles survive while XX animals with the null *Sxl* allele die.** In the pathway shown in **Figure 16.32** there must be a bifurcation of function of Sxl. As shown in the diagram below, Sxl affects the production of Tra, but Sxl independently affects dosage compensation in a way that does not depend on Tra.

* Female Dosage Compensation () = each X chromosome is transcribed at a rate = n;

** Male Dosage Compensation () = each X chromosome is transcribed at a rate = 2n.

f. The *tra-2* gene is transcribed in both males and females. In females *tra-2* works together with the functional Tra protein to affect the splicing of the *doublesex* transcript into the Dsx-F transcription factor which leads to phenotypically female animals. In the absence of the Tra protein the *doublesex* transcript is spliced in the Dsx-M form which results in animals that are phenotypically male. **Null mutations in *tra-2* should have the same effect as null mutations in *tra* - XX animals should be transformed into sterile males because the primary transcript of the *Dsx* gene will be spliced into the Dsx-M form.**

g. **You would predict from Figure 16.33 that the normal XY males or the sterile XX males carrying null alleles of *tra* would have normal male courtship behavior since the male-specific form of the Fru protein is made in the absence of Tra.** Note that this answer is correct only if all aspects of male courtship depend on the Fru-M protein - there could be other pathways downstream of Sxl that are independent of Tra.

Chapter 17   Somatic Mutation and the Genetics of Cancer

Synopsis:

This chapter describes characteristics of the eukaryotic cell cycle and the cancer phenotype. Genetic analysis of the regulation of the cell cycle has lead to our understanding of the genetic basis of cancer. Cancer is an uncontrolled growth of cells in which the cell cycle is no longer correctly regulated (**Feature Figure 17.4**). There are genes whose products are necessary for the cells to proceed in the cell cycle and genes whose products monitor progress of the cell cycle, stopping the cycle if there is something amiss. Signals for beginning cell division are relayed through a series of steps (signal transduction). A series of mutations, each one either inherited or newly arising in a somatic cell, lead to a cell dividing out of control. Inherited predisposition to particular cancers is due to a mutation in one of the genes controlling cell cycle growth.

Significant Elements:

After reading the chapter and thinking about the concepts, you should be able to:

♦   Describe the role of cyclins and cyclin-dependent kinases in controlling timing and the sequence of the phases of cell cycle (**Figures 17.16, 17.18 and 17.19**) – G_1, S, G_2 and M (**Figure 17.13**).

♦   Describe the role of growth factors and the pathway of signaling that leads to the start of the cell cycle (**Figures 17.2** and **17.3**).

♦   Distinguish between oncogenes (**Table 17.3**) and tumor suppressor genes (**Table 17.4**) based on the behavior of cells containing a mutation in a gene (**Figure 17.10**).

♦   Understand that many abnormalities must accumulate in a cell before it can become cancerous (**Feature Figure 17.4**).

♦   Explain inherited predisposition to a non-scientist.

Problem Solving Tips:

♦   Think about gene regulation, cell signaling and how they all might fit together in a system for regulating the cell cycle.

♦   Remember that there will be regions of genes coding for interacting proteins where mutations will affect the ability of the proteins to interact.

♦   Use your knowledge of molecular techniques (hybridizations, PCR, transcription analysis, etc) to design experiments.

- Multiple occurrences of cancer and/or early onset of cancer are suggestive indicators of inherited predisposition alleles, but further experiments are necessary to identify the mutant alleles especially if that type of cancer is very prevalent in the population.

Solutions to Problems:

Vocabulary

17-1. a. **7**; b. **6**; c. **8**; d. **2**; e. **1**; f. **9**; g. **3**; h. **5**; i. **4**.

Section 17.1 – Initiation of Cell Division

17-2.

a. Molecules outside the cell that regulate cell cycle include **hormones and growth factors** (see **Figure 17.2**).

b. The whole of Chapter 17 is a discussion of the molecules inside the cell that regulate the cell cycle! These include **cyclins, cyclin dependent protein kinases, any molecule in the signal transduction pathway, receptors or molecules that transmits a second signal** (**Table 17.1**).

17-3.

a. RAS is inactive when it is bound to GDP (RAS-GDP) (**Figure 17.3d**). When RAS is bound to GTP it is activated and in turn activates three protein kinases (the MAP kinase cascade). This cascade activates a transcription factor which causes cells to divide. Therefore, **a RAS mutant that stays in the GTP-bound state is permanently activated and will cause the cell to continue dividing**.

b. This second RAS mutant is blocked in the RAS-GDP form at the restrictive temperature. Therefore, **under the restrictive conditions the cells will not divide**.

17-4. c, e, a, b, d (**Figure 17.3**).

Section 17.2 – Cancer: A Failure of Control Over Cell Division

17-5.

a. T antigen binds to p53 and prevents it from acting in cell cycle regulation by binding a transcription factor. By supplying excess p53 from the high level promoter there is enough now in the cell to bind all the T antigen (quenching, as described in Figure 16.13) and still have enough unbound p53 to regulate the cell cycle. Thus, **the effect of the T antigen is minimized**.

b. Mutants 1 and 2 have the phenotype seen in cells with the non-elevated levels of p53 in the presence of T antigen. These mutants can no longer rescue the cells from the effect of T antigen, so they must **decrease the ability of p53 to function in cell cycle control**. The further presence of the T1 antigen blocks the ability of the mutant p53 to function even more.

c. Mutant 3 still rescues the cells from T antigen so this mutation does not seem to alter the ability of the p53 protein to bind the T antigen, nor does it alter the ability of the p53 to regulate the cell cycle. Therefore, this mutation must be **in a functional domain other than those that bind the T antigen and the transcription factor**.

17-6. For an overview, see <u>Feature Figure 17.4</u>. The four main characteristics of cancer are: **uncontrolled growth; genomic instability; potential for immortality; and the disruption of local tissues and the ability to invade distant tissues.**

17-7.

a. Amplification of a specific DNA sequence can be observed using hybridization to DNA from tumor and normal cells. **Use two different probes- one representing the specific sequence that you are analyzing and the other representing an unamplified control sequence.** Look for an increase in the signal of the amplified region in tumor cells while the level of the control region does not change.

b. To find gross rearrangements look for **alterations in the chromosomal banding patterns in a karyotype analysis**.

17-8. These 'conflicting views' can be easily reconciled. We know that **some of the environmental agents that are implicated in increased cancer risk cause increased level of mutations,** so this is consistent with the fact that mutations in genes are necessary to cause cancer. **The inherited mutations that lead to predisposition to cancer inactivate one allele of a gene (often a tumor suppressor gene) that inhibits cell growth**. This inherited mutation is just one of several mutations that must occur within a cell to lead to cancer. The environmental and inherited factors are therefore both affecting the same thing – the genes that are important for regulating the cell cycle.

17-9. You need to do epidemiological studies of the colon cancer profile looking for a correlation between diet and the incidence of cancer. To assess **the role of diet, studies can be set up within the recent immigrant population vs. the United States-based native Indian population examining the effect of a Westernized diet vs. a diet resembling that of the ethnic group. Also, the effects of the same diets on non-Indian Westerners should be examined.**

To assess **the role of genetic differences, you need to keep other factors, for example, diet, as constant as possible. You could look at the incidence in Indians and Americans who have similar diets**. This is often done by studying cancer rates in immigrants. For example, people who moved from India to Toronto could be compared with Toronto natives.

17-10. For the background on positional cloning see **<u>Figure 11.17</u>**. First, you must localize the trait to a specific region of the genome. Then you must sort through the candidate genes in that region to decide which one (or two) to focus your studies on. Lastly, you should sequence your clone and predict the protein structure and function from the predicted amino acid sequence. Step d can actually fit in anywhere after step b is done (after you have the clones). You should also look for or create mutations (knock-out mice, for example) in the gene in other species in order to see what mutant phenotype is produced. **The order of the steps is: e, b, a or d, c, f.**

17-11. Steps d, a and c must be done in that order. First you must localize the human homolog of the yeast gene in the human genome and find nearby markers that can be easily followed. Next, see if that region of the genome is linked to a hereditary predisposition to any sort of cancer. If you do find families with a predisposition that is linked to that region of the genome, then you must isolate the human gene and examine that gene in the affected and unaffected members of these families. **The order of the steps is: d; a; b, c.**

17-12. Protooncogenes are genes that code for proteins that stimulate cell division. Therefore, oncogenic mutations in such genes usually increase the level of expression of the gene product. Increasing the level of expression of the protooncogene results in cells that divide too much or at the wrong time, increasing the probability of cancer. A deletion of a protooncogene will decrease the level of expression, not increase it. Therefore, **choice c. will not be associated with cancer**. All of the rest of the changes will increase the probability that the affected cell, or its mitotic descendants, will become cancerous.

17-13.

a. The fact that everyone develops the disease at an early age is often a clue that an inherited mutation is an important factor in the development of colon cancer in this family. However, it is not an absolute predictor. Likewise, the appearance of so many affected individuals in one family is not a necessarily a predictor of an inherited basis for the prevalence of colon cancer in this family either. Both of these may indicate that there is an important environmental factor that contributes to development of the disease. Examination of the pedigree suggests that

predisposition to colon cancer in this family could be an autosomal dominant trait. If this is true, then individuals II-2, and either I-1 or I-2 must have the mutation, but not express it.

b.   **Individuals I-1, I-2 and II-2 are not among the high coffee consumers. Perhaps the predisposition to colon cancer is a combination of a particular genotype and the environmental factor of consumption of the special coffee**. Thus, the interaction of a certain genotype and an environmental factor results in the phenotype of colon cancer. In this scenario, the genotype alone is not enough to predispose you to cancer, as seen in individuals I-1, I-2 and II-2. On the other hand, note that a substantial number of individuals who drink coffee are not affected (for example III-8, III-9, III-10, III-11, III-12) so coffee alone does not appear to have a major effect on the early onset of colon cancer.

17-14. If there is a dominant predisposing allele segregating, **individual II-2 carries the mutant allele** since male III-2 has colon cancer and thus has the mutant, predisposing allele. Two individuals in the second generation carry the mutation, so one of the generation I parents must have had the mutant allele and passed it on to individuals II-2 and II-3. Thus, **either I-1 or I-2 also carries the mutation**. Note that if only one member of generation 2 inherited the mutant allele (developed early onset colon cancer) then it is possible that the mutation arose in development of a germ cell in a generation I individual. In this case, the somatic tissues of that parent would <u>not</u> contain the mutation.

17-15. Technique d will be most useful in detecting a specific point mutation in the p53 gene. Unlike techniques b and c, hybridization with ASOs is a high throughput method that looks directly at the DNA change. Method a will give you a lot of DNA, but you still have to assay for the DNA sequence at the nucleotide of interest by sequencing the DNA that was amplified. If the mutation is one in which a restriction site is altered, you could use restriction enzyme digestion followed by Southern blot.

17-16. a. **F**; b. **F**; c. **F**; d. **T**; e. **T**; f. **F**; g. **T**; h. **T**; i. **F**; j. **F**.

17-17. The genesis of the chronic myelogenous leukemia (CML) in <u>problem 17-13</u> is a chromosome rearrangement. A reciprocal translocation between non-homologous chromosomes is found in the cancerous cells of this patient while normal cells do not contain this rearrangement. Presumably translocation locates a protooncogene near a strong promoter creating a gain-of-function mutation. **If one PCR primer binds to one of the chromosomes at one side of the translocation while the other primer binds to the other chromosome on the other side of the breakpoint then your PCR**

primers will span the translocation. This would amplify a PCR fragment only if there were still cells in the blood that had the translocated chromosomes (**Figure 13.20**).

17-18.

a.   **All of these are potential oncogenes** since their normal function is to promote cell division and growth. None of them could be tumor suppressor genes.

b.   **If adding a phosphate inhibited the phosphatase then all the actions of KinaseA would ensure that the transcription factor was activated**. By contrast, if adding a phosphate activated the phosphatase then KinaseA would simultaneously add and remove phosphates from the transcription factor.

c.   The phosphatase gene is likely to be **a tumor suppressor** since its protein product inhibits the production of growth factors.

d.   In the table homozygous means homozygous for the mutation, heterozygous means one mutant allele and one wild type allele, E = excessive cell growth and D = decreased cell growth.

Mutation	homozygous	heterozygous
i.	E	
ii.	D	
iii.	D	
iv.	D	
v.	E	E
vi.	E	E
vii.	E	E
viii.	D	D
ix.	E	

17-19. Both. Genome and karyotype instability are seen in many types of cancer (**Feature Figure 17.4b2**). **These instabilities can be caused by somatic or germline mutations in genes such as** *p53* **or the genes for DNA repair enzymes** that are defective in xeroderma pigmentosum. However, **genome and karyotype instability can then result in additional problems that can contribute to cancer progression**. For instance, **mutations in DNA repair enzymes lead to a high rate of mutation and such mutations might inactivate a tumor suppressor gene or activate a protooncogene**. High rates of chromosome aberrations are also mutagenic and aneuploid cells might lose functional copies of tumor suppressor genes.

17-20.

a.   A **tumor suppressor gene** - most inherited tumors are due to the inheritance of an inactive allele of a tumor suppressor gene. The inheritance is dominant because it is highly likely that the wild type allele of the tumor suppressor gene obtained from the non-affected parent will be inactivated or lost in some of the progeny's cells.

b. As implied in part a, the allele inherited from the affected parent is a **loss-of-function** mutation that inactivates the *NF1* gene.

c. **RAS-GDP**. If neurofibromin is a tumor suppressor protein its presence would slow down or restrict the cell cycle. This could be done if neurofibromin favors the formation of the inactive form of RAS, RAS-GDP.

d. **ii, iv, v** and **vi** could all be "second hits" that would inactivate or cause the loss of the functional allele of NF1 inherited from the normal parent. The allele from the afflicted parent is already inactivated or deleted, so additional mutations in that allele would have no effect. Mitotic chromosome non-disjunction or chromosome loss (v) might produce an aneuploid cell without a functional copy of the NF1 gene. Mitotic recombination in a heterozygote could produce a cell homozygous for the inherited NF1 mutation, but the recombination event would have to occur between the NF1 gene and the centromere of the chromosome carrying it (**Figure 5.24**).

e. These patients **probably inherited a loss-of-function NF1 allele <u>and</u> another mutant gene on the same chromosome**. For instance, **both the affected parent and the child could have a chromosome with a deletion that removes NF1 as well as another gene needed for proper facial development**. Haploinsufficiency for the latter gene would then cause the additional phenotype. Other neurofibromatosis patients would have an NF1 gene inactivated in other ways.

f. The neurofibromas in these patients are **sporadic**. In other words, these patients inherit wild-type alleles of NF1 from both parents. Then **there must be a clone of cells in which one of the copies of NF1 is inactivated by some rare event then the other allele of NF1 is inactivated or lost**. These tumors are rare because two rare events must happen in the same line of cells. The inherited disease requires just one rare event as the patient starts out with a defective allele.

Section 17.3 – The Normal Control of Cell Division

17-21. See **Figure 17.13** for an overview.

a. **M** (mitosis)

b. **M**

c. **S** phase

d. **G$_1$** phase

17-22. See **Figure 17.15 Plate mutagenized haploid yeast cells at 30°C (permissive temperature) and replica plate colonies onto two sets of plates- one incubated at 30°C and the other at 23°C.** Genes necessary for the cell cycle are essential, so cold-sensitive mutants defective in cell cycle genes should die at the non-permissive temperature (23°C). Those mutants that are cold-sensitive can be studied more extensively by growing a liquid culture of the strain at permissive temperature, shifting to the non-permissive temperature and watching the cells microscopically to see if they stop with the same phenotype. If they have the same phenotype when they stop dividing, this indicates the stage in the cycle at which they stop.

17-23. There are **three complementation groups** and therefore three genes. Mutant 1 does not complement mutants 4 or 5, so these three mutations define one complementation group. Mutant 2 does not complement mutant 8, so these make up a second complementation group. Mutant 3 does not complement mutants 6, 7, or 9 so these are in a third complementation group.

17-24.

a. Although condensation begins during prophase, chromosomes are most visible during mitosis. Mitosis is 1/24 of the cycle, so 1/24 or **4% of the population of cells are in M.**

b. **The remaining 96% (23/24) of the cells are in interphase** (G1, S and G2 combined).

17-25. There is **cyclical regulation of transcription** that allows certain genes to be expressed only at certain points in the cell cycle. There is also **cyclical regulation of translation** for certain mRNAs and **cyclical control of posttranslational modifications** that lead to proteins that are active at certain points in the cell cycle and inactive at other times.

17-26.

a. **False** - CDKs (<u>c</u>yclin <u>d</u>ependent <u>k</u>inases) function in CDK/cyclin complexes; only when the CDKs are complexed are they are active kinases (**Figure 17.16**).

b. **True** - their appearance and disappearance activates the correct kinases and regulates the phosphorylation/dephosphorylation (activation/deactivation) of other proteins.

c. **False** - Checkpoint monitoring proteins check for aberrant cell cycle events (**Figure 17.21**).

17-27. a. **2**; b. **3**; c. **1** (**Figure 17.21**).

17-28.

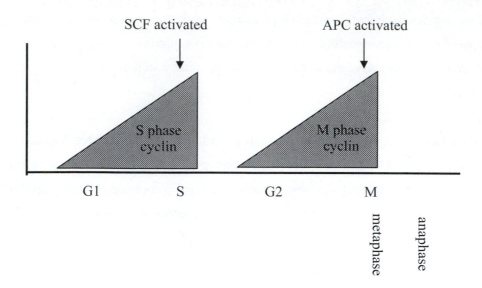

17-29.

As explained in Chapter 4 kinetochores that are not properly attached to spindle fibers generate molecular checkpoint signals (Figure 4.11). The M phase checkpoint ensures that cohesin remains intact until all the chromosomes are connected to the spindle (Figure 17.21b). **In the M phase checkpoint molecules made by unattached kinetochores prevent the anaphase promoting complex (APC) from being activated** (see below). If the APC is not active then M phase cyclins remain at high levels and cohesin remains intact. In other words the cells stay in prometaphase or metaphase while the checkpoint is signaling.

As seen in problem 17-12 **the APC must become activated at the beginning of anaphase to destroy M phase cyclin, allowing cells to leave M phase. The activated APC adds ubiquitin to protein substrates. When this happens the ubiquinylated proteins are rapidly destroyed by the proteosome. One simple hypothesis is that cohesin is also targeted by the APC since it must be destroyed at the beginning of anaphase**, thus ensuring that sister chromatids separate at the beginning of anaphase.

Activated APC does in fact lead to cohesin degradation and thus sister chromatid separation, but the actual situation is slightly more complicated than you could have predicted from the given information. The APC-directed degradation of cohesin turns out to be indirect. Active APC adds ubiquitin to another protein called Securin which is found in a complex with a proteolytic enzyme named Separase during M phase. When Securin and Separase are together in the complex the Separase is non-functional. When Securin is destroyed then Separase acts to cleave cohesin and separate the sister chromatids.

Chapter 18   Using Genetics to Study Development

Synopsis:

This chapter examines how the single cell of a fertilized egg, or zygote, differentiates into hundreds of different types of cells. This chapter presents some general conclusions that provide a framework for how genes help control cellular differentiation during development – that is how they direct most of the changes that turn the fertilized egg into a functioning adult. Much of this understanding has been elucidated from research on model organisms that serve as prototypes for research in developmental genetics. You will see how researchers extrapolate the insights from research in model organisms to all living forms. For instance the genetic analysis of body plan development in *Drosophila* and the relevance of results from this analysis to the study of body plan specification in mammals, including humans.

Significant Elements:

After reading the chapter and thinking about the concepts, you should be able to:

♦ The advantages of using the four model organisms that are discussed in this chapter - *S. cerevisiae*, *C. elegans*, *D. melanogaster* and mouse.

♦ Understand the differences between mosaic and regulative determination.

♦ Describe how careful analysis of various types of genetic mutants helps researchers understand the function of the normal protein in development.

♦ Explain the differences between different sorts of loss-of-function mutations: null mutations hypomorphic mutations, conditional mutations and dominant-negative mutations (**Figure 18.7**).

♦ Explain how to make knock-out mice (**Figure 18.5**).

♦ Understand how to use RNA interference to create phenocopies that mimic loss-of-function mutations (**Figure 18.8**).

♦ Discuss gain-of-function mutations and mutations causing ectopic gene expression.

♦ Understand how the developmental phenotype, the nature of the protein encoded by the mutant gene, when and in what tissues the gene is expressed, where the protein product is found in the cell and in the animal, what cells and tissues are affected by the loss of gene function.

♦ How proteins are localized using various techniques like antibodies and GFP tagging (**Figure 18.13**).

♦ Describe the potential uses of stem cells and human cloning – understand the differences between reproductive cloning and therapeutic cloning (**Genetics and Society**).

Solutions to Problems:

Vocabulary

18-1. a. **11**; b. **4**; c. **8**; d. **7**; e. **12**; f. **9**; g. **13**; h. **3**; i. **1**, j. **5**; k. **6**; l. **10**; m. **2**.

Section 18.1 – Model Organisms

18-2.

a. **Mutations in the gene that affected heart development would probably cause lethality early in embryonic or fetal development**. It would not be ethical to study such mutations in humans, **so you would study this gene in a model organism**. The chosen model organism must have to have a heart and a homologous (actually orthologous) gene. **Possible model organisms would be *Drosophila* or mice; interesting work on the genetics of heart development is also being done in zebrafish**. Gene knockouts or knockdowns can be made by various techniques in all these organisms. *Drosophila* and zebrafish have the advantage that early development occurs externally of the mother. Zebrafish are partially transparent so that hearts can be observed to some extent in live animals. Mice have the advantage that their heart development is more similar to that of humans than the other model organisms – for instance early development occurs internally, in the womb, in both organisms. *C. elegans* does not have a heart thus removing it from consideration as a model organism. Remarkably, the development of their pharynx (an organ used to grind food) mimics that of the human heart, so relevant research is also done on this organism.

b. **The search for genes involved in heart development would be easier in model systems that can be subjected to large-scale genetic screens. It is much easier and cheaper to conduct such screens in *Drosophila*** than in mice. Genetic screens for mutations affecting heart development have also been done in zebrafish, as have screens for pharynx development defects in *C. elegans*.

18-3.

a. In *C. elegans,* **laser ablation at this early stage of development would almost certainly be lethal**, since you would destroy a large proportion of the cell types that would eventually develop from the descendants of the ablated cell, and there are no other cells that could replace these. In **mice, the loss of one out of four early embryonic cells would have no effect**, since the descendants of the cells that were not destroyed could take the place of the descendants of the cell that was destroyed.

b. Separating the four cells would be **lethal to *C. elegans*** since none of these cells would be able to develop into a complete organism. In mice it is **possible that the separated cells could develop into a mouse**.

c. Fusing two four-cell embryos **in *C. elegans* would likely be lethal**, since proper development depends upon signals between the various cells and this would almost certainly be disrupted by the fusion event. **In mice, such a fusion would be tolerated giving rise to a chimeric animal** if the two embryos had different genotypes. Such fusion is in fact an important step in the procedures for constructing gene knockouts in mice (**Figure 18.5 step e**).

Sections 18.2 and 18.3 – Using Mutations to Dissect Developmental Pathways

18-4.

a. **The *rg$^{\gamma3}$* allele** causes the more severe phenotype (very rough eyes) and is therefore the stronger allele.

b. Since the ***rg^{41}*** phenotype is more wildtype (less severe) this allele directs either the production of more Rugose protein, or the production of the same amount of a more active Rugose protein than the *rg$^{\gamma3}$* allele.

c. **Muller's test compares the phenotype associated with homozygosity for a new, recessive mutation with that associated with heterozygosity for the mutation and a deletion of the gene. If the phenotypes are identical in these two genotypes then the new mutation is a null mutation. If the mutation/deletion has a more severe phenotype then the new mutation is not null but hypomorphic.** The answers to parts a and b above establish that *rg^{41}* **is a weak allele that makes at least some Rugose protein**, so it cannot be a null (no activity) allele. This leaves *rg$^{\gamma3}$* as the only candidate for a null allele. Muller's test suggests that *rg$^{\gamma3}$* **is in fact a null allele** because the phenotype of a *rg$^{\gamma3}$* homozygote is the same as that of *rg$^{\gamma3}$ / Df(1)JC70*. We know that the deletion must be null for the *rg* gene because the gene is missing. If *rg$^{\gamma3}$* encoded some Rugose protein then you would expect that *rg$^{\gamma3}$ / Df(1)JC70* would make half the amount of gene product of *rg$^{\gamma3}$* homozygotes thus giving a more severe phenotype.

18-5. You could see if any mRNA for *rugose* was produced by **making mRNA preparations from homozygotes for the new null allele and then analyzing these preparations on Northern blots.** Another way to measure the amount of *rugose* mRNA is a technique called **RT-PCR** (reverse transcriptase PCR), in which you would use reverse transcriptase to make cDNA copies of the mRNAs in these cells, and then amplify the *rugose* cDNA by PCR. If you failed to detect *rugose* mRNA in these animals although it detected it in wild type controls, the mutation is null because the gene is not transcribed. In some cases, like the presence of a nonsense mutation early in the open reading frame, **the mutation could be null but the gene would still be transcribed**. To check such cases, you could **analyze protein extracts from the homozygous mutant animals by Western blot using the antibody against the Rugose protein as a probe**. If the protein could not be detected, or if the size of the protein was much shorter than normal, then the mutation is likely to be null.

18-6.

a. The rough eye phenotype suggests that **the Rugose protein is made in the eye imaginal disks during larval development** - this is the only time and place that eye development would be impacted. It also appears that **the Rugose protein plays a role in embryonic development** since 35% of the embryos fail to hatch into larvae. Thus the Rugose protein is also likely to be found in at least some tissues during embryonic development. It seems that the role of Rugose protein in embryos is not absolutely essential, since the majority of embryos do develop into larvae and then adult flies with defective eye surfaces. One scenario is that there is another gene that serves a related function in embryos and the loss of Rugose reduces the total function to a level near a threshold that allows embryo survival.

b. You could see if there was *rugose* gene mRNA in a particular tissue by using a technique like *in situ* **hybridization** as shown in **Figure 18.12**. You could also **dissect the tissue and analyze the** *rugose* **mRNA by Northern blot or by RT-PCR** (see <u>Problem 18-5</u> above). You could **detect the presence of the Rugose protein in the tissue by staining the tissue with fluorescent antibody** to this protein (as in **Figure 18.13b**). **The expression of a gene in a particular tissue suggests, but does not prove, that the protein product of the gene plays an important or essential role in that tissue**. This role can be demonstrated only if the loss of the gene product in a mutant animal shows a defect in that tissue; in other words, you have to analyze the mutant phenotype.

18-7.

a. The $w^+ rg^+ / w\ rg$ genotype gives wild type eyes. The mitotic crossover between the centromere and the rugose gene will give two types of recombinant cells: $w^+ rg^+ / w^+ rg^+$ which are wild type like the non-recombinant cells, and $w\ rg / w\ rg$ which give rise to white, rugose patches of eye tissue.

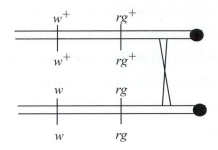

b. Expose larvae with developing eye imaginal disks to X-rays in order to induce the mitotic recombination in animals of the genotype shown in part a. These animals would have to be females in order to have two X chromosomes that could recombine. **As a result of the mitotic crossover developing ommatidia in the eye would be simultaneously homozygous for the mutations in *rugose* and *white*, while adjacent ommatidia would be heterozygous for the wild type and mutant alleles of both gene.** The former ommatidia would be white whereas the latter would be red. **If the red ommatidia are abnormal even though their genotype predicts a normal structure, then the lack of rugose in the adjacent white ommatidia affects the red ommatidia.** This could occur, for example, if the rugose protein was a ligand excreted by some ommatidial cells that binds to receptors in other ommatidia. See **<u>Figure 18.14</u>** for a similar example.

c. If *rugose* function was needed for embryonic development then animals homozygous for a null allele of the gene would die as embryos and never develop eyes. In this case you could take female larvae of the genotype shown in the answer to part a and subject them to X-rays. The adult animals would have patches of white eye tissue homozygous for the *rg* null mutation. **If these patches were normal in appearance, then rugose does not have an important role in eye development. If the white patch is abnormal, then rugose is important for eye development.** Another possible outcome is you would never see any white patches. If you conducted enough control experiments to ensure that enough mitotic recombination occurred so that you should have detected white patches, then their absence suggests that any cell homozygous for the *rugose* null mutation dies.

18-8.

a. **Mutagenize worms containing the *myo-2::GFP* transgene and look for fluorescent patterns indicating defects in pharyngeal structure.**

b. **These worms will ultimately die because they cannot ingest food.**

c. **Transform worms containing *myo-2::GFP* with a DNA construct that expresses the wild-type *pha-4* gene in ectopic locations** - that is, in locations where the pharynx does not normally develop. For example, you could place the *pha-4* gene under the control of a heat shock promoter that would express *pha-4* throughout the entire animal if the animal was grown at high temperature. You **then look for green fluorescence at locations other than the pharynx. If you see ectopic green fluorescence it means that expression of the *pha-4* gene is sufficient to elicit the development of pharyngeal tissue**, even in locations where the pharynx is not normally found. When this experiment was actually performed, *pha-4* was in fact established to be a master regulator of pharyngeal development.

18-9. Mutate possible regulatory DNA elements. This could be done by deleting various DNA sequences near the 5'-end of the *myo-2* gene. Then transform these deleted *myo-2::GFP* transgenes back into the worm. Then look to see if and how the normal pattern of green fluorescent protein expression in the pharynx is affected in these worms.

18-10. You want to determine if gene X is important in specification of the pharynx so you want to disrupt expression there; however you want to maintain normal expression of gene X in all other tissues. Therefore **express a double-stranded RNA (RNAi) corresponding to gene *X* in the pharynx.** One way to do this is to put gene *X* between two promoters for *myo-2* so that gene *X* is transcribed in opposite directions (**Figure 18.8a** and the figure on the left below). Alternatively you can place two copies of gene *X* oriented in opposite directions downstream of a single promoter for *myo-2*; when these genes are transcribed, you will get a double-stranded hairpin RNA (the figure on the right below).

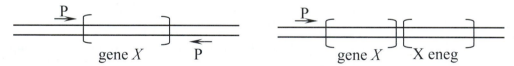

Transform these constructs into worms containing *myo-2::GFP* and examine the pattern of green fluorescence for indications of the effects of RNAi for gene *X* on pharyngeal development.

18-11. Make DNA constructs that place a wild type genomic copy of gene *X* adjacent to *myo-2::GFP*. You then transform these constructs into worms that are homozygous for a null allele of gene *X* (and that did not contain any *GFP* source). The constructs form extrachromosomal arrays as described. Pharyngeal cells containing the arrays would be wild type for gene *X* and express GFP. Pharygeal cells that had lost the arrays would be homozygous mutant for gene *X* and would not express GFP.

18-12. The knockout methodology removes an allele of the normal gene and replaces it with your altered version. This technique must be used anytime you want to examine the effects of a reduced level of function. In the add-on strategy the normal alleles of the gene are still present as well as the altered or extra version. This technique can be used when you want to increase the level of expression or when you want to express the gene in a different tissue (ectopic expression).

a. **Knockout**, because the human mutation is likely to be a loss-of-function. You knock the gene out and look for an effect in mice that are homozygous for this knockout mutation.

b. **Add-on**, because the mutation is dominant and is therefore likely to be gain-of-function. The wild type mouse gene is mutated it *in vitro* with the corresponding amino acid change found in the human protein. This mutant gene is then transformed into mice without knocking out the endogenous alleles, and you look for effects in any mouse with the added-on mutant gene.

c. **Add-on**. Make a construct in which the coding sequence of the gene was placed downstream of a promoter that turns on gene expression in other tissues, and then add this construct on to the mouse genome.

d. Deletion of the gene is by definition a **knockout** experiment. You look for possible dominant effects in a mouse in which one homolog of the gene was normal and the other was knocked out.

e. You would construct a gene **knockout** and then cross the heterozygous knockout mice to make mice that are homozygous for the knockout.

f. A dominant-negative mutation the inactive protein "poisons" the normal protein. You would have to express the dominant-negative allele of the gene in mice in an **add-on** experiment (**Figure 18.7**).

g. This is actually an **add-on** experiment: Make a construct that expresses a double-stranded RNA corresponding to the gene; (see problem 18-10). You expect the effect of this add-on to be equivalent to a knockout, or more precisely knockdown, experiment since you are trying to achieve a loss of function of the gene.

h. Perform an **add-on** experiment in which you transform mice with recombinant DNA constructs containing a reporter gene like *GFP* downstream of the promoter you want to investigate. By altering, adding and deleting different portions of the promoter region in the construct you can see which tissues express the reporter. Problem 18-9 discusses a similar scenario in *C. elegans*.

i. You could do **either a knockout or add-on** experiment in all cases. The knockout experiment would involve constructing mice with a knockout of the candidate gene. You then see if mice that are heterozygous or homozygous for the knockout express an abnormal developmental phenotype similar to that seen in the original mutant. The add-on experiment for a recessive mutant would involve taking a wild-type copy of the gene and adding it to the genome of a mouse homozygous for the recessive mutation. You then ask whether the wild-type transgene "rescues" the effect of the mutation leading to animals with a wild type phenotype. If this is the case then the candidate gene is the gene whose mutation causes the phenotype. The add-on experiment for a dominant mutation would involve taking a copy of the gene that contained the polymorphism you detected and adding it on to the genome of a wild type mouse. If mice with the transgene showed the phenotype this implicates the added gene in the production of the phenotype.

18-13.

a. **If the mutation was due to an insertion of the transgene, the MMTV *c-myc* gene should segregate with the phenotype.** That is, all subsequent animals that had the limb deformity should have the *c-myc* fusion and vice versa. The presence of the *c-myc* fusion could be recognized by Southern or PCR analysis.

b. **Clones containing the *c-myc* fusion could be identified by hybridization of MMTV sequences versus a library of genomic clones produced from the cells of the mutant mouse.** The DNA surrounding the MMTV *c-myc* fusion in this clone would be the gene of interest.

c. **The sequence of the gene into which the MMTV *c-myc* fusion inserted could be analyzed in the *ld* mutant to determine if there were mutations in the gene.** Alternatively a clone containing only the wild-type copy of the gene into which the MMTV *c-myc* fusion inserted could be injected into a mouse heterozygous for the *ld* mutation. If the homozygous *ld / ld* progeny that had also received the wild-type transgene are phenotypically wild-type, the transgene rescues and therefore must correspond to the *ld* gene.

Section 18.4 – Body Plan Development in *Drosophila*

18-14. Statement e is the only statement which is not true. Instead, the Bicoid protein activates the zygotic transcription of the *hunchback* gap gene.

18-15.

a. The **promoter** sequence itself is recognized by RNA polymerase. The region also contains **binding sites for transcription factors such as Bicoid** that ensure that the *hb* gene is transcribed only in zygotic nuclei in the anterior part of the egg. The promoter region also must have **binding sites for other transcription factors that ensure the *hb* gene is transcribed in the proper cells in the mother** so that it is deposited uniformly in the egg before fertilization.

b. The coding region contains codons that specify **the amino acids in Hunchback that comprise DNA binding domains and domains involved in the transcriptional regulation of gap and pair rule genes**.

c. The sequence could be needed for **translational repression carried out by Nanos protein**. This repression, prevents translation of the maternally supplied *hb* mRNA in the posterior portion of the embryo.

18-16. The gap and pair rule genes act before cellularization so the gene products can freely diffuse in the syncytium in the embryo. The protein products of most gap and pair rule genes are transcription factors. In contrast, **the segment polarity genes act after cellularization of the embryo,** when intrasegmental patterning is determined by the interaction between ligands made in one type of cell and receptors found on nearby cells. Most of the segment polarity genes therefore encode proteins involved in signal transduction, only a few of which are transcription factors.

18-17. The cytoplasm from the anterior of a wild type embryo could be injected into the anterior end of a *bicoid* mutant embryo to see if there was rescue of the mutant phenotype. You need a control that would indicate that the physical act of injection does not cause rescue. The control would be injection of cytoplasm from a *bicoid* mutant embryo into another mutant embryo. **Purified *bicoid* mRNA injected into the anterior end of a *bicoid* mutant embryo would be a more definitive experiment** indicating that *bicoid* alone is sufficient to rescue. Finally, **purified *bicoid* mRNA could be injected into the posterior end of a wild-type embryo.** If *bicoid* is an anterior determinant, there should be two anterior ends developing.

18-18. These results indicate that 1) maternally-supplied *hunchback* mRNA is completely dispensable and 2) that the function of the Nanos protein is only needed to restrict the translation of the maternally-supplied *hunchback* mRNA in the posterior of the egg. Development is fine if there is no maternally-supplied *hunchback* mRNA, so if there is no *hunchback* mRNA in an embryo, Nanos is not needed and can be defective (mutant) without showing any effect. Too much maternally-supplied *hunchback* mRNA swamps out the Nanos protein supplied even by wild-type mothers. The result is too much Hunchback protein in the posterior of the egg and this prevents proper abdominal development. This is a peculiar situation: the fly doesn't really need *hunchback* maternal mRNA at all, but it makes it anyway. Because it makes this unnecessary *hunchback* maternal mRNA it now needs *nanos* and the other posterior group genes to prevent its translation in the posterior of the egg. Evolution does not always come up with the simplest solution to a problem, just one that happens to work. Note that Hunchback is still needed for proper development even if the maternal hunchback mRNA is not because *hunchback* must still be transcribed from zygotic nuclei so it can perform its role as a gap gene.

18-19.

a   Absence of *knirps* function has no effect on Hunchback protein distribution but does affect Kruppel protein localization. More specifically, these results show that **the wild-type Knirps protein is needed to restrict the posterior limit of the zone of Kruppel expression.** This suggests that the wild-type Knirps protein negatively regulates the expression of Kruppel, at least in this part of the embryo.

b. **Hunchback protein would be seen throughout the embryo** because there is no Nanos protein to inhibit its translation (see problem 18-18 above).

18-20.

a. You expect that **there would be an anterior-directed transformation of the segments in which *Abd-B* is expressed most heavily.** Here the likely result would be that the segments A5-A8 would all be transformed to look like A4 which is the next most anterior segment in which *Abd-B* is not expressed.

b. This is probably due to the fact that the 3 genes are not expressed at the same levels in all the A5-A8 segments. It is known that the expression of *Ubx* and *Abd-A* decreases in successively posterior segments. **Evidently segment identity is determined by integrating the information provided by the levels of expression of the various homeotic genes.**

c. Animals deleted for all 3 genes of the BX-C would have **posterior segments that resembled those of the adjacent anterior segment in which none of the genes were expressed.** In this case you expect segments T3 and A1-A8 to all resemble T2 (**Figure 18.27**).

d. In wild type flies the wing is in segment T2 while the haltere is in segment T3 (**Figure 18.27**), so this would be a transformation from an anterior structure to a posterior structure. Since posterior-to-anterior transformations (T3 to T2) are associated with loss-of-function mutations of the *Ubx* gene, **it is likely that *contrabithorax* mutations are gain of function mutations in which the *Ubx* gene is expressed ectopically in T2.** Remember that *Ubx* is ordinarily expressed in T3 but not in T2. **Most gain-of-function mutations are dominant.** The majority of *contrabithorax* mutations are chromosomal rearrangements that place the *Ubx* gene near an enhancer element that turns on gene expression in T2. This situation is similar to that of gain-of-function mutations in *Antennapedia* that place a leg on the head of the fly (**Figure 8.31d**).

e. **Expression of the genes of the BX-C seems to inhibit the expression of *Antp*. The BX-C proteins have homeodomains and act as DNA-binding transcription factors.** In fact, it is also likely that the more posteriorly-expressed genes of the BX-C inhibit the expression of the more anterior genes; that is Abd-B inhibits expression of *Ubx* and *Abd-A* as well as *Antp* while the

Abd-A protein inhibits transcription of *Ubx* as well as *Antp*. Segment identity is mostly determined by the most "anterior" of the BX-C protein that is expressed in that segment.

Section 18.5 – Genes Help Control Development

18-21. A mutation in the genes encoding a maternally supplied component which affects early development must be a mutation in the mother's genome. If the mutation affecting early development is in a gene whose transcription begins after fertilization then the mutation must be in the genome of the zygote (these are thus sometimes called "zygotic genes"). You would need two different kinds of genetic screens to make mutations either in the mothers' genome or the zygotes' genome.

18-22.

a. At least partially, **mosaic development results from the differential distribution of molecules in the two daughter cells derived from the first division of the zygote. Since these cells are molecularly different from each other their descendants can have different fates**; thus the descendants of one cell cannot replace the descendants of the other cell. This means that molecules that help determine cell fates <u>cannot</u> be differentially distributed to daughter cells during the first several rounds of mitotic divisions of the zygote in mammals and other organisms with regulative development.

b. The PAR proteins are present in unfertilized eggs so they must be encoded by maternal genes. Therefore you will more likely find mutants in the *par* genes by doing a screen for **maternal-effect** genes rather than a screen for zygotic genes.

c. It appears that the PAR-2 protein excludes the PAR-3 protein from the posterior cortex while the PAR-3 protein excludes PAR-2 from the anterior cortex. **The entry of the sperm into the egg must set up of chemical reactions inside the fertilized zygote that favor nearby PAR-2 binding and disfavors nearby PAR-3 binding**. This distinction is then made clearer and is maintained by the fact that each protein mutually excludes the other protein from its side of the embryonic cortex.

18-23.

a. The fact that *glp-1* mRNA is found in all four cells but the GLP-1 protein is found only in ABa and ABp suggests that there is some kind of translational regulation. Perhaps a factor necessary for the translation of the *glp-1* mRNA is found only in the ABa and ABp cells, or a factor inhibiting the translation of this mRNA is found in the other two cells (EMS and P_2). These possibilities are real because, as seen in <u>Problem 18-22,</u> the two cells derived from the first mitotic division are dissimilar from each other since they have different PAR proteins. One simple model is that ABa and ABp are the descendants formed by the second round of mitosis of either the original anterior or the original posterior cell. Although the data presented does not let you decide, these two cells are actually descendants of the anterior cell. Thus **the presence of PAR-2 and absence of PAR-3 from these cells indirectly dictates their ability to translate** ***glp-1*** **mRNA into GLP-1 protein**.

b. APX-1 is located only at the region of the P_2 cell membrane that is in contact with ABp and thus with the GLP-1 protein. It appears that the stable localization of APX-1 at the membrane requires an interaction with GLP-1. **Such an interaction could occur through the extracellular domains of both proteins.**

c. The changes in cell fate that occur in response to the interaction between APX-1 and GLP-1 occur within the ABp cell, which now develops differently from that of ABa. This implies that the signal transduction pathway leading to this change in cell fate occurs within ABp. The signal transduction pathway is initiated by a receptor within the ABp cell. This **receptor is the GLP-1 protein. Thus APX-1 would be the ligand.**

d. **(i)** If the P_2 cell were ablated with a laser there would be no signaling from this cell to ABp. Since this signaling is necessary for ABp to have a different fate from ABa **the ablation of P_2 would make ABp and its descendants would have the same fate shown by ABa and its descendants. (ii) A null mutation of** ***apx-1*** **would have the same effect** - ABp and its descendants would have the same fate shown by ABa and its descendants. **(iii) Same. (iv) Same**.

Chapter 19   Variation and Selection in Populations

Synopsis:

This chapter involves the study of how genetic laws impact the genetic makeup of a population. Mendelian principles are the basis for the Hardy-Weinberg law which allows one to calculate allele and genotype frequencies from one generation to the next. The Hardy-Weinberg law can be used only if other forces are not acting on the allele frequency. Those forces include selection, migration, mutation, and population size.

Population geneticists try to determine the extent to which a trait is determined by genetic factors and how much is determined by environmental factors. Knowing if a trait is largely determined by genetic factors introduces the possibility for animal and plant breeders to select and maintain populations with desired traits.

Significant Elements:

After reading the chapter and thinking about the concepts, you should be able to:

♦ Determine allele frequencies in a population given the frequencies of genotypes.

♦ Determine genotype frequencies in a population given the frequencies of alleles.

♦ Determine genotype and/or allele frequencies in the next generation given the genotype or allele frequencies in the present generation.

♦ Determine if a population is in equilibrium.

♦ Determine allele and genotype frequencies after migration has occurred.

♦ Describe how heritability of a trait can be determined.

Problem solving tips:

♦ p and q represent allele frequencies for a gene with two alleles; the sum of p + q must equal 100% of the alleles (or gametes) in the population. In other words, $p + q = 1$.

♦ p^2, 2pq, q^2 represent genotype frequencies; p^2 corresponds to one homozygous genotype, 2pq represents the heterozygous genotype and q^2 is the other homozygous genotype.

♦ Determining allele and genotype frequencies can be done two slightly different ways. One method involves converting the initial numbers of each genotype to frequencies and then doing all calculations as frequencies. In this case the frequency of the p allele = the frequency of the p/p homozygotes + 1/2 the frequency of the heterozygotes. The frequency of the q allele = the frequency of the q/q homozygotes + 1/2 the frequency of the heterozygotes. If the problem asks for numbers of individuals then the genotype frequencies will be multiplied by the number of individuals in the population, as in <u>problem 19-2 solution #1</u>. The problems in this chapter will be solved using this method. Alternately the problems can be solved by calculating the numbers of each allele present in the population. When doing this the total must be divided by the total number of alleles, as in <u>problem 19-2 solution #2</u>.

- When any values of p and q that sum to 1 are binomially expanded ($p^2 + 2pq + q^2$) the resulting genotype frequencies are those of a population in Hardy-Weinberg equilibrium.

- The binomial expansion is the equivalent of a Punnett square. If p (or the A allele) = 0.7 and q (or the a allele) = 0.3 then the binomial expansion gives the genotype frequencies in the next generation as 0.49 AA + 0.42 Aa + 0.09 aa. You can use a Punnett square to generate the same genotype frequencies in the F_1 generation:

	0.7 A	0.3 a
0.7 A	0.49 AA	0.21 Aa
0.3 a	0.21 Aa	0.09 aa

- If the allele frequencies are calculated correctly then p + q MUST sum to 1. If the genotype frequencies are calculated correctly than $p^2 + 2pq + q^2$ MUST sum to 1.

- Once a population is at equilibrium allele and genotype frequencies do not change.

- If a genotype is selected against or if populations are combined (by migration) or if there is significant mutation (usually together with selection) allele frequencies will change.

- Selection for one genotype and selection against another genotype are balanced at a particular allele frequency (equilibrium frequency).

- Genetic drift is most often seen in small populations.

- Polygenic traits are controlled solely by alleles of two or more gene; multifactorial traits include polygenic traits and traits that are influenced by both genes and environment.

- Genetic and environmental contributions to a phenotype are sorted out by setting up conditions in which the genetic background is consistent (to analyze environmental contributions) or conditions in which the environment is constant (to analyze genetic contributions).

Solutions to Problems:

Vocabulary

19-1. a. **3**; b. **5**; c. **8**; d. **7**; e. **6**; f. **1**; g. **9**; h. **2**; i. **4**.

Section 19.1 – the Hardy-Weinberg Law

19-2. Solution #1:

a. Calculate genotype frequencies:

G^GG^G = 120 / 200 = **0.6**; G^GG^B = 60 / 200 = **0.3**; G^BG^B = 20 / 120 = **0.1**.

b. Frequency of G^G = 0.6 + 1/2 (0.3) = **0.75**; frequency of G^B = 0.1 + 1/2 (0.3) = **0.25**.

c. Binomially expand p and q: $(0.75\ G^G)^2$ + 2 (0.75 G^G) (0.25 G^B) + (0.25 G^B)2 = **0.563** G^GG^G +

0.375 G^GG^B + **0.063** G^BG^B = 1.

Solution #2:

a. The genotype frequencies are calculated as in Solution #1 part a above.

b. The allele frequencies are determined by totaling all alleles within each genotype.

G^GG^G = 120 individuals with 2 G^G alleles = 240 G^G alleles

G^GG^B = 60 individuals with one G^G allele = 60 G^G alleles

G^GG^B = 60 individuals with one G^B allele = 60 G^B alleles

G^BG^B = 20 individuals with two G^B alleles = 40 G^B alleles

There are 300 G^G alleles/400 total alleles so the frequency of G^G **(p) = 0.75**. There are 100 G^B

alleles/400 total alleles so the frequency of G^B **(q) = 0.25**.

c. The expected frequencies can be calculated using the allele frequency and the terms of the Hardy-

Weinberg law, p^2 + 2pq + p^2 = 1; **0.5625** G^GG^G + **0.375** G^GG^B + **0.0625** G^BG^B = 1.

19-3. For each population determine the allele frequency Use the genotype frequencies to calculate
the allele frequency. Binomially expand the allele frequencies to generate the genotype frequencies of
the next generation (the Hardy-Weinberg genotype frequencies).

a. The frequency of A = 0.25 + 1/2 (0.5) = 0.5; the frequency of a = 0.25 + 1.2 (0.5) = 0.25.

   Binomially expanding A (p) and a (q) gives genotype frequencies of 0.25 AA + 0.5 Aa + 0.25 aa.

   Because these genotype frequencies (the Hardy-Weinberg equilibrium) are the same as those of

   the original population the original **population is in equilibrium**.

b. The frequency of A = 0.1 + 1/2 (0.74) = 0.47; the frequency of a = 0.16 + 1/2 (0.74) = 0.53.

   Binomially expanding A and a gives genotype frequencies of 0.221 AA + 0.498 Aa + 0.281 aa.

These are the Hardy-Weinberg genotype frequencies for p = 0.47 and q = 0.53. They are NOT the same as the original genotype frequencies, so **population b is not in equilibrium**

c. **Population c is not in equilibrium** - the Hardy-Weinberg genotype frequencies are 0.61 AA + 0.34 Aa + 0.05 aa.

d. **Population d is not in equilibrium** – the Hardy-Weinberg genotype frequencies are 0.50 AA + 0.41 Aa + 0.08 aa.

e. **Population e is in equilibrium** – the Hardy-Weinberg genotype frequencies are 0.81 AA + 0.18 Aa + 0.01 aa.

19-4.

a. There are 60 D^+D^+ flies with normal wings and 90 DD^+ flies with Delta wings. The frequencies of these two genotypes are 60/150 = 0.4 D^+D^+ and 90/150 = 0.6 DD^+. The allele frequency for **D** = 0 DD homozygotes + 1.2 (0.6) = **0.3**; allele frequency for D^+ = 0.4 + 1/2 (0.6) = **0.7**.

b. The following frequencies of F_1 zygotes will be produced: 0.49 D^+D^+ + 0.42 DD^+ + 0.09 DD = 1. The homozygous DD zygotes do not live, so the viable progeny are 0.49 + 0.42 = 0.91 (91%) of the original zygotes. If there are 160 viable adults in the F_1 generation, then there must have been 160 / 0.91 = 176 F_1 zygotes.

c. As noted in part b above, the viable progeny are only 91% of the F1 zygotes (0.49 D^+D^+ + 0.42 DD^+ = 0.91. This equation must be normalized before you can calculate expected numbers of progeny. Therefore, 0.49 D^+D^+ / 0.91 = 0.54 D^+D^+ = proportion of the viable offspring with the D^+D^+ genotype and 0.42 D^+D / 0.91 = 0.46 D^+D = proportion of the viable offspring with the D^+D genotype. Thus, the numbers of viable individuals with these genotypes are: 0.54 D^+D^+ x 160 = **86 D^+D^+** and 0.46 D^+D x 160 = **74 D^+D**.

d. **No**. The lethality of the DD genotype means **that there are genotype-dependent differences, violating one of the basic tenets of the Hardy-Weinberg equilibrium**. Therefore this population will never achieve equilibrium.

19-5.

a. The frequency of M = 0.5 + 1/2 (0.2) = 0.6; the frequency of N = 0.3 + 1/2 (0.2) = 0.4. The expected genotype frequencies in the next generation are calculated using these allele frequencies in the binomial expansion $p^2 + 2pq + p^2$ = 1: 0.36 MM + 0.48 MN + 0.16 NN = 1. These Hardy-Weinberg genotype frequencies are not the same as the genotype frequencies of the initial population, therefore **the initial population is not in equilibrium.**

b. The **genotype frequencies in the F$_1$ will be: 0.36 _MM_ + 0.48 _MN_ + 0.16 _NN_ = 1**. From this the **allele frequencies in the F$_1$ generation** can be calculated: _M_ = 0.36 + 1/2 (0.48) = **0.6** and _N_ = 0.16 + 1/2 (0.48) = **0.4**.

c. **The same as in part b.**

19-6.

a. Convert the genotypes into frequencies by dividing each by the total number (= 1480).

$Q^FQ^FR^CR^C$ 0.137

$Q^FQ^GR^CR^C$ 0.068

$Q^GQ^GR^CR^C$ 0.068

$Q^FQ^FR^CR^D$ 0.251

$Q^FQ^GR^CR^D$ 0.126

$Q^GQ^GR^CR^D$ 0.126

$Q^FQ^FR^DR^D$ 0.112

$Q^FQ^GR^DR^D$ 0.056

$Q^GQ^GR^DR^D$ 0.056

Calculate the allele frequencies for Q and R genes separately. In this population the frequencies of the Q gene are 0.5 Q^FQ^F + 0.25 Q^FQ^G + 0.25 Q^GQ^G = 1. The allele frequencies are Q^F = 0.625 and Q^G = 0.375. The expected genotype frequencies in the next generation (Hardy-Weinberg equilibrium) is therefore 0.39 Q^FQ^F + 0.47 Q^FQ^G + 0.14 Q^GQ^G = 1. **The population is <u>NOT</u> in equilibrium for the _Q_ gene.** The genotype frequencies for the R gene in this population are 0.27 R^CR^C + 0.5 R^CR^D + 0.22 R^DR^D = 1. The R^C allele frequency = 0.52 and the R^D allele frequency = 0.48. The expected genotype frequencies in the next generation (Hardy-Weinberg equilibrium) is 0.27 R^CR^C + 0.5 R^CR^D + 0.23 R^DR^D =1. In this case the observed genotype frequencies are very close to the Hardy-Weinberg genotype frequencies, so **the population <u>IS</u> in equilibrium for the _R_ gene.**

b. the fraction that will be Q^FQ^F in the next generation is the expected genotype frequency in part a: **0.39 for the Q^FQ^F genotype.**

c. The fraction that will be **R^CR^C in the next generation** will again be the expected frequency calculated based on the allele frequency: **0.27.**

d. This probability is <u>not</u> influenced by allele frequencies calculated in part a above. Instead this is a standard probability question starting with parents of specific genotypes as discussed in <u>Chapter 2</u>

(see problem 2-4). This cross is $Q^F Q^G R^C R^D$ and $Q^F Q^F R^C R^D$. The probability is the product of the individual probabilities for each of the genes. There is a 1/2 chance that the female will contribute the Q^G allele and a 1/1 chance that the male will contribute the Q^F allele = 1/2 chance of a $Q^F Q^G$ child. Both parents are heterozygous for the R gene, so there is a 1/4 chance the child will be homozygous $R^D R^D$. There is a 1/2 chance the child will be male. The overall probability is (1/2)(1/4)(1/2) = **1/16 that the child will be a $Q^F Q^G R^D R^D$ male**.

19-7. If a population is in Hardy-Weinberg equilibrium then the allele frequency p is squared to give the genotype frequency p^2. Thus, **each different allele frequency of p has a different p^2 value and a different set of genotype frequencies at equilibrium.**

19-8. The 3:1 ratio is seen when two heterozygous individuals are crossed. This ratio is not relevant for a population where the crosses are of many different sorts – some will be homozygous dominant x homozygous dominant, others homozygous dominant x heterozygous, some homozygous dominant x homozygous recessive, some heterozygous x homozygous recessive and others homozygous recessive x homozygous recessive. **The ratio of wild type : mutant progeny in the population will depend on specific allele frequencies in that population.**

19-9.

a.   The frequencies of the genotypes in the sailor population are: MM = 324/400 = 0.81, MN = 72/400 = 0.18 and NN = 4/400 = 0.01. The frequency **of the N allele = 0.01 + 1/2 (0.18) = 0.1**.

b.   In order to calculate the allele and genotype frequencies in the children you must calculate the allele frequencies of each original population and then combine them in the correct proportions. From part a the frequency of N = 0.1 so the frequency of M = 0.9. In the Polynesian population the allele frequencies are N = 0.94 and M = 0.06. When these two populations mix randomly then 40% of the M and N alleles will come from the sailor population (400/1000) and the remaining 60% will be provided by the Polynesians. Therefore the frequency of N in the mixed population = 0.4 (0.1) + 0.6 (0.94) = 0.04 + 0.564 = 0.604. The frequency of M in the mixed population = 0.4 (0.9) + 0.6 (0.06) = 0.36 + 0.036 = 0.396. The genotype frequencies in the next generation will be: $(0.604)^2$ NN + 2 (0.604) (0.396) MN + $(0.396)^2$ MM = 0.365 NN + 0.478 MN + 0.157 MM = 1. If there are 1,000 children then there will be **478 children with the MN genotype**.

c.   The observed genotype frequencies in the children are 100/1000 NN + 850/1000 MN + 50/1000 MM = 0.1 NN + 0.85 MN + 0.05 MM. The allele frequencies among the children therefore are N = **0.1 + 1/2 (0.85) = 0.525** and M = 0.05 + 1/2 (0.85) = 0.475.

19-10.

a. If a population is in Hardy-Weinberg equilibrium then $\sqrt{q^2} = q$. If 1/250,000 people are affected by the autosomal recessive disorder alkaptonuria then the genotype frequency of the recessive homozygote $(q^2) = 4 \times 10^{-6}$. Therefore $q = \sqrt{4 \times 10^{-6}} = 0.002$.

b. Remember that the population is in Hardy-Weinberg equilibrium, that $q = 0.002$ and p (the normal allele) $= 0.998$. Therefore **the frequency of carriers (heterozygotes) = 2 (0.998) (0.002) = 0.004 = 4 \times 10^{-3}. The ratio of carriers / affected individuals = 4 \times 10^{-3} / 4 \times 10^{-6} = 1,000:1**.

c. If the unaffected woman has an affected child then she must be a carrier, as was the father of the child. She has remarried. If her new husband is homozygous normal she can never have an affected child. However if the new husband is a carrier or is affected himself then she can have an affected child. The total probability of an affected child is the sum of each of these individual probabilities. **The probability that she will have an affected child by this marriage** = [(0.004 probability that her new husband is a carrier) (0.5 probability that he passes on his mutant allele) (0.5 probability that she passes on her mutant allele)] + [(4 \times 10^{-6} probability that her new husband is affected) (1 probability that he passes on a mutant allele) (0.5 probability that she passes on her mutant allele)] = [0.001] + [2 \times 10^{-6}] = 0.001002 = **0.001**. As you can see, the probability that her second husband is affected is insignificant. Therefore the probability that this woman will have an affected child with her second husband is only dependent on the frequency of heterozygotes in the population.

d. **No - if one of the genotypes is selected against then the frequency of p and q will change each generation**. The population will never reach equilibrium (see problem 19-4 and **Figure 19.8**).

19-11.

a. If a gene has three alleles then $p + q + r = 1$. The genotype frequencies of a population at Hardy-Weinberg equilibrium would therefore be represented by the binomial expansion of the allele frequencies: $(p + q + r)^2 = (1)^2 = p^2 + 2pq + q^2 + 2pr + r^2 + 2qr = 1$. As discussed in the Problem Solving Tips at the beginning of this chapter, a Punnett square will give the same genotype frequencies:

	p	q	r
p	p^2	pq	pr
q	pq	q^2	qr

r	pr	pr	r^2

b. In the Armenian population the allele frequencies are $I^A = 0.36$, $I^B = 0.104$ and $i = 0.536$. If the population is in Hardy-Weinberg equilibrium then the binomial expansion of these allele frequencies gives: $0.13\ I^A I^A + 0.075\ I^A I^B + 0.011\ I^B I^B + 0.111\ I^B i + 0.287\ ii + 0.386\ I^A i = 1$. Therefore the frequencies of the four blood types in this population are **0.516 A, 0.122 B, 0.075 AB and 0.287 O**.

19-12.

a. When considering an X linked gene, women have two alleles. Therefore the **allele and genotype frequencies in women are** calculated exactly as we have been doing – **$p + q = 1$ and $p^2 + 2pq + q^2 = 1$**. However men are hemizygous (have only one allele for genes on the X chromosome) so the **men's allele frequencies and genotype frequencies are the same – $p + q = 1$**.

b. If 1/10,000 males is a hemophiliac then the allele frequency of the mutant allele (q) = 1×10^{-4}. Therefore the frequency of affected females = $q^2 = 1 \times 10^{-8}$. If there are 1×10^8 women in the United States then **only one of them should be afflicted with hemophilia**.

19-13.

a. Remember that colorblindness is an X-linked recessive trait. Therefore boys are hemizygous – the allele frequency for **$C = 8324/9049 = 0.92$** and for **$c = 725/9049 = 0.08$**.

b. If the population is in Hardy-Weinberg equilibrium then the girls should have the same allele frequencies as the boys. The genotype frequency of colorblindness (cc) in the girls = 40/9072 = 0.0044 and if the girls are in equilibrium then the allele frequency of $c = \sqrt{0.0044} = 0.066$. This does not equal 0.08 (from part a).

    Alternately if the population is in equilibrium then the allele frequency of c in the boys should accurately predict the genotype frequencies in the girls. Thus c^2 should equal the frequency of colorblind girls - $(0.08)^2 = 0.0064$ which does <u>not</u> equal the cc genotype frequency of 0.0044.

    Therefore **this sample does not demonstrate Hardy-Weinberg equilibrium**.

c. Based on this information the frequency of the $c^p c^p$ genotype in girls = $3/9072 = 3.3 \times 10^{-4}$ and **the frequency of $c^p = \sqrt{3.3 \times 10^{-4}} = 0.018$**. The frequency of the $c^d c^d$ genotype in the girls = $37/9072 = 0.0041$ and **the frequency of $c^d = \sqrt{0.0041} = 0.064$**. The **frequency of the C allele is** $1 - (0.018 + 0.064) = 0.918$.

d. The genotype frequencies among the boys are the same as the allele frequencies in the girls calculated in part c. **Thus in boys $C = 0.918$ (normal vision), $c^d = 0.064$ (colorblind) and $c^p =$ 0.018 (colorblind). In the girls the genotype frequencies are: $CC = 0.843$ (normal vision), $Cc^d = 0.118$ (normal vision), $Cc^p = 0.033$ (normal vision), $c^pc^p = 3.3$ x 10-4 (colorblind), c^dc^d $= 0.004$ (colorblind) and $c^dc^p = 0.002$ (normal vision).**

e. These results make it much more likely that **the population is in equilibrium. As seen in part c, the allele frequency of C is the same in boys and girls and the allele frequency of c in the boys is the same as the total frequencies of $c^d + c^p$ in girls.** Likewise the frequencies of genotypes with normal vision (0.918 predicted in boys : 0.92 observed in boys and 0.996 predicted in girls vs 0.996 observed in girls) vs colorblind vision (0.082 predicted in the boys vs 0.08 observed and 0.004 predicted in the girls vs 0.004 observed) are the same in boys and girls.

19-14. The genotype frequency of the recessive, non-eye rolling phenotype (*ugh ugh*) is 410/500 = 0.82 for the French population and 125/200 = 0.625 for the Kenyan population. If each of these groups is in equilibrium then the allele frequency for *ugh* in the French group is $\sqrt{0.82} = 0.906$ and in the Kenyan group the *ugh* allele frequency = $\sqrt{0.625} = 0.791$. If the two groups married and had children at random then the French group would provide 500/700 = 71.4% of the total ugh alleles and the Kenyans would provide 200/700 = 28.6% of these alleles (see problem 19.9). Therefore the frequency of the *ugh* allele in the mixed group = 0.714 (0.906) + 0.286 (0.791) = 0.873. The *Ugh* allele frequency = 1 – 0.873 = 0.127. The genotype frequencies in the children will be 0.762 *ugh ugh* + 0.222 *Ugh ugh* + 0.016 *Ugh Ugh* = 1. If there are 1,000 children then there will be 762 *ugh ugh* children that will <u>not</u> express the eye rolling trait and **238 *Ugh* - children that will roll their eyes.**

19-15.

a. Diagram the cross:

$vg^+ vg^+$ x $vg\ vg$ → F$_1$ $vg^+ vg$ → F$_2$ 1/4 $vg^+ vg^+$: 1/2 $vg^+ vg$: 1/4 $vg\ vg$ (3 wild type : 1 vestigial). If the vestigial F$_2$ flies are selected against then the remaining F$_2$ genotypes are 1/3 $vg^+ vg^+$ and 2/3 $vg^+ vg$. Therefore **the genotype frequencies in the F$_2$ are 0.33 $vg^+ vg^+$ and 0.67 $vg^+ vg$; the allele frequencies in the F$_2$ for vg^+ = 0.33 + 1/2 (0.67) = 0.67 and for vg = 1/2 (0.67) = 0.33.**

b. The expected **genotype frequencies in the F$_3$ progeny are 0.449 $vg^+ vg^+$ + 0.442 $vg^+ vg$ + 0.109 $vg\ vg$ = 1, or 0.891 wild type and 0.109 vestigial.**

c. If the F_3 vestigial flies are selected against then the altered F_3 genotypic ratio becomes 0.449 vg^+ vg^+ + 0.442 vg^+ vg = 0.891. In order to calculate allele frequencies this equation must be normalized (see <u>problem 19-4c</u>), becoming 0.504 vg^+ vg^+ + 0.496 vg^+ vg = 1. The F_3 allele frequencies: vg^+ = 0.504 + 1/2 (0.496) = 0.752 and vg = 1/2 (0.496) = 0.248. The genotype frequencies in the F_4 generation will be: 0.566 vg^+ vg^+ + 0.373 vg^+ vg + 0.062 vg vg = 1. The genotype frequencies in the F_4 generation will be: 0.566 vg^+ vg^+ + 0.373 vg^+ vg + 0.062 vg vg = 1. Therefore the **F_4 allele frequencies are vg^+ = 0.566 + 1/2 (0.373) = 0.753 and vg = 0.062** + 1/2 (0.373) = **0.247**.

d. **If all of the F_4 flies are allowed to mate at random** then there is no selection and **the population will be in Hardy-Weinberg equilibrium**. Therefore the F_5 genotype and allele frequencies will be the same as those in the F_4 generation in part c above: **0.566 vg^+ vg^+ + 0.373 vg^+ vg + 0.062 vg vg = 1; vg^+ = 0.753 and vg = 0.247**.

19-16.

a. Convert the genotypes to frequencies: 60/150 t^+t^+ = 0.4 t^+t^+ and 90/150 t^+t = 0.6 t^+t. The **allele frequencies are: t^+ = 0.4 + 1/2 (0.6) = 0.7 and t = 1/2 (0.6) = 0.3**.

b. First determine the frequencies of the three genotypes if all lived, then remove the inviable mice from your calculations and normalize the genotype frequencies in order to calculate the allele frequencies. The expected genotypes frequencies of the zygotes in the next generation are: 0.49 t^+t^+ + 0.42 t^+t + 0.09 tt = 1. However the tt zygotes die, so the remaining genotype frequencies are: 0.49 t^+t^+ + 0.42 t^+t = 0.91. When this is normalized it becomes 0.538 t^+t^+ + 0.462 t^+t = 1. If 200 progeny mice are scored there will be **108 normal mice and 92 tailless mice**.

c. The Dom 1 population has 64 members and the genotype frequencies are 0.25 t^+t^+ and 0.75 t^+t. Thus the allele frequencies are t^+ = 0.625 and t = 0.375. The Dom 2 population has 84 members and the genotype frequencies are 0.571 t^+t^+ and 0.429 t^+t. The allele frequencies here are t^+ = 0.785 and t = 0.215. If the populations interbreed randomly then the Dom 1 parents will provide 64/148 = 0.432 of the alleles (gametes) found in the next generation and Dom 2 will provide 0.568 or 56.8% of the alleles. Therefore the combined gamete frequencies are: t^+ = 0.432 (0.625) + 0.568 (0.785) = 0.716 and t = 0.432 (0.375) + 0.568 (0.215) = 0.284. The genotype frequencies in the next generation will be $(0.716)^2$ t^+t^+ + 2(0.716)(0.284) t^+t + $(0.284)^2$ tt = 1. Of course the *tt* zygotes die (see part b above) so the normalized genotype frequencies of the two viable genotypes are **0.558 t^+t^+ and 0.442 *tt*.**

Section 19.2 – Causes of Allele Frequency Changes

19-17. A fully recessive allele is not expressed in a heterozygous organism, so there is no selection against the heterozygotes. **Selection against the homozygous recessive genotype will decrease the frequency of the recessive allele in the population, but it will never totally remove it, as the recessive allele is hidden in the heterozygote (<u>Figure 19.8</u>). In addition, a recessive allele sometimes confers an advantage when present in the heterozygote**, as seen for the sickle cell allele in areas where malaria is prevalent. Finally, **mutation can produce new recessive alleles in the population**.

19-18. The farther from equilibrium, the greater the Δq, so the population with an allele frequency of 0.2 will have the larger Δq.

19-19.

a.   The allele frequency of $b = \sqrt{0.25} = 0.5$ so the allele frequency of **$B = 0.5$**.

b.   To calculate Δq, first determine q in both generations. For tank 1 (and all three tanks), q = 0.5 (see part a) and q in the next generation (q') = $\sqrt{0.16} = 0.4$. Therefore **Δq for tank 1 is q'- q = 0.4 - 0.5 = -0.1**. The same calculations are carried out for the other two tanks: **q for all tanks = 0.5; q' for tank 2 = 0.5 so Δq = 0; q' for tank 3 = 0.55 and Δq = 0.05.**

	Tank 1	Tank 2	Tank 3
b. Δ*q*	**-0.1**	**0.0**	**0.05**
c. w_{Bb}	**1.0**	**1.0**	**1.0**
d. w_{bb}	**<1.0**	**1.0**	**>1.0**

c.   **If the fitness of the *BB* genotype = 1 (w_{BB} = 1) and the *b* allele is totally recessive then the fitness of the *Bb* genotype (w_{Bb}) = 1 also (see row c. in the table above).**

d.   In all three tanks the original frequency of the *b* allele (*q*) = 0.5. **In tank 1** the frequency of *b* in the progeny (*q'*) = 0.4 so the frequency of the *bb* (small tail) males has decreased. Therefore the fitness of the small tailed males decreased, so **w_{bb} <1.0. In tank 2** *q'* = 0.5 (the *b* allele frequency remained the same) so **w_{bb} = 1.0. In tank 3** *q'* = 0.55 (the *b* allele frequency increased) so **w_{bb} >1.0.**

19-20. The equilibrium frequency will be different for the two populations. Equilibrium frequency is a balance between selection and mutation and the selection is very different in the two populations.

19-21.

a.   The affected genotype dies before reproductive age. Therefore the **fitness value (w) = 0 and the selection coefficient (s) = 1 for the affected genotype. There is no selection pressure against the carrier or the homozygous normal genotypes, so for both of these $w = 1$ and $s = 0$.**

b. $\overline{w} = p^2 \, w_{RR} + 2pq \, _{Rr} + q^2 \, w_{rr} = (0.96)^2 \times 1.0 + 2 \, (0.96) \, (0.04) \times 1.0 + (0.04)^2 \times 0 = 0.9984$ (See equation 21.4a).

$\Delta q = -s_{rr} \, p \, q^2 / \overline{w} = -1 \, (0.96) \, (0.04)^2 / 0.9984 = \textbf{-1.54} \times \textbf{10}^{\textbf{-3}}$ (see equation 21.7).

c.   If the mutation rate from CF^+ to CF^- is 1×10^{-6} then the expected evolutionary equilibrium frequency ($\hat{q}$) of the CF^- allele is:

$\hat{q} = \sqrt{(\mu / s \, p)} = \sqrt{[(10^{-6}) / (1 \times 0.96)]} = \sqrt{[10^{-6} / 0.96]} = \textbf{1.02} \times \textbf{10}^{\textbf{-3}}$. **This number ($1.02 \times 10^{-3}$) is smaller than the observed q which is 0.04.** This implies that there must be some heterozygote advantage to the mutant CF^- allele.

d.   Interestingly one recent study suggests the hypothesis that ***CF^+/CF^-* heterozygotes may be better able to survive outbreaks of cholera.** This is possible because people with cholera have diarrhea that pumps water and chloride ions out of the small intestine. The CFTR protein encoded by the *CF* gene is a chloride ion channel. CF^+/CF^- heterozygotes thus lose less water than CF^+/CF^+ individuals when infected with cholera. Thus the heterozygotes are less likely to die of dehydration.

Section 19.3 – Analyzing Quantitative Variation

19-22.

a.   Using genetic clones, **only environmental effects contribute to variation**.

b.   Monozygotic twins are genetically identical so they can be thought of as genetic clones whereas dizygotic twins are genetically different. **Comparison of MZ and DZ twins provides an assessment of the effect of genes versus environment.**

c.   Cross-fostering is removing offspring from a mother and placing with several different mothers to randomize the effects of different mothering environments. This is done to **reduce environmental effects** when determining heritability of a trait.

19-23. High heritability indicates that the phenotypic differences observed are due in large part to genetic differences. **Choice b** would be true.

19-24.

a. Calculations of heritability vary in different environments, so **by controlling environmental similarity we can get a better estimate of the effect of genetic factors in that population**.

b. Concordance values for **MZ twins would allow a more accurate estimation of heritability**, as MZ twins have a genetic similarity of 1 but an environmental similarity about equal to either non-twin siblings or DZ twins.

19-25.

a. The table shows the average differences for each category. If a trait has a high heritability the MZ (monozygotic) twins would have small average differences compared with DZ (dizygotic) twins or siblings. You also expect that MZ twins raised together would have about the same (small) average differences as MZ twins raised apart since the environment would not contribute much. Thus, the table shows that **Height has the highest heritability** because both categories of MZ twins have about the same average differences, and these are much lower than the average differences for DZ twins or siblings. The table also implies that **Weight has the lowest heritability** since the average differences for MZ twins raised apart are much higher than for MZ twins raised together; in fact, the average differences for MZ twins raised apart is almost the same as for DZ twins or siblings. IQ is somewhere in between, since the average differences for MZ twins raised apart is apparently significantly less than the differences for DZ twins or siblings. However, the data shows that environment is a very strong influence on the phenotype of IQ scores since there is considerably more variation among MZ twins raised apart than for MZ twins raised together.

b. **The data from the CDC does not affect the conclusions from part a**. During the 42-year period of the study, there can have been very little change in the genetic composition of such a large group of people (all 15 year-old boys in the United States), yet there have been very significant changes in the height and weight of this population. The difference of roughly 15 pounds indicates that environment (most probably, diet) plays a critical role in determining the phenotype of weight. The data from the CDC does seem to show that the environment also plays more of a role in determining the phenotype of height than would have been inferred from the data in part a.

19-26. Heritability values depend on the frequency of the trait in the population, on the total amount of phenotypic variation in the population and on environmental factors that could influence the trait. Comparing frequency and total phenotypic variation for the trait between the two populations could provide evidence that heritability values for the two populations would be expected to be different.

19-27.

a. **There is a founder effect of the descendants coming from a small number of individuals,** so whatever recessive alleles were present in the population are more likely to be combined. Therefore the frequency of some alleles and genotypes can be higher in that population. Other alleles may not have been included in that original gene pool.

b. **An advantage to studying the Finnish population is that there is genetic homogeneity and probably fewer genes** (potential modifiers) that may affect the trait and therefore can be more easily dissected. **A disadvantage is that some mutations that are present in general population may not be found in this small, inbred population** and therefore will not be identified in studies of Finns.

19-28. The second trait, because the greater number of loci increases the potential for the accumulation of new mutations that will affect the trait. The first trait will reach a selective plateau more quickly.

19-29.

a. The relationship can be expressed with the formula **$2n +1$ where n = number of genes.**

b. For one gene one of the extreme phenotypes (either homozygous genotype) will be found with a frequency of 1/4 in the F2 generation. When you know the ratio of one of the extreme phenotypes in the progeny you can us **$(1/4)^n$** to calculate the number of genes. In this case **$(1/4)^n$** = 1/256, so $(1/4)^4$ = 1/256 and **n=4.**

19-30.

a. Remember that A', B', C' and D' each add 2 cm to leaf length, while A, B, C and D add 4 cm each to leaf length. Each allele is incompletely dominant and the alleles of all four genes have additive effects. Thus the genotype $A'A'$ $B'B'$ $C'C'$ $D'D'$ will give the shortest leaves, 16cm long. The phenotype will increase by an additional 2 cm with each non-prime allele that is added to the genotype. Therefore there are 9 possible phenotypes (2n + 1 as in <u>problem 19-19a</u>). The probability of each allele = (1/2) and there are a total of 8 alleles controlling the phenotype.

Therefore the probability of any single genotype is $(1/2)^8 = 1/256$. The simplest, although quite time consuming, way to calculate the frequency of each phenotype is to set up the Punnett square of all possible genotypes, as shown in **Figure 3.22**. In this case there are 16 different genotypes of gametes: A'B'C'D', AB'C'D', A'BC'D', A'B'CD', A'B'C'D, ABC'D', AB'CD', AB'C'D, A'BC'D, A'B'CD, A'B'CD, A'BCD, AB'CD, ABC'D, ABCD', ABCD. **The frequency of each phenotype is: 16 cm (8 prime alleles), 1/256; 18 cm (7 prime alleles), 8/256; 20 cm (6 prime alleles), 28/256; 22 cm (5 prime alleles), 56/256; 24 cm (4 prime alleles), 70/256; 26 cm (3 prime alleles), 56/256; 28 cm (2 prime alleles), 28/256; 30 cm (1 prime allele), 8/256; and 32 cm (0 prime alleles or 8 non-prime alleles), 1/256.**

b. If the allele frequencies for gene A are $A = 0.9$ and $A' = 0.1$ then **the genotype frequencies are AA (p^2) = 0.81, AA' ($2pq$) = 0.18 and $A'A'$ (q^2) = 0.01. For gene B the genotype frequencies are BB = 0.81, BB' = 0.18 and $B'B'$ = 0.01. For gene C the genotype frequencies are CC = 0.01, CC' = 0.18 and $C'C'$ = 0.81. For gene D the genotype frequencies are DD = 0.25, DD' = 0.5 and $D'D'$ = 0.25.**

c. When the frequency of the alleles in the population is not uniform as it is in part a, you have to use the allele or genotype frequencies to determine the probability of obtaining a particular genotype, in this case AA BB CC DD. The probability of this genotype is the product of the appropriate genotypes for the individual genes calculated in part b above. Therefore **the probability of AA BB CC DD = (0.81) (0.81) (0.01) (0.25) = 0.0016.**

Chapter 20   Evolution at the Molecular Level

Synopsis:

This chapter takes a look at the evolution of the information of life, combining some of our recent knowledge of the structure and function of nucleic acids and the organization of information in the genomes with fossil record and paleontological evidence. Much of the information is speculative and hypotheses are proposed that are thought-provoking. Genome analysis provides information that drives the development of new hypotheses in this field.

Significant Elements:

After reading the chapter and thinking about the concepts, you should be able to:

♦   discuss the merits of DNA and RNA as information carrying molecules

♦   distinguish between synonymous and non-synonymous changes in DNA sequence

♦   describe how duplications can occur

♦   derive a molecular clock value and understand how to use it

♦   think creatively

Problem Solving Tips:

♦   Synonymous base changes in DNA do not change the amino acid that is encoded. Non-synonymous base mutations will result in a different amino acid being present in the protein.

♦   Duplications can occur by unequal crossing-over between repeated sequences or transposition of DNA.

♦   Molecular clock rates can be established by dividing the number of base changes by the length of time when the two organisms diverged (according to fossil or paleontological evidence).

Solutions to Problems:

Vocabulary

20-1. a. **4**; b .**6**; c. **5**; d. **2**; e. **1**; f. **7**; g. **3**.

Section 20.1 – The Origin of Life on Earth

20-2. The fact that **the same genetic code directs the cellular basis of life in varied organisms** supports the unity of life hypothesis.

20-3. Statements a. and c.

20-4.

a.   RNA is a **relatively unstable** molecule, so would not be as good a storage information molecule as DNA. RNA **can be degraded by chemical or enzymatic hydrolysis** (RNases). RNA is also **not easily compacted**.

b.   Because **RNA has only 4 "letters" in its alphabet, there is less variety than can be achieved in proteins**.

20-5.

a.   The **enzyme consists of an RNA molecule**.

b.   The **enzyme has both an RNA and a protein component**.

20-6. The differences should affect many processes. The **majority of the alterations should be effects on brain structure and function** since the differences between humans and chimps are primarily in the cognitive functions. One way that many functions can be affected is by **alterations in master regulatory genes** (changing function or creating new regulators) that control the expression of many other genes.

Section 20.2 – The Evolution of Genomes

20-7.

a.   The rates of non-synonymous substitutions differ for these three genes because there is **a different constraint on the function of each of the proteins**. The function of the histones is very fixed in different organisms, so there is little room for variation, whereas growth hormone may have evolved to interact with other proteins that also have evolved.

b.   The rates of **synonymous substitutions** are more constant between different genes because these base changes **do not affect function of the gene product**.

20-8. The immunoglobulin genes encode subunits of antibodies and **a great diversity of antibodies is needed in the immune response**. The changes that occur are therefore evolutionarily advantageous and would be maintained rather than being selected against.

20-9. The maintenance at a relatively high level suggests that there is **some benefit to the *CF* allele in the heterozygous state**. Recent evidence indicates that the allele may be somewhat protective against certain diseases. The *CF* allele may be similar to the sickle cell allele that protects heterozygous individuals against malaria.

20-10. A single copy of a gene can be duplicated if it is surrounded by **small repetitive sequences and these sequences misalign during meiosis and crossing-over occurs**. The gene that lies between the misaligned sequences will be duplicated (**Figure 13.29**).

20-11. Because the three color vision genes are cross-hybridizing, it is **likely that they arose by duplication events and then diverged evolutionarily to function as different color receptors**.

20-12. In transposition, **a copy is transposed to a new location in the genome and therefore can be found at a very different location than the original copy. Unequal crossing-over produces extra copies adjacent to the original copy**.

20-13.
a. The *A* allele in humans and the *M* allele in mice are 600 base pairs different and mice and humans are 60 million years apart in evolutionary time. The clock rate we can derive from this information is that over 1 million years, 10 bp on average were changed. The *A* allele has 3000 base differences compared to *Xenopus* (*X* allele). Using the mouse-human numbers, we estimate that **frogs and humans are 300 million years apart. That means that frogs and mice are separated evolutionarily by 240 million years**.

b. **Two duplications** must have occurred to lead to alleles *B* and *C* in humans. Because the *C* allele has 300 bases different from *A* the ***A - C* duplication occurred 30 million years ago** and has undergone base changes over that time period. The *B* allele has fewer differences compared with *C* so was duplicated and evolved more recently than *C*. Using our clock rate, **the *B* allele arose by duplication 1 million years ago**.

c. **The duplication of *B* resulted in a copy on a different chromosome** (or far enough away on the same chromosome so that it is unlinked). The duplication **could have been a transposition** type of event. Because ***C* is a tandem copy, it could have arisen by a misalignment of repeated sequences** around the original copy followed by a crossing over.

20-14. Chromosome mutations have the potential for much more dramatic effects. For example, regulatory regions of one gene get juxtaposed to regions of other genes; several genes can be lost by deletion.

20-15. This aberrant phylogeny of the glucose-6-phosphate isomerase could indicate that **this particular gene was introduced from a different species**.

Section 20.3 – The Organization of Genomes

22-16. The **plasmid could have been introduced** (via natural transformation or conjugation- see chapter 14 for more information on these processes) **from another species in the wild**.

20-17. a. **exons**; b. **genes**.

20-18.

a. **In vertebrates there are four *HOX* gene family clusters** compared with one cluster in *Drosophila*, for example. This fits well with the tetraploidization hypothesis since the two doublings should lead to four copies of the genes.

b. The variation within the four gene family clusters **could be the result of duplications within the gene family in certain lineages**.

20-19. LINES or SINES can mediate genome rearrangements by unequal crossing-over between elements and they can contribute regulatory elements adjacent to a gene.

20-20. The size of dinucleotide repeat sequences can change **by unequal crossing-over between the repeated sequences within the element**. There is also some variation that can occur by slippage during DNA replication.

20-21.

a. **SINES or LINES**.

b. **centromeric satellite DNA**; although we don't know the precise function, it appears to be needed for proper centromere function in higher eukaryotes.

20-22. a. **2**; b. **4**; c. **1**; d. **3**.

Section 20.4 – Rapid Evolution in the Immune Response and in HIV

20-23. The highly active antiretroviral therapy (HAART) involves simultaneous delivery of three or four different anti-HIV drugs. Two of the drugs block the function of the HIV reverse transcriptase through different mechanisms and the third blocks the functioning of the viral protease critical to the viral reproduction. The focus is on indinavir sulfate, a viral protease inhibitor that has several possible side effects. These include increased risk of kidney stones, redistribution of fat deposits in the body and acute hemolytic anemia.

a.   **The side effects suggest that the protease inhibitor is not completely specific to the HIV protease**, or else the drug would not have these effects on what appear to be processes unrelated to the viral infection. **Presumably the HIV protease may have distant evolutionary relationships with other proteases found in normal cells**, so the drug could be affecting the performance of these other proteases as well. However the existence of side effects is not proof that the effects are due to the ation of the drug as a protease inhibitor. It could be that the body is mounting some kind of response (like an immune response) to the presence of the drug that does not depend on its function as a protease inhibitor, or the drug could be metabolized to some ther molecule that is toxic and that does not function as a protease inhibitor.

b.   **Protease cleavage could be involved in the generation of a hormone like cortisol that could affect fat distribution.** The insulin resistance would be secondary to the lack of such a hormone. **Many hormones are generated from prohormones by protease cleavage.** Protease cleavage is unlikely to be involved in the generation of insulin. Perhaps protease activity is needed for the production of proper amounts of the receptor for insulin or some aspect of cellular signaling downstream of the binding of insulin to its receptor.

c.   **Perhaps the indinavir sulfate affects the processes needed for energy metabolism, such as the function or the integrity of the mitochondria. This hypothesis could be tested by analyzing the energy content of the cells from patients treated with the drug.** Specific aspects of energy generation such as mitochondrial performance can also be examined in patients treated with the protease inhibitor.

Chapter 21   Systems Biology and the Future of Medicine

Synopsis:

Systems biology is a newly emerging field that attempts to define all the components of a biological system and understand how they function in conjunction with one another. This relies on the tools and strategies of both genomics – the global studies of genomes - and proteomics – the global analysis of all of the proteins in a particular cell type or organism.

The fundamental questions of systems biology include:

1. What are the elements of the system (proteins, cells, etc)?
2. What are the physical associations among the elements?
3. How do other biological systems that connect to the system under study change with perturbations of the original system?
4. How do the system's elements, their associations and their relations to changes in the biological context explain its emergent properties?

Systems biology is a developing science. Systems biologists must practice cross-disciplinary biology with teams of biologists, computer scientists, chemists, engineers, mathematicians and physicists. These teams must develop new, high-throughput measuring instruments to generate global data sets and then develop the computational tools to analyze and annotate this data.

There are two fundamental types of biological information: the digital information in the DNA of the genome and environmental signals. The genomic information includes the genes that encode proteins and short (10-15bp) cis-control elements. Some of these cis-control elements are called enhancers and were discussed in Chapter 16.

Gene regulatory networks receive diverse inputs of information, integrate and modify those inputs, and then output the transformed information to various protein networks (**Figure 21.7**). The interaction between a transcription factor and its cis-control element is the fundamental linkage in gene regulatory networks.

Protein networks are sets of interacting proteins that execute a particular biological function. These networks channel biological information and use it to execute particular function.

Significant Elements:

♦   After reading the chapter and thinking about the concepts, you should be able to understand the tools and technologies of systems biology and what types of data they collect. These tools include:

- the mass spectrometer allows identification of the types and amounts of specific proteins in complex mixtures of proteins; protein-protein interactions can be analyzed with affinity capture/mass spectrometry;

- the ICAT (isotope-coded affinity tags) technique allows the analysis of changing patterns of protein expression in two different cellular states;

- protein-protein interactions can be studied using the yeast two-hybrid system and affinity capture/mass spectrometry;

- protein arrays are another technique that allow the study of protein-protein interactions;

- ChIP/chip analysis is used to identify protein-DNA interactions.

Solutions to Problems:

Section 21.1 – What is Systems Biology?

21-1. The fundamental concepts that define a biological system are: 1. **the elements of the system** – proteins, cells, etc.; 2. **the physical associations among the elements**. These are often represented in a graphic representation of a network with nodes representing individual proteins and connections or edges representing physical interactions between the proteins (**Figure 21.2**); 3. **the biological context of the system** – within individual cells types (prostate cells) or within an entire organism (yeast or mice or humans); 4. **how the association of the system's elements and their relation to changes in the biological context explain its emergent property** (the property that arises from the operation of the entire system, like the ability of the heart to pump blood).

21-2.

a. **Engineers are critical in developing new high throughput measuring instruments** necessary to study emergent properties in systems biology.

b. **Mathematicians and computer scientists are necessary to develop new computational tools to organize, annotate, analyze, integrate and model the accumulated data.**

Section 12.2 – Biology as an Informational Science

21-3. Digital genomic information consists of two types of DNA sequences. These are the **genes that encode proteins and untranslated RNAs** (rRNAs, tRNAs, miRNAs, etc) **and the short DNA sequences that make up the control elements adjacent to the genes**. These cis-control elements are 6-15 bases long.

21-4. Gene regulatory networks examine the interactions of transcription factors with their cognate cis-control elements. These networks receive information from different sources and then they ingrate and modify this input and send the transformed information to various protein networks. Protein networks are sets of interacting proteins and other molecules such as metabolites that execute a particular biological function. Protein networks channel biological information and use it to execute particular functions such as signal transduction, physiological responses, development and metabolism – networks are dynamic. **A catalog of cis-control elements would be part of a gene regulatory network** as it could define specific transcription factors involved in a process such as gut development (Figure 21.7). To gain a further understanding of the network you need to understand **what environmental information affects the network via signal transduction pathways and other inputs, how the network integrates and modifies these inputs and the output of the transformed information to various protein networks**.

Section 21.3 – The Practice of Systems Biology

21-5. The sequences of the entire genomes of humans and other model organisms like *E. coli*, yeast and mice provide a genetics parts list of all the genes and the proteins they encode. The study of human genome defined the basic blocks of DNA, proteins and other molecules that systems biology hopes to fit together into networks. These blocks include cis-control elements, the discovery and definition of many transcription factors (**Figure 21.5**) and regulatory cascades where transcription factors affect many different genes (**Figure 21.6**). This **also defined many of the complex molecular machines (Figure 21.3) and protein networks, and enforced the idea that biological information is hierarchic** – it starts with digital DNA, progresses through mRNA, protein, molecular machines, networks, cells, networks of cells and tissues to individual organisms, populations of organisms to ecosystems.

The Genome Projects also drive the development of powerful computational tools which make it possible to acquire, store, analyze, integrate, display and model biological information. The high-throughput platforms for genomics and proteomics enable the acquisition of global data sets of differing types of biological information. Scientists can now gather global data sets from the genetic and environmental perturbations (experimental manipulations) of the biological systems of the model organisms and thus learn how to do systems biology in more complex organisms. The use of comparative genomics allows scientists to begin to determine the logic of life for individual organisms, (*e.g.* what type of energy generating systems does an organism use?) and to determine how that logic has changed in different evolutionary lineages.

21-6. a. **True**; b. **True**; c. **False**; d. **False**; e. **True** (see p446) – there are about 6,000 genes in the yeast genome and <u>very little alternate mRNA splicing</u>, protein processing and protein modification increase the number of variations of these gene products; f. **True**; g. **True**; h. **False**; i. **False**; j. **True**; k. **True**.

21-7. You should ask your friend lots of questions: **What exactly is the biological system that he is perturbing? Does he know all of the elements in this system? Some of the cytokines may be the output of the system, but does he know all of the outputs? Or all of the inputs? If he is measuring protein levels, he has only one type of biological information** and systems biology cannot be done with just a single type of information – this is closer to discovery science because systems biology requires the integration of different types of information. **Also your friend is not looking at the other systems whose behaviors will be altered by the knockout perturbations** – there are no contextual studies. You could argue that **your friend's research is not systems biology**.

21-8. ICAT (isotope coded affinity tags) allows quantification of changes in protein concentrations in different cellular states. Of the 30 proteins whose levels changed under the conditions of +galactose and -galactose only half of them showed a corresponding change in mRNA levels. How can the levels of the remaining 15 proteins change if their mRNA levels are constant in the 2 different conditions? It is possible that **the regulation of concentration is at the post-transcriptional level**. For example **perhaps the mRNA is altered so the level of translation changes, or perhaps the protein is modified by a chemical reaction. It is also possible that this group of proteins is involved in protein-protein interactions with one or more of the proteins whose level does fluctuate**. In this model the interacting partner's level decreases so there is less to interact with and the non-interacting subunits are destabilized and degraded.

<u>Section 21.4 – A Systems Approach to Disease</u>
21-9. a. **True**; b. **True**; c. **True**; d. **True**; e. **True**.